IUTAM Symposium on Computational Approaches to Multiphase Flow

T0191762

FLUID MECHANICS AND ITS APPLICATIONS
Volume 81

Series Editor: R. MOREAU
MADYLAM
Ecole Nationale Supérieure d'Hydraulique de Grenoble
Boîte Postale 95
38402 Saint Martin d'Hères Cedex, France

Aims and Scope of the Series

The purpose of this series is to focus on subjects in which fluid mechanics plays a fundamental role.

As well as the more traditional applications of aeronautics, hydraulics, heat and mass transfer etc., books will be published dealing with topics which are currently in a state of rapid development, such as turbulence, suspensions and multiphase fluids, super and hypersonic flows and numerical modelling techniques.

It is a widely held view that it is the interdisciplinary subjects that will receive intense scientific attention, bringing them to the forefront of technological advancement. Fluids have the ability to transport matter and its properties as well as transmit force, therefore fluid mechanics is a subject that is particulary open to cross fertilisation with other sciences and disciplines of engineering. The subject of fluid mechanics will be highly relevant in domains such as chemical, metallurgical, biological and ecological engineering. This series is particularly open to such new multidisciplinary domains.

The median level of presentation is the first year graduate student. Some texts are monographs defining the current state of a field; others are accessible to final year undergraduates; but essentially the emphasis is on readability and clarity.

For a list of related mechanics titles, see final pages.

IUTAM Symposium on Computational Approaches to Multiphase Flow

Proceedings of an IUTAM Symposium
held at Argonne National Laboratory, October 4–7, 2004

Edited by

S. BALACHANDAR
University of Florida, Gainesville, Florida, U.S.A.

and

A. PROSPERETTI
The Johns Hopkins University, Baltimore, Maryland, U.S.A.

 Springer

A C.I.P. Catalogue record for this book is available from the Library of Congress.

ISBN-13 978-90-481-7241-2
ISBN-10 1-4020-4977-3 (e-book)
ISBN-13 978-1-4020-4977-4 (e-book)

Published by Springer,
P.O. Box 17, 3300 AA Dordrecht, The Netherlands.

www.springer.com

Printed on acid-free paper

Printed in the Netherlands.

Table of Contents

PART II: LATTICE-BOLTZMANN AND MOLECULAR DYNAMIC SIMULATIONS

PART III: FULLY-RESOLVED MULTI-PARTICLE SIMULATIONS

PART IV: FREE SURFACE FLOWS, DROPS AND BUBBLES

Preface

This volume contains a large fraction of the papers presented at a symposium on *Computational Approaches to Disperse Multiphase Flow*, sponsored by the International Union of Theoretical and Applied Mechanics and generously supported by the Office of Basic Energy Sciences of the US Department of Energy. The symposium, which attracted about 90 participants from fifteen different countries, was held at Argonne National Laboratory on October 4–7, 2004. There were 48 oral presentations and an additional 17 poster papers.

Together with experiment and theory, computation has been for a long time an integral component of multiphase flow research. A striking feature common to most papers presented at the symposium was the power, maturity and sophistication reached by this approach.

A few papers conclusively demonstrate that, for some problems, computing is the *only* means by which key physical phenomena can be elucidated. A prime example is the analysis of Leonardo's paradox, i.e., the instability of the rectilinear path of an ascending bubble. The explanation of the phenomenon rests on computations in which the bubble shape is constrained to remain spheroidal with a varying eccentricity – not a situation amenable to experiment, but a key step in understanding the physics. Another case in point is the study of the detailed action of surfactants at the surface of a rising bubble. While the general physical mechanism at work has been known for some time, this is the first visualization of the microphysical processes acting at the bubble surface and their impact on the local flow field.

It is also interesting to realize that computation may be the key to interpreting experimental results. A striking example is offered by the study of the sudden exposure of a liquid drop to an incident Mach-3 flow. High-speed video recordings of experiments present an extremely complex interaction of shocks and drop deformation which it would be next to impossible to unravel without simulation results, themselves rendered possible by very sophisticated numerics.

New algorithms and new problems begin to leave their mark. The progress of lattice-Boltzmann methods was demonstrated by many papers in which they are used for modeling over a range of scales, from single particles to suspensions and fluidized beds. Several papers described new methods to simulate thousands of extended particles in Navier–Stokes flows, simulations which could hardly have been imagined just a few years ago. At the other end, molecular dynamics proves useful to elucidate phenomena, such as contact line motion and metastable nano-bubbles, which have eluded a full understanding for decades. Interesting problems and methods also arise at the intermediate scale, between micro and macro, a good example being the direct numerical simulation of the Brownian motion of particles.

Many of the papers contained in this volume address more traditional topics in computational multi-phase flow in which research continues apace: point-particle models in turbulence, both DNS and LES, stochastic methods, free-surface flows treated with diffuse- or sharp-interface algorithms. While this type of studies has been ongoing for some years, progress is still rapid at the level of both algorithms and physical understanding.

A special lecture, which unfortunately could not be included in this volume, was given by Gad Hetsroni. His presentation ranged from some historical notes on boiling, dating back to the Bible and Homer, to reflections on scientific progress in boiling research, to recent statistics on papers published in the *International Journal of Multiphase Flow*. The opening lecture, which could not be included here either, was given by John Hinch, who discussed of the scaling of velocity fluctuations induced by sedimenting particles. The closing lecture, included in this volume, was given by Daniel Joseph, who presented numerical results on particle migration in Poiseuille flows. He also gave a personal summary of the current status of high-end scientific computating in multiphase flow as presented at the symposium and conclued on a high note as to the bright future for *Computational Multiphase Flow*.

The symposium made two points abundantly clear. In the first place, computational multiphase flow suscitates a strong interest in the fluid mechanics community – a heartening corollary being the high quality of much of the work in this field. Secondly, it was extremely gratifying – and perhaps even somewhat surprising – to gain such a palpable appreciation of the maturity of the field, its impressive developments, and the level of complexity and detail that progress in hardware and algorithms currently permit. It is hard to imagine that an external observer coming to the meeting with misgivings about the usefulness of computing would have left nurturing the same doubts.

Together with the enhanced appreciation of the role of computing, it is wise to always keep firmly in mind the other two legs of progress in science – theory and experiment. This is all the more true in multiphase flow in which we are still faced by problems of such magnitude and complexity that it would be unrealistic to imagine solving by computing alone.

In conclusion, we wish to express our gratitude to the International Union of Theoretical and Applied Mechanics for sponsoring this symposium, and to Dr. Timothy Fitzsimmons, Materials Sciences and Engineering Division, Office of Basic Energy Sciences, US Department of Energy, for the Department's generous support.[1]

Urbana and Baltimore   *S. Balachandar*
November 2005   *A. Prosperetti*

[1] The preparation of this volume has been supported in part by the ASCI Center for the Simulation of Advanced Rockets through DOE subcontract B341494 (S.B.), and by NSF grant CTS-0210044 (A.P.).

PART I
POINT PARTICLE APPROACH

An Updated Classification Map of Particle-Laden Turbulent Flows

Said Elghobashi

Department of Mechanical and Aerospace Engineering, University of California, Irvine, CA 92697, USA

1 Introduction

The classification map of particle-laden turbulent flows shown in Figure 1, originally proposed by Elghobashi [4] and slightly modified by Elghobashi [5], was based on the experimental and direct numerical simulation (DNS) data available at the time. However, recent DNS results of particle-laden isotropic turbulence [8] now provide sufficient details to update a small region in the map. The purpose of this short paper is to describe the updated map.

It should be noted that the map covers a wide-range of particle-laden turbulent flow regimes, most of which are not fully understood at present. The statements made by Elghobashi [5] about the challenges facing the attempts of numerically predicting particle-laden turbulent flows are still valid today. Thus, it is expected that the map will be further updated when new *reliable* data become available.

2 Description of the original map

Figure 1 shows the classification map of Elghobashi [5]. The quantities appearing on the dimensionless coordinates are defined below.

ϕ : volume fraction of particles, $\phi = NV_p/V$

N : total number of particles in the flow

V_p : volume of a single particle

V : total volume occupied by particles and fluid

d : diameter of particle

τ_p : particle response time $= \rho_p d^2/(18\rho_f \nu)$ for Stokes flow, and

τ_k : Kolmogorov time scale $= (\nu/\varepsilon)^{1/2}$

In the above definitions, ρ is the material density and the subscripts p and f denote respectively the particle and carrier fluid. ν is the kinematic viscosity of the fluid, and

3

S. Balachandar and A. Prosperetti (eds), Proceedings of the IUTAM Symposium on Computational Multiphase Flow, 3–10.

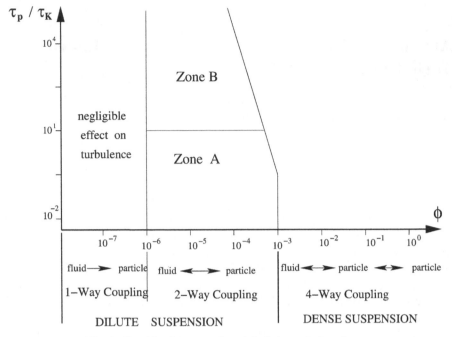

Fig. 1. Classification map of particle-laden turbulent flows.

ε is the dissipation rate of turbulence kinetic energy. For very small values of ϕ ($\leq 10^{-6}$), the particles have negligible effect on turbulence, and the interaction between the particles and turbulence is termed as **one-way coupling**. This means that particle dispersion depends on the state of turbulence but due to the negligible concentration of the particles in this regime, the momentum exchange between the particles and the turbulence has an insignificant effect on the flow. In the second regime, $10^{-6} < \phi \leq 10^{-3}$, the momentum exchange between the particles and turbulence is large enough to alter the turbulence structure. This interaction is called **two-way coupling**. Now, in this regime and for a given value of ϕ, there are two zones (A and B), depending on the ratio τ_p/τ_k, where the transition from A to B occurs at about $\tau_p/\tau_k = 10$. In zone A, the particle Reynolds number, R_p is ≤ 1, but within the range $0.01 \leq (\tau_p/\tau_k) \leq 10$ and for a fixed ϕ, our recent DNS results [8] show that the effects of the particles on the turbulence vary significantly as a function of (τ_p/τ_k), as depicted in Figure 2 and discussed later. In zone B, as τ_p increases (e.g. by increasing the particle diameter) for the same ϕ, the particle Reynolds number increases, and at values of $R_p \geq 400$, vortex shedding takes place resulting in enhanced production of turbulence energy. In the third regime, because of the increased particle loading, $\phi > 10^{-3}$, flows are referred to as dense suspensions. Here, in addition to the two-way coupling between the particles and turbulence, particle/particle collision takes place, hence the term **four-way coupling**. As ϕ approaches 1, we obtain a **granular flow** in which there is no fluid.

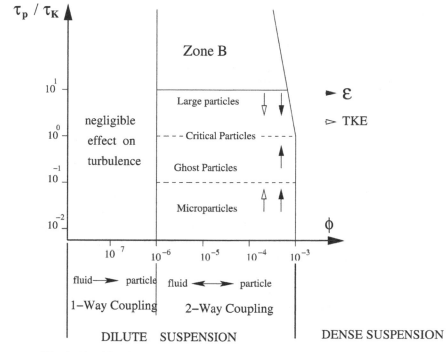

Fig. 2. Classification map of particle-laden turbulent flows. Details of Zone A.

Table 1. Flow parameters (dimensionless) at initial time ($t = 0$), injection time ($t = 1$), and for case A at time $t = 5$.

t	u_0	ε	l	λ	η	Re_l	Re_λ	l/η	τ_k	τ_l
0.0	0.0503	7.4×10^{-4}	0.0684	0.0345	0.00202	150	75	33.8	0.177	1.36
1.0	0.0436	9.8×10^{-4}	0.0685	0.0259	0.00188	129	49	36.4	0.154	1.57
5.0	0.0233	2.0×10^{-4}	0.0891	0.0305	0.00280	90	31	31.9	0.338	3.83

The line separating the two-way and four-way coupling regimes is inclined to indicate the tendency of particle-particle collision to take place at higher values of τ_p/τ_k, thus transforming the two-way to four-way coupling regime even for $\phi < 10^{-3}$.

The dispersion of particles and their preferential accumulation in unconfined homogeneous turbulent flows with one-way coupling are reasonably understood [1, 6–8]. On the other hand, flows in the two-way or four-way coupling regimes are still challenging and require more reliable, detailed experimental and numerical studies to improve their understanding.

We restrict the present discussion to isothermal incompressible flows without phase changes (e.g. vaporization) or chemical reaction. Also, the effects of particle-particle or particle-wall collisions are not considered here.

Table 2. Particle properties (dimensionless) at injection time ($t = 1$) with $\phi = 10^{-3}$ and $\phi_m = 1.0$ (for $\rho_p/\rho = 1000$).

Case	τ_p	τ_p/τ_l	τ_p/τ_k	d	d/l	d/η	$d\ (\mu m)$	M_c	M_r/M_c	$\mathrm{Re}_{p,max}$	v_t/u_0^*
A	–	–	–	–	–	–	–	–	0	–	
B	0.0154	0.0098	0.1	0.80×10^{-4}	0.00117	0.043	30	80×10^6	46.7	0.11	0.0
C	0.0385	0.0245	0.25	1.26×10^{-4}	0.00185	0.067	47	80×10^6	11.8	0.31	0.0
D	0.1540	0.0979	1.0	2.53×10^{-4}	0.00369	0.134	94	80×10^6	1.5	1.34	0.0
E	0.7700	0.4895	5.0	5.66×10^{-4}	0.00825	0.300	211	10.6×10^6	1.0	5.33	0.0
F	0.0385	0.0245	0.25	1.26×10^{-4}	0.00185	0.067	47	80×10^6	11.8	0.32	0.25

3 Recent DNS results of particle-laden isotropic turbulence

Ferrante and Elghobashi [8] performed DNS of particle-laden isotropic turbulence with higher resolution ($\mathrm{Re}_\lambda = 75$) and a considerably larger number (80 million) of particles in comparison to the previous DNS studies [2, 3, 7, 10, 11]. We studied six cases to understand how particles with different inertia, τ_p, modify the decay rate of isotropic turbulence in both zero- and finite-gravity conditions. The Lagrangian equation of particle motion included only the forces due to Stokes drag and buoyancy. The fluid velocity at the particle location was computed via a fourth-order accurate three-dimensional Hermite cubic interpolation polynomial. The flow parameters are shown in Table 1, and the particle properties are shown in Table 2. These two tables, which are copied from [8], include more information than required for the present paper. The definitions of the additional quantities therein are available in [8] and thus will not be repeated here. Case A represents the particle-free flow, whereas cases B-E represent particle-laden flows with different inertia particles in zero gravity, and case F represents the particle-laden flow in *finite gravity*. It is important to note that all five cases (B-F) of particle-laden turbulence have the same volume fraction of particles, $\phi = 10^{-3}$, and the same mass loading ratio $\phi_m = 1.0$ (for $\rho_p/\rho = 1000$), and thus the differences between the resulting modifications of turbulence in these cases are only due to the different values of (τ_p/τ_k). We changed the particle diameter for each case to obtain a different ratio (τ_p/τ_k) of the particle response time to the Kolmogorov time scale at the injection time, e.g. $\tau_p/\tau_k = 0.1$ in case B, and $\tau_p/\tau_k = 5.0$ in case E. The effects of gravity are studied in case F where $\tau_p/\tau_k = 0.25$ (as in case C) and $v_t/u_0^* = 0.25$, where v_t is the terminal velocity ($v_t = g\,\tau_p$) of the particle and u_0^* is the rms velocity of the surrounding fluid at the injection time, and gravity is in the negative x_3 direction. However, the non-zero gravity effects are *not included* in the map since that would require a third axis for the ratio (v_t/u_0^*), rendering the current two-dimensional map three-dimensional. At present, the experimental and numerical data necessary for that extension are not available.

More details about the governing equations and numerical solution method are given in [8].

E(t)/E(0)

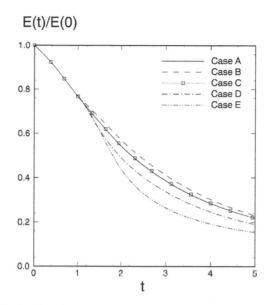

Fig. 3. Time development of the turbulence kinetic energy.

4 Time evolution of turbulence kinetic energy

Figure 3 shows the temporal evolution of the turbulence kinetic energy (TKE)normalized by its initial value, $E(t)/E_0$, for the zero gravity cases (A-E). The microparticles (case B), with $\tau_p/\tau_k < 0.25$, initially ($1 < t \leq 2.1$), reduce the decay rate of TKE resulting in TKE being larger than that of case A at all times, whereas particles with higher inertia (critical particles, case D, and large particles, case E), $\tau_p/\tau_k > 0.25$, initially enhance the TKE decay rate considerably resulting in TKE being smaller than that of case A at all times. Figure 3 also shows that particles with $\tau_p/\tau_k = 0.25$ (case C) keep TKE nearly identical to that of case A at all times, with a percentage difference smaller than 0.6%. Thus we name the particles in case C '*ghost' particles*, since their effects on the turbulence cannot be detected by TKE's temporal behavior, $E(t)$. However, as we will discuss later (Figure 4), these 'ghost' particles do modify the spectrum $E(\kappa)$ of TKE. Figure 3 shows that at time $t = 5$, in comparison to TKE in case A, TKE in case B is larger by more than 5%; TKE in case C is nearly identical; TKE in case D is smaller by about 13%, and TKE in case E is smaller by about 30%.

The effects of gravity on the time evolution of TKE are described in [8].

4.1 Energy spectrum

Figure 4 shows the three-dimensional energy spectra $E(\kappa)$ for the five cases A-E at time $t = 5$. Microparticles (case B) increase $E(\kappa)$ relative to case A at wavenumbers $\kappa \geq 12$, and reduce $E(\kappa)$ relative to case A for $\kappa < 12$, such that $\int E(\kappa)d\kappa \equiv$ TKE

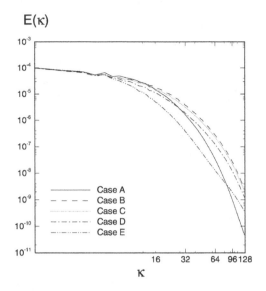

Fig. 4. Three-dimensional spatial spectrum of energy $E(\kappa)$ at $t = 5.0$.

in case B is larger than in case A as shown in Figure 3. For 'ghost' particles (case C), although $E(t)$ is nearly identical to that of case A at all times (Figure 3), it is clear in Figure 4 that the energy spectrum in case C differs from that in case A such that its integral, TKE, is nearly identical to that of case A. Figure 4 shows that ghost particles reduce $E(\kappa)$ relative to that of case A for $\kappa < 15$ and increase it above that of A for $\kappa \geq 15$. Critical particles (case D) increase $E(\kappa)$ above that of case A for $\kappa \geq 27$ and reduce it for smaller wavenumbers. In this case (D) the modulation of $E(\kappa)$ is such that its integral, TKE, is smaller than in case A (Figure 3). Large particles (case E) contribute to a faster decay of TKE by reducing the energy content at almost all wavenumbers, except for $\kappa > 87$ where a slight increase of $E(\kappa)$ occurs. The physical mechanisms responsible for the above observations are discussed in detail in [8] and thus will not be repeated here. Only a brief description of the different cases relevant to the map is given below.

4.2 Microparticles ($\tau_p/\tau_k \ll 1$)

Microparticles (case B) behave almost like flow tracers because their response time, τ_p, is much smaller than the Kolmogorov time scale, τ_k, but since their material density, ρ_p, is much higher than that of the carrier fluid, ρ, they cause the fluid to behave like a 'heavy gas' [9]. Our DNS results [8] show that the microparticles increase both TKE and ε relative to their values in single-phase flows.

4.3 Critical particles ($\tau_p/\tau_k \approx 1$)

We label the particles in case D ($\tau_p/\tau_k = 1$) '*critical*' particles because of their property of maximum preferential accumulation in comparison to other particles. The main characteristic of these critical particles is that they are ejected from the cores of the smallest vortical structures (of a size nearly equal to the Kolmogorov length scale) and remain orbiting around their peripheries, leaving these structures free of particles. Since these structures are subjected to the largest strain rates, they control the viscous dissipation rate, ε, of TKE. Consequently, the dissipation rate, ε, in this case is nearly identical to the single-phase case A. However, due to their significant inertia, the critical particles' two-way coupling force reduces TKE relative to that of case A.

4.4 'Ghost' particles ($0.1 < \tau_p/\tau_k < 0.5$)

It is clear from Figure 3 and the above discussion that in comparison to the particle-free flow (case A), microparticles (case B, $\tau_p/\tau_k = 0.1$) reduce the decay rate of TKE, and critical particles (case D, $\tau_p/\tau_k = 1.0$) enhance that rate. These two opposing effects in cases B and D lead us to search for particles which have a '*neutral*' effect on that decay rate. More specifically, we searched for particles whose τ_p is in the range $0.1 < \tau_p/\tau_k < 1.0$ and which maintain the decay rate of TKE as that of the particle-free flow (case A). Our DNS results show that particles with $\tau_p/\tau_k = 0.25$ (case C) satisfy this condition at all times, as shown in Figure 3. As mentioned above, these particles are '*ghost*' particles because their presence in the flow cannot be detected by examining only the temporal development of TKE. It is important to emphasize that the value of $\tau_p/\tau_k = 0.25$ is not universal but depends on Re_{λ_0}, ϕ_m and the magnitude of the gravitational acceleration (zero in our case). However, the significance of this finding is that dispersed particles are capable of modifying the turbulence energy spectrum (Figure 4) in such a unique way that the amount of energy gained by the turbulence at high wave numbers balances exactly the amount of energy lost at low wave numbers, with the net result of retaining the integral of the spectrum equal to that of the particle-free flow at all times (Figure 3).

4.5 Large particles ($\tau_p/\tau_k > 1$)

Large particles (case E) here denote particles whose response time, τ_p, is larger than the Kolmogorov time scale, τ_k. Because of their large τ_p, large particles do not respond to the velocity fluctuations of the surrounding fluid as quickly as microparticles do but rather 'escape' from their initial surrounding fluid, 'crossing' the trajectories of the fluid points [6, 12]. Whereas microparticles remain 'trapped' in the vortical structures of their initial surrounding fluid, large particles are 'ejected' from these structures. The net result is the reduction of $E(\kappa)$ in case E at nearly all wave numbers relative to case A (Figure 4), thus reducing both TKE and ε relative to the single-phase flow.

5 The updated map

The results discussed above provide new information about the behavior of the turbulence kinetic energy (TKE) and its dissipation rate (ε) in Zone A of the classification map in Figure 1. Figure 2 shows the new details of Zone A in the updated map which includes three regions:

- Microparticles ($\tau_p/\tau_k \leq 0.1$) cause both TKE and its dissipation rate, ε, to be larger than in the single-phase flow.
- Ghost particles ($0.1 < \tau_p/\tau_k < 0.5$) modify the energy spectrum $E(\kappa)$ in a way such that TKE is *unchanged* but ε is larger than that of the single-phase flow.
- Critical particles ($\tau_p/\tau_k \approx 1$) reduce TKE but keep ε *unchanged* relative the single-phase flow.
- Large particles ($\tau_p/\tau_k > 1$) reduce both TKE and ε relative to their values in the single-phase flow.

References

1. Ahmed, A. and Elghobashi, S., 2001, Direct numerical simulation of particle dispersion in homogeneous turbulent shear flows, *Phys. Fluids* **13**, 3346–3364.
2. Boivin, M., Simonin, O. and Squires, K., 1998, Direct numerical simulation of turbulence modulation by particles in isotropic turbulence, *J. Fluid Mech.* **375**, 235–263.
3. Druzhinin, O. and Elghobashi, S., 1999, On the decay rate of isotropic turbulence laden with microparticles, *Phys. Fluids* **11**, 602–610.
4. Elghobashi, S., 1991, Particle-laden turbulent flows: Direct simulation and closure models, *Appl. Sci. Res.* **48**, 301–314.
5. Elghobashi, S., 1994, On predicting particle-laden turbulent flows, *Appl. Sci. Res.* **52**.
6. Elghobashi, S. and Truesdell, G., 1992, Direct simulation of particle dispersion in decaying isotropic turbulence, *J. Fluid Mech.* **242**, 655–700.
7. Elghobashi, S. and Truesdell, G., 1993, On the two-way interaction between homogeneous turbulence and dispersed solid particles, Part 1: Turbulence modification, *Phys. Fluids* **A5**, 1790–1801.
8. Ferrante, A. and Elghobashi, S., 2003, On the physical mechanisms of two-way coupling in particle-laden isotropic turbulence, *Phys. Fluids* **15**, 315–329.
9. Saffman, P., 1962, On the stability of laminar dusty gas, *J. Fluid Mech.* **13**, 120–128.
10. Squires, K. and Eaton, J., 1990, Particle response and turbulence modification in isotropic turbulence, *Phys. Fluids* **A 2**, 1191–1203.
11. Sundaram, S. and Collins, L., 1999, A numerical study of the modulation of isotropic turbulence by suspended particles, *J. Fluid Mech.* **379**, 105–143.
12. Yudine, M., 1959, Physical considerations on heavy-particle diffusion, *Adv. Geophys.* **6**, 185–191.

On Fluid-Particle and Particle-Particle Interactions in Gas-Solid Turbulent Channel Flow

Kyle D. Squires[1] and Olivier Simonin[2]

[1]*Mechanical and Aerospace Engineering Department, Arizona State University,
Tempe, AZ 85287-6106, USA; e-mail: squires@asu.edu*
[2]*Institut de Mécanique des Fluides de Toulouse, UMR CNRS-INPT-UPS 5502, Allée du
Professeur Camille Soula, 31400 Toulouse, France; e-mail: simonin@imft.fr*

Abstract. Large-Eddy Simulation (LES) and Discrete Particle Simulation (DPS) are used to
highlight effects of fluid-particle and particle-particle interactions on dispersed-phase transport in fully-developed turbulent channel flow. A range of particle Stokes numbers in the
simulations are considered that lead to strong changes in particle response. In the absence of
inter-particle collisions, the calculations illustrate the characteristic build-up of particles in the
near-wall region. While mean shear in the carrier and dispersed phase velocities is an important effect in wall-bounded flows, LES/DPS results show that the particle velocity fluctuations
in the wall-normal direction are controlled primarily by the drag force and in equilibrium with
the corresponding components of the fluid-particle velocity correlation. Inter-particle collisions provide a redistribution mechanism that reduces the strong anisotropy of the particle velocity fluctuations and substantially elevates cross-stream transport. Spatial properties of the
particle velocity field are examined using two-point correlations. The correlation functions
are discontinuous at the origin and are consistent with a partitioning of the particle velocity by
inertia into a spatially-correlated contribution and random component that is not correlated in
space. Perspectives and implications of these findings are also discussed.

1 Introduction and overview

Turbulent flows laden with dense particles or droplets occur in a large number of
engineering and environmental systems. Examples include coal combustors, chemical reactors, semiconductor processing devices, pneumatic transport and processing
systems, and atmospheric dispersion of pollutants. The flows within these systems
are complex due to the presence of a wide range of turbulent scales in the continuum,
in addition to the dispersed particulate phase.

The presence of a dispersed phase of heavy particles introduces several new parameters over those used to characterize single-phase turbulent flows. The relevant
timescales, for example, include the particle response time, the inter-particle collision time, and for wall-bounded flows a timescale that could be used to characterize
particle-wall collisions. The values of these timescales compared to the appropriate
fluid flow timescales has generally been thought to indicate the relative importance

*S. Balachandar and A. Prosperetti (eds), Proceedings of the IUTAM Symposium on Computa-
tional Multiphase Flow, 11–20.*

of a given effect. The complexity of gas-solid turbulent flows, however, limits the utility of simple scaling arguments that might be applied to predict the dominance of a particular phenomena in a new flow regime.

The complex features of multi-phase flows in general and particle-laden flows in particular motivates the application of numerical simulations that enable detailed investigation. Further, for the practical applications in which these flows are encountered statistical models that require substantial empirical input will continue to form the basis for engineering prediction. This further motivates numerical simulation strategies that can be used not only to study fundamental processes but also to supply results for evaluation of engineering turbulence models.

The focus of the present contribution is on the application of computations to study dispersed-phase transport in wall-bounded turbulent shear flows. The main objective is to highlight specific aspects of fluid-particle and particle-particle interactions that illustrate the complex features of dispersed-phase motion in gas-solid flows. The computational approach is based on Large-Eddy Simulation (LES) of fully-developed turbulent channel flow for the gas and Discrete Particle Simulation (DPS) of the dispersed phase. Following an overview of the simulations, single-point statistical measures of the particle motion are presented. Spatial characteristics from recent investigations of the particle velocity field are then summarized. The paper is concluded with a summary and perspectives developed from these studies.

2 Approach

2.1 Particle equation of motion

The computations consider dilute gas-solid flows for which the dispersed-phase volume fraction $\alpha_p = n_p m_p / \rho_p$, is negligible where n_p is the particle number density, m_p the particle mass, and ρ_p the particle density. The particle diameter d_p is small compared to the smallest turbulent length scales of the undisturbed fluid flow, though owing to the large particle density ρ_p relative to the fluid value ρ_f, the particle response time is large compared to the Kolmogorov timescale of the undisturbed flow.

The effect of particle momentum exchange on properties of the fluid flow are neglected and the volume force induced by the surrounding fluid flow on the particles reduces to the drag. The equation of motion for a single particle is written as

$$\frac{dv_{p,i}}{dt} = -\frac{3}{4}\frac{\rho_f}{\rho_p}\frac{C_D}{d_p}|\mathbf{v}_r|v_{r,i}\,, \tag{1}$$

where $v_{p,i}$ is the ith component of the particle velocity and $v_{r,i}$ is the particle relative velocity,

$$v_{r,i} = v_{p,i} - \widetilde{u}_{f,i}\,, \qquad C_D = \frac{24}{\mathrm{Re}_p}(1 + 0.15\mathrm{Re}_p^{0.687})\,, \qquad \mathrm{Re}_p = \frac{|\mathbf{v}_r|d_p}{v_f}\,, \tag{2}$$

where ν_f is the molecular viscosity of the fluid and $\tilde{u}_{f,i}$ is the undisturbed fluid velocity at the location of the particle. As shown in (2), the correlation for the drag coefficient from Schiller and Nauman [5] is introduced to extend the Reynolds number range of the drag force.

2.2 LES and DPS of turbulent channel flow

The fully-developed particle-laden turbulent flow between plane, parallel walls is predicted using Large-Eddy Simulation (LES) for the carrier phase and Discrete Particle Simulation for the dispersed phase. The Reynolds number based on the friction velocity u_τ and channel halfwidth δ is $Re_\tau = 180$. The dimensions of the channel are $4\pi\delta$ in the streamwise (x or x_1), $4\pi\delta/3$ in the spanwise (z or x_3), and 2δ in the wall-normal (y or x_2) directions. Periodic boundary conditions are applied to the dependent variables in the streamwise and spanwise dimensions and no-slip boundary conditions to the fluid velocity at the channel walls. The subgrid stress arising from the filtering of the Navier–Stokes equations is closed using an eddy viscosity model.

The equations governing the fluid flow are solved using a fractional step method on a staggered mesh comprised of 64^3 cells. Spatial derivatives are approximated using second-order accurate central differences. The grid spacing is uniform in the x and z directions with the corresponding grid spacings in wall units $\Delta x^+ = 35$ and $\Delta z^+ = 12$, respectively. The wall-normal mesh is clustered near the solid surfaces and stretched away from the wall using a hyperbolic tangent function. The discretized system is advanced in time using an implicit/explicit scheme (Crank–Nicholson and second-order Adams–Bashforth).

The fluid flow is not influenced by momentum exchange with the particles and the (undisturbed) fluid velocity $\tilde{u}_{f,i}$ required in (1) is the value interpolated to the particle position that is computed in the LES, representing the spatially-filtered (volume averaged) solution of the Navier–Stokes equations. The neglect of turbulence modulation is a strong assumption though not incompatible with the present approach of using the simulations to isolate effects.

The influence of subgrid-scale transport on particle motion is not considered, which should be a reasonable assumption given the filtering by particle inertia of the smaller-scale, high-frequency components of the subgrid fluid velocity (e.g., see [10]). However, the neglect of subgrid transport restricts the parameter range of the current calculations to moderate Reynolds numbers for which there is a relatively weak effect of the unresolved motions on the resolved scales. In other regimes such as very small particle response times, the errors introduced by transporting the particulate phase by a filtered fluid velocity should be significant and will require models of the subgrid velocities on particle motion. In boundary layers, subgrid modeling of the fluid velocities viewed by the particles will depend on, among other factors, the distance from the wall and for small Stokes numbers be important to problems of particle deposition and wall collisions.

Statistics are presented for three particle Stokes numbers, $St = \tau_{ps}/(\delta/u_\tau)$, where τ_{ps} is the Stokes relaxation timescale of the particle. For all simulations

Table 1. Particle parameters for turbulent channel flow, $Re_\tau = 180$. The particle diameter for each case is one viscous unit, $d_p^+ = 1$. The Stokes time constant in viscous units, $\tau_{ps}^+ = \tau_{ps} Re_\tau$ where τ_{ps} is the Stokes response time.

ρ_p/ρ_f	527	2106	8424
St	0.1625	0.65	2.60
τ_{ps}^+	29	117	468

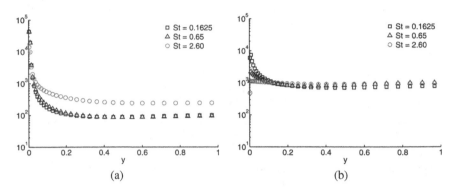

(a)   (b)

Fig. 1. Mean number density for St = 0.1625, St = 0.65, and St = 2.60; (a) without inter-particle collisions, (b) including particle-particle collisions.

the particle diameter was specified as one viscous unit and therefore the variation in the Stokes number is achieved via a variation in the density ratio, as summarized in Table 1. The particle response times are chosen so that the lightest particles (St = 0.1625) follow reasonably well the scales of carrier-phase motion resolved in the LES while the particles with St = 2.60 are the most sluggish in their response to the turbulent fluid velocity fluctuations.

Properties of the dispersed phase are obtained by following the trajectories of 1×10^5 particles, corresponding to an average number density of 950 particles per unit volume, equivalent to a dispersed-phase volume fraction of 8.5×10^{-5}. A particle is assumed to contact the smooth channel walls when its center is one radius from the wall. Elastic rebound is imposed for particles colliding with the wall. For computations that account for inter-particle collisions, binary, elastic collisions are assumed between particles. The algorithm for collision detection is described in [8].

3 Results

3.1 Single-point statistics

Shown in Figure 1 are profiles of the mean number density, cases without inter-particle collisions illustrated in Figure 1a and including particle-particle collisions in Figure 1b. The flows without inter-particle collisions exhibit an accumulation of

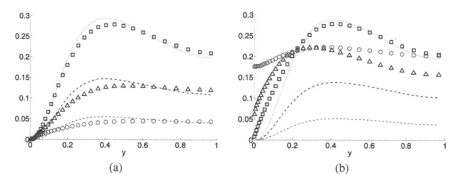

Fig. 2. Wall-normal components of the fluid-particle ($R_{fp,22}$, shown by lines) and particle ($R_{p,22}$, shown by symbols) velocity correlations; (a) without inter-particle collisions, (b) including particle-particle collisions. $----$, ○ St = 2.60; $—\cdot—$, △ St = 0.65; $\cdots\cdots$, □ St = 0.1625.

particles in the near-wall region, a well-known effect observed in many previous computations that do not include inter-particle collisions (e.g., see [4, 6, 7]). Changes in wall-normal ("radial") transport across the channel induced by inter-particle collisions leads to a more uniform number density profile, as shown in Figure 1b. For the smallest Stokes number St = 0.1625, the number density peaks near the wall, as observed in the profile without inter-particle collisions, though is less pronounced compared to the non-colliding case. Increases in the Stokes number lessen the non-uniformity in the distribution and Figure 1b shows that the number density profile is nearly uniform for St = 2.60.

Shown in Figure 2 are profiles of the wall-normal components of the particle kinetic stress $R_{p,22} = \langle v'_{p,2} v'_{p,2} \rangle_p$ where $\langle \cdot \rangle_p$ indicates an average over the dispersed phase. Also plotted are the corresponding components of the fluid-particle velocity correlation, $R_{fp,22} = \langle \tilde{u}'_{f,2} v'_{p,2} \rangle_p$. For the flows without inter-particle collisions depicted in Figure 2a, the particle kinetic stress and fluid-particle correlation are nearly equal. This equivalence in the flows without collisions reflects the fact that the particle velocity fluctuations in the wall-normal direction (and spanwise direction, not shown) are controlled by the drag force and in local equilibrium with the turbulent fluid flow.

The effect of inter-particle collisions is to re-distribute the particle velocity variance amongst the three velocity components. In wall-bounded flows the anisotropy of the particle velocity fluctuations in the near-wall region can be very large, especially in flows without inter-particle collisions (e.g., see [9]). Consequently, substantial effects are observed in Figure 2b in the near-wall region with large increases in the wall-normal kinetic stress, $R_{p,22}$, compared to the simulations that do not include particle-particle collisions. Further, Figure 2b also shows that the equilibrium between the particle fluctuations and fluid-particle correlation that is observed in the flows without inter-particle collisions is disrupted, i.e., $R_{p,22}$ and $R_{fp,22}$ are no longer equal.

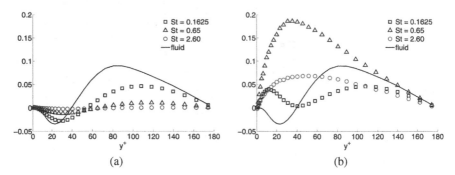

Fig. 3. Turbulent transport of the wall-normal particle velocity variance by the wall-normal particle velocity, $\langle v'_{p,2}v'_{p,2}v'_{p,2}\rangle_p$. The solid line in both frames is the corresponding component of the fluid velocity triple correlation; (a) without inter-particle collisions, (b) including particle-particle collisions.

Figure 3 displays the profiles of the wall-normal turbulent transport of the wall-normal particle velocity variance, $\langle v'_{p,2}v'_{p,2}v'_{p,2}\rangle_p$. The wall-normal gradient of $\langle v'_{p,2}v'_{p,2}v'_{p,2}\rangle_p$ appears in the transport equation for the wall-normal velocity variance and therefore the behavior of the triple correlation provides insight into the changes observed in the particle velocity fluctuations. Figure 3a shows that in the absence of inter-particle collisions, the triple velocity correlations are not large and, consequently, effects of turbulent transport are weak (which is also consistent with the relatively good agreement between $R_{p,22}$ and $R_{fp,22}$ discussed above). Figure 3b shows that in flows which include inter-particle collisions, turbulent transport of the wall-normal velocity variance is substantially increased over that observed in flows without particle-particle collisions. Thus, the elevated levels in the wall-normal particle velocity variance from the near-wall region are transported by the fluctuating particle velocity to the core region of the channel. Similar features were also observed by Caraman et al. [1] in experimental measurements of particle-laden pipe flow.

3.2 Spatial properties of the particle velocity field

The results summarized above highlight aspects of dispersed-phase transport that are efficiently studied using numerical simulation techniques such as Large-Eddy Simulation and Discrete Particle Simulation. The statistics highlighted in the previous sections are single-point measures and are useful for understanding the effects of fluid-particle and particle-particle interactions on dispersed-phase transport. Additional measures for assessing dispersed-properties require information at more than a single point in the flow. Of particular interest to the present effort is the spatial structure of the particle velocity field.

For decreasing Stokes numbers, particle motion follows more closely that of the underlying carrier flow. While that notion is intuitive, also important to recognize

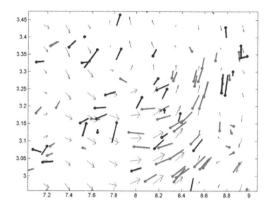

Fig. 4. Fluid and particle velocity vectors in a portion of the centerplane of turbulent channel flow. Fluid velocity vectors shown in blue, particle velocity streaks for St = 0.65 shown in red and for St = 2.60 in black.

is that neighboring particle velocities will be correlated in space through the interactions with the same local fluid flow. This effect is illustrated in Figure 4 which shows fluid and particle velocity vectors for St = 0.65 and St = 2.60 in a portion of the centerplane of the channel flow calculations described previously. For the lower Stokes number the particle velocities follow somewhat closely the fluid vectors and also exhibit, at least visually, a correlation in space. In contrast, for larger-inertia particles, neighboring particle velocities should become uncorrelated since these particles maintain stronger connection (memory) to their interactions with very distant, and independent, turbulent eddies. This effect is also somewhat apparent for St = 2.60 for which Figure 4 shows that the velocities of neighboring particles appear less correlated spatially than for St = 0.65 and the fluid. Of particular interest is the impact of the structural features indicated by Figure 4 on the spatial correlation of the particle velocity field.

Spatial correlations of the streamwise particle velocity with separations in the streamwise and spanwise directions are shown in Figures 5a and 5b, respectively, for St = 0.65. The correlations are in the plane $y^+ = 90$ and from simulations without inter-particle collisions. The correlations shown in Figure 5 are normalized (to unity) by the corresponding single-particle velocity variance in the plane.

The general behavior of the spatial correlations is similar to the simulations of particle-laden isotropic turbulence reported by Fevrier et al. [2] and measurements in turbulent channel flow reported by Khalitov and Longmire [3]. Figure 5 shows that the correlations for the lightest particles, St = 0.1625, are the closest to those of the turbulent fluid flow, consistent with the low-inertia particles being the most responsive to the local fluid velocity. For larger Stokes number the spatial correlations exhibit greater departures from those of the carrier phase. As also observed by Fevrier et al. [2] and Khalitov and Longmire [3], the correlations are discontinuous at the origin. Fevrier et al. [2] have shown that this feature is consistent with a partitioning of the velocity by particle inertia into a component that is spatially-correlated and

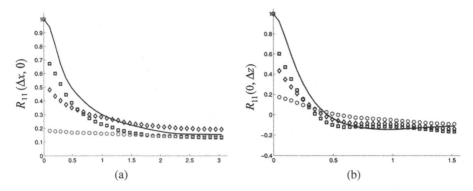

Fig. 5. Streamwise (in (a)) and spanwise (in (b)) correlation functions of the streamwise velocity in the plane $y^+ = 90$. Normalization is by the single-particle streamwise velocity variance in the corresponding plane. Correlations shown from simulations without inter-particle collisions. ——— fluid; □ St $= 0.1625$; ◇ St $= 0.65$; ○ St $= 2.60$.

another component that is random and uncorrelated in space. The spatial correlations show that the relative contribution of the random-uncorrelated velocity accounts for a larger fraction of the particle fluctuating motion for increasing Stokes number.

4 Summary and perspectives

The present computations show that there can be a substantial effect of inter-particle collisions on dispersed phase transport. The simulations show that re-distribution of the particle kinetic energy by particle-particle collisions substantially increases wall-normal ("radial") transport. With increases in the Stokes number, particle collisions have an increasingly significant effect with wall-normal profiles of the mean number density becoming more uniform across the channel. In the calculations that include colliding particles, the equilibrium between the particle-particle and fluid-particle velocity correlations is disrupted, with increases in the wall-normal particle fluctuating velocities compared to the cases without colliding particles. In general, the agreement (or lack of) between the particle velocity variance and corresponding component of the fluid-particle correlation is a useful diagnostic that could be used in many simulations of gas-solid flows to assess the characteristics of the particle fluctuating motion and its equilibrium with the local turbulent fluid flow.

That inter-particle collisions can cause substantial changes in dispersed-phase statistics even in relatively dilute regimes is consistent with the increase of the particle concentration near the walls of the channel and the very strong anisotropy of the particle velocity fluctuations in the near-wall region. The relatively larger concentration of particles near the wall ensures higher collision rates in the region where the redistribution effect of inter-particle collisions will be strongest. Effects of two-way coupling, neglected in the present study, might be expected to damp the smaller scales of the fluid motion, which would result in an effectively larger particle Stokes number. If true, then effects of inter-particle collisions might become even more pronounced.

Simulation techniques such as LES/DPS offer a useful tool that complements modeling strategies that solve an averaged set of equations, e.g., as currently employed in most practical engineering applications, and fully-resolved simulations of the flow around a few particles. Fully-resolved simulations can be, and currently are by a few groups, used to study specific aspects of the details of fluid-particle and particle-particle interactions. Such simulations should benefit LES/DPS approaches by improving the empirical input inherent in such computations. Two examples are the parameterization of particle dynamics, i.e., the forces acting on a particle, and subgrid modeling treatments in regimes that include effects of turbulence modulation. Formulation of more accurate force models for quantities such as drag, particle rotation, shear-induced lift, etc., for use in LES/DPS approaches could conceivably be proposed based on results from fully-resolved computations. Two-way coupling requires accurate subgrid models that are capable of taking into account the distortion of the fluid flow by the particles. Fully-resolved calculations of particle-laden flows would yield a database that could be used to test modeling hypotheses.

A wider range of applications for LES/DPS methods will also be possible with further improvements in subgrid models for the fluid. At the relatively low Reynolds numbers considered these modeling errors are not significant. In other regimes, e.g., substantially higher Reynolds numbers, subgrid modeling of wall-bounded flows remains one of the principle problems in LES. Improvements to subgrid modeling for fluid flow prediction will obviously enhance the accuracy of two-phase flow predictions using LES/DPS methods.

The correlations of the particle velocity fields in channel flow are consistent with the particle velocity being comprised of a spatially correlated contribution and a random component that is not correlated spatially. As discussed in [2], an important consequence is that in the large-inertia limit, the particle velocity distribution cannot be assumed to correspond to a spatially-continuous velocity field. The implications of these findings, as well as those reported in [2] and [3], also impact Eulerian-based prediction of dispersed two-phase flows using techniques that attempt to resolve the spatially-variable and time-dependent motions of the particulate phase. Such simulation strategies should recognize that only the spatially-correlated motions of the particle velocity field are computed from a set of field equations, given the fact that the random-uncorrelated particle velocity component is not differentiable. It is important, therefore, to account in Eulerian-based prediction for the influence of the random component of the particle velocity on the correlated motions. The equations governing the correlated part of the particle velocity are developed in [2].

References

1. Caraman, N., Borée, J. and Simonin, O., 2003, *Physics of Fluids* **15**, 3602–3612.
2. Fevrier, P., Simonin, O. and Squires, K.D., 2005, *Journal of Fluid Mechanics* **533**, 1–46.
3. Khalitov, D.A. and Longmire, E.K., 2003, Effect of particle size on velocity correlations in turbulent channel flow, FEDSM03-45730.

4. Rouson, D.W.I. and Eaton, J.K., 2001, *Journal of Fluid Mechanics* **424**, 149–169.
5. Schiller, L. and Nauman, A., 1935, *V.D.I. Zeitung* **77**, 318–320.
6. McLaughlin, J.B., 1989, *Physics of Fluids* **1**, 1211–1224.
7. Marchioli, C. and Soldati, A., 2002, *Journal of Fluid Mechanics* **468**, 283–315.
8. Vance, M.W. and Squires, K.D., 2002, An approach to parallel computing in an Eulerian–Lagrangian two-phase flow model, FEDSM 2002-31225.
9. Wang, Q. and Squires, K.D., 1996, *Physics of Fluids* **8**, 1207–1223.
10. Yamamoto, Y., Potthoff, M., Tanaka, T. and Tsuji, Y., 2001, *Journal of Fluid Mechanics* **442**, 303–334.

Simulation of Particle Diffusion, Segregation, and Intermittency in Turbulent Flows

Michael W. Reeks

School of Mechanical & Systems Engineering, The University of Newcastle upon Tyne, UK;
e-mail: mike.reeks@ncl.ac.uk

Key words: particles, turbulent structures, segregation, kinematic simulation

1 Introduction

It is by now well known that turbulence, contrary to traditionally held views, can demix a suspension of particles, segregating the particles into regions of high strain rate. The process depends upon the ratio of the particle response time to the timescale of the turbulent structures in the flow (i.e. the Stokes number), the maximum segregation occurring when this ratio ~ 1. There have been numerous simulations and PIV based measurements that have demonstrated this phenomenon in various types of flows from simple homogeneous turbulent flow to turbulent boundary layers [1]. In this regard, there have been four major areas of study where segregation and demixing have played a vital role: (a) the deposition of particles in turbulent boundary layers [2]; (b) droplet/particle coalescence in clouds which controls droplet size and distribution and eventual rain fall [3]; (c) particle agglomeration in turbulent boundary layers which leads to the formation of ropes; (d) two-way coupling between the dispersed and continuous phases and in particular the generation and dissipation of turbulence [4].

The work described here is part of a long term program on the formulation of a PDF approach for modelling dispersed flows [5] that will take account of these important effects of turbulent structures in a statistical manner. So for instance in a continuum description of the dispersed phase, we would be addressing the way particle interactions with turbulent structures influence the constitutive relations of the dispersed phase, two-way coupling between phases and agglomeration where the effect of of segregation is the most influential. The particular focus of the current work is on a PDF formulation for two particle dispersion and follows on from early work in developing a PDF approach for single particle transport [7]. Of particular importance in the statistical formulation of this approach was the compressibility of the particle flows along a particle trajectory i.e. the local divergence of the underlying

S. Balachandar and A. Prosperetti (eds), Proceedings of the IUTAM Symposium on Computational Multiphase Flow, 21–30.

particle velocity field. It has already been shown by Maxey [6] that the enhancement of gravitational settling Δv_s in homogeneous stationary turbulence due to the biasing of the particle motion in the direction of gravity is formally given by

$$\Delta v_s = -\int_0^t \langle \mathbf{u}'(\mathbf{x}, t) \nabla \cdot \mathbf{v}(\mathbf{x}, t \mid s) \rangle ds, \tag{1}$$

where $\mathbf{u}'(\mathbf{x}, t)$ is the carrier flow fluctuating velocity and $\nabla \cdot \mathbf{v}(\mathbf{x}, t \mid s)$ is the divergence of the particle velocity field $\mathbf{v}(\mathbf{x}, t)$ along a particle trajectory at time s which passes through $\mathbf{x}$ at time t. The biasing of the particle trajectory is reflected in the value of the compressibility though whether it represents a negative contribution or $+ve$ contributions depends upon the *preferential sweeping* of the trajectories themselves. We note that the result does not explicitly contain gravity and is a formal result that applies to turbulence in general whether gravity is present or not. So it has a finite value in inhomogeneous turbulence especially in near wall turbulence contributing to the net drift velocity of particles towards the wall and at equilibrium to the build up of concentration. It also contributes to the particle diffusion coefficient so that in strictly inhomogeneous flows the formula is different from the well-known Taylor formulation for homogeneous stationary turbulence.

The basis of this paper is to highlight the role played by simulation in the validation of statistical models as e.g. PDF approach. In particular, I will illustrate by way of example, the value of simple kinematic simulations (with structures) in contrast to the the more familiar role that DNS has played. Whilst DNS has played and continues to play an important role in identifying turbulent structures and linking them with particular mechanisms, the drawback is that whilst DNS represents low Reynolds number real turbulence especially near wall turbulence, we have no real control over the type of structure present in the flow, neither its density or scale: in fact we have to identify the structures rather than control them. There is a clear need for kinematic simulations of turbulence which can be easily implemented, containing structures which we can control and in turn use to replicate the influence of real structures and scales of those in real turbulence. Such simple random flows can be used to identify single effects which might otherwise go unnoticed in more complex flows, their influence masked by other scales and other effects.

Such an example of a simple flow is that of a random array of counter rotating vortices which I have used to examine and quantify the demixing process and which in turn has revealed some important features that may well be present in real turbulent flows [8]. In particular I will examine here the statistics of the compressibility or more precisely $\ln(J(t))$ where $J(t)$ measures the fractional change in an elemental volume of particles along a particle trajectory at time t. $\ln[J(t)]$ is related to the compressibility $\nabla \cdot \mathbf{v}(t)$ along a particle trajectory by the equation

$$d \ln[J(t)]/dt = -\nabla \cdot \mathbf{v}(t). \tag{2}$$

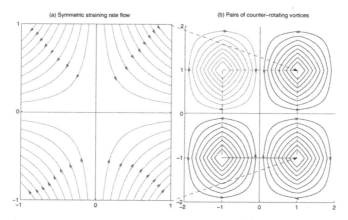

Fig. 1. Generation of instantaneous flow field: horizontal and vertical axes are scaled by a factor L.

2 Diffusion in a simple homogeneous turbulent flow

2.1 Description of random flow field

We consider dispersion in a simple ihomogeneous turbulent flow field composed of pairs of counter rotating vortices which are periodic in both the x-, y-directions with the same periodicity. Each lattice cell (the basic periodic element) contains a pair of counter-rotating vortices in both the x, y orthogonal directions and is constructed from a linear symmetric straining flow field in the manner shown in Figure 1.

So starting from an initial symmetric straining flow pattern of width $2L$ in both the x, y-directions (see Figure 1(a)), this pattern is repeated *front to back* in both the x, y-directions with a strain rate S drawn from a uniform distribution $[0, S_0]$. We note that each quadrant of this straining rate pattern in Figure 1(a) is a quadrant of one of the two pairs of counter-rotating vortices formed within the lattice cell in Figure 1(b). As shown in Figure 2(a), the flow velocity u_x in the x-direction has a linear saw tooth profile $U(x)$, with a slope of constant magnitude S but with a change in sign across the y-centre line of a vortex (the line running in the y-direction passing through the centre of the vortex) where the maximum and minimum values $\pm SL/2$ of $U(x)$ are located: across the x-centre line, u_x changes to $-U(x)$ as shown in Figure 2(b), consistent with the change in direction of the streamlines shown in Figure 1(a). The flow velocity u_y in the y-direction at (x, y) is $-U(y)$ to preserve continuity of flow through out. This cellular flow pattern of counter-rotating vortices so formed, persists for a fixed life-time selected from an exponential distribution with a decay time of S_0^{-1}, at the end of which time, a fresh flow field is generated with new values of the life-time and S and the origin of the pattern at the same time shifted by random displacements in both the x and y-directions, drawn independently from a uniform distribution $[0, 2L]$. This makes the average flow homogeneous with zero mean in the x- and y-directions. The important feature of this randomized flow field is that the equations of motion of an individual particle in both the x, y-directions are linear

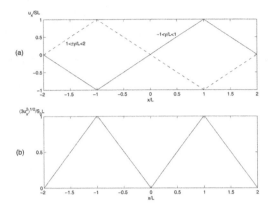

Fig. 2. (a) Flow in x-direction within vortex; (b) rms velocity in x-direction.

and independent of one another (other than through the maximum length of time a particle can experience a particular value of the straining of the flow in either the x- or y-directions before it changes sign). With respect to the centre (stagnation point) of a symmetric straining flow pattern (see Figure 1(a)), the flow velocity within that flow region is given by

$$u_x = +Sx; \quad u_y = -Sy \quad (-L \leq x \leq L, \ -L \leq y \leq L). \tag{3}$$

The flow field so generated turns out to be homogeneous and stationary but not isotropic. It has the interesting property that the Lagrangian fluid point rms velocity (along its trajectory) is different from its value at a fixed point (Eulerian).

2.2 Solutions of particle equations of motion

Based on Stokes drag, the particle equation of motion is

$$\ddot{x}_i + \tau_p^{-1} \dot{x}_i + (-1)^{i+1} \tau_p^{-1} S x_i = 0 \tag{4}$$
$$-L \leq x_i \leq L \ (i = 1, 2)(x_1 = x, x_2 = y),$$

where x_i is measured from a stagnation point. For convenience we normalize x_i on L and express the particle response time τ_p and strain rate S in units of S_0^{-1} so in this case S is drawn from a uniform distribution $\wp [0, 1]$. We note that when the strain rate in the x, y-directions the x, y equations are those of a damped simple harmonic oscillator and as a consequence there are two types of motion, namely heavily damped for $\tau_p S < 0.25$ and lightly damped for $\tau_p S > 0.25$. In this case τ_p corresponds to the particle Stokes number. This has some bearing on the motion of the particle within these vortices. For heavily damped motion, a particle remains trapped within a vortex but approaches the extremity (the stagnation region) in a manner which decreases exponentially with time. On the other hand for lightly damped motion the particle can escape the vortex or *overshoot* into an adjacent vortex, though

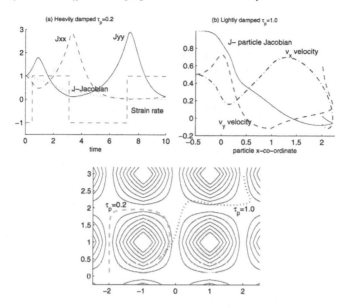

Fig. 3. Particle trajectories in counter-rotating vortices: (a) heavily damped $\tau_p = 0.1$; (b) lightly damped $\tau_p = 1$.

in time it ends up in the stagnation region between the vortices. Both these types of behaviour are illustrated in Figures 3(a) and (b) for particle response times $\tau_p = 0.1$ (heavily damped) and $\tau_p = 1$ (lightly damped). Figure 3 also shows the corresponding values of components J_{11} and J_{22} of the the unit deformation tensor $\mathbf{J}$ given by $\partial \mathbf{x}(\mathbf{x_0},t)/\partial \mathbf{x_0}$ where $\mathbf{x_0}$ is the position of the particle at some initial time say $t = 0$. In this case we have set $J_{ij}(0) = \delta_{ij}$. The equations of motion for J_{ij} are obtained from the particle equations of motion by partial differentiation wrt $x_{0,i}$ giving

$$\ddot{J}_{ij} + \tau_p^{-1}\dot{J}_{ij} + (-1)^{i+1}\tau_p^{-1}SJ_{ij} = 0 \tag{5}$$

for which we choose the initial conditions $\dot{J}_{ij} = \partial u_i/\partial x_j = (-1)^{i+1}\tau_p^{-1}S\delta_{ij}$. This means that together with the initial conditions on J_{ij} that $J_{ij}(t) = 0$ for $i \neq j$ and that

$$J(t) = |\mathbf{J}| = J_{11}J_{22}. \tag{6}$$

Note that in this linear system the equation of motion for $J(t)$ is the same as that for x_i and that the components of $\mathbf{J}$ are dependent on the position of the particle only through the time at which the strain rate experienced by the particle changes sign. Figures 3(a) and (b) show the corresponding values of $J(t)$ along the heavily damped and lightly damped (b) trajectories. In both cases the value of J approaches zero as $t \to \infty$. However in the lightly damped case $J(t)$ passes through zero at intermediate times as the particle oscillates backwards and forwards across a stagnation line. In so doing the value of J oscillates from $+ve$ to $-ve$, with the corresponding elemental volume rotating through $180°$ as it passes through zero volume. Each time

$J(t)$ passes through zero, the corresponding particle concentration becomes infinite instantaneously. This raises the possibility that such events may occur in real turbulent flows and that the process of particle dispersion could be a highly intermittent process associated with large deviations in the particle concentrations.

2.3 Concentration and particle diffusion coefficients

Figure 4 shows the dispersion of particles released from a single point into the flow. In particular Figures 4(a)–(c) show the positions of 10^5 particles after 100 time steps from their release into a single realisation of the flow (a time step corresponding to the lifetime of each flow pattern). Each particle was released from a single point with a Gaussian distribution of velocities with an rms the same as the long term value for the particles. The cases in Figures 4(a)–(c) refer to particles with normalized response times (Stokes numbers) of $\tau_p = 0.1$, 1.0 and 10 respectively. All three cases show a segregation of particles being most marked for the case of $\tau_p = 1$. The role of particle response time here determines the degree to which particles can segregate into the stagnation regions of the flow pattern as the pattern and hence stagnation regions both shift in position from one time step to the next. This is most effective for the case of $\tau_p = 1$, but in all cases even with $\tau_p = 1$ the segregation does not align with the location of the stagnation lines at any instant of time. Even so the pattern of segregation gets stringier as time progresses, the rate at which that happens being greatest for τ_p 1 ($\tau_p = 1$ for the three cases considered here).

In contrast Figure 4(d) shows the positions together with the corresponding concentration contours superimposed for a particle with $\tau_p = 1$ released with the same distribution of velocities as in the previous cases from the same position but in 10^5

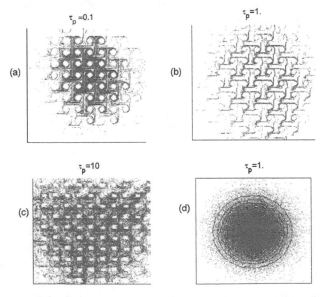

Fig. 4. Dispersion of particles in random flow field: (a) $\tau_p = 0.1$; (b) $\tau_p = 1$;(c)$\tau_p = 10$.

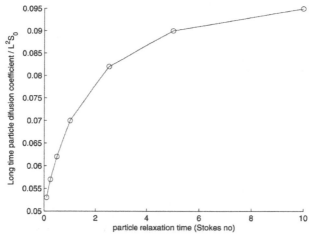

Fig. 5. Long-time particle diffusion coefficients.

separate realisations of the flow. In this case there is no evidence of segregation – the result being the superposition of all the concentration patterns in Figure 4(b) uniformally shifted over a lattice spacing.

Figure 5 shows the values of the particle diffusion coefficient as a function of time based on the displacements for the case in Figure 4(d). The long term values exhibit no intermediate peaking of the particle diffusion coefficient as with the particle segregation: the particle diffusion coefficient increases with increasing particle response time reaching an asymptotic limit for very large response times which is greater than that of the fluid point (passive scalar). This latter result is consistent with an early analysis [11]and those of the simulation of particles in DNS homogeneous isotropic turbulence [12].

2.4 Statistics of the compressibility and deformation

In contrast to the values for the long term particle diffusion coefficient, Figure 6 shows the corresponding values of the quantity $\frac{1}{2}t^{-1}\langle \ln |J(t)| \rangle$ for the same particle response time as in Figure 4. This quantity may be interpreted as the net time averaged normal strain rate associated with each axis or more succinctly the net Liapounov exponent of the particle flow. In all three cases of the particle response time the values of these exponents reach a constant negative value consistent with a steady average value of the compressibility in the long term (after many particle relaxation times and flow integral time scales) (in the case of $\tau_p = 10$ it is a very small value not distinguishable from zero in Figure 6). However the most important feature is that the maximum value of the compressibility in these three cases is for $\tau_p = 1$ consistent with particle segregation behaviour shown in Figures 4(a)–(c). The implication of the constant value of the compressibility is that in this simple flow field (in which the particles are entirely transported by the underlying flow velocity) the accumulation of particle would continue indefinitely: this of course ignores

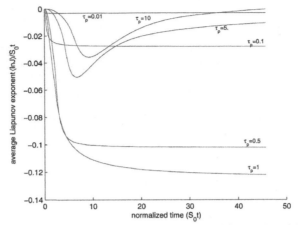

Fig. 6. Net compression (Liapunov exponent) as a function of time and particle response time.

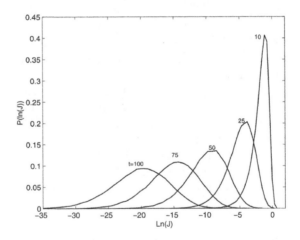

Fig. 7. Evolution of the distribution of $\ln J(t)$ for $\tau_p = 1$.

the absence of molecular or Brownian diffusion but nevertheless suggests that the concentration will be extremely high before Brownian diffusion can counteract the accumulation process.

Figure 7 shows the distribution of $\ln |J(t)|$ for $\tau_p = 1$ and how it evolves with time. For large times the distribution is very close to Gaussian with a mean which is $-ve$ and $\propto$ time as is also the a variance

$$\sigma^2 = \left\langle \left(\ln |J| - \langle \ln |J| \rangle \right)^2 \right\rangle = 2Dt, \tag{7}$$

so that the dispersion of $\ln |J|$ may be accurately described by a Gaussian convection diffusion process

$$\frac{\partial P}{\partial t} + c \frac{\partial P}{\partial \phi} = D \frac{\partial^2 P}{\partial \phi^2}, \tag{8}$$

where $\phi = \ln|J|$ and $P(\phi)$ the pdf for the occurrence of ϕ and $c = (d/dt)\ln J$.

3 Summary and conclusions

The way particles interact with turbulent structures, particularly in regions of high vorticity and strain rate, has been simulated in a simple incompressible random flow of counter rotating vortices which has a periodic persistent structure for a fixed time interval period of time but is randomly shifted from one time interval to the next. A particular focus of the simulation has been the statistical properties of $\ln|J(t)|$ along a particle trajectory at time t, where $J(t)$ is the fraction value of an elemental volume of particles compared to its initial value. The compressibility of the the particle velocity field $\nabla \cdot \mathbf{v}(t)$ is related to $\ln|J(t)|$ by Equation (2). Depending on the particle response time, the particle motion is identical to that of a lightly or heavily damped harmonic oscillator. For a lightly damped flow the particles overshoot the stagnation lines and pass into the neighbouring vortices. In so doing the value $J(t)$ passes through zero and the concentration instantaneously becomes infinite. The possibility of this occurring in real turbulent flows and with it occurrence of intermittency in the fluctuations of the particle concentration was mentioned. Particles released into a single realisation of the flow field segregated without alignment with the instantaneous flow pattern, the maximum segregation occurring for particle response time (Stokes number) ~ 1. Noticeably the segregation progresses without reaching any state of equilibrium and would continue indefinitely for all particles response times, the particle response time being responsible for the rate at which the segregation occurs. This pattern of behaviour was also revealed in the plot of $\frac{1}{2}t^{-1}\langle \ln|J(t)|\rangle$ versus time t which approaches a constant $-ve$ value as $t \to \infty$, corresponding to a compression rather than a dilation. The statistics of $\ln J(t)$ revealed a Gaussian convection diffusion process in the limit of $t \to \infty$. for the particle response times of $\tau_p = 0.1, 1, 10$ considered.

It is clear that the type of flow field considered both in the Eulerian and Lagrangian simulations exhibits the maximum degree of segregation, compressibility intermittency. Ways of controlling the compressibility would be to introduce free vortices into the flow which on their own would introduce a dilatation and so reduce the compressibility/segregation: regions of isolated strain rate would have the reverse effect. The possibility exists of creating a more turbulent like flow filed by introducing admixtures of counter-rotating vortices, free strain regions and and vortices.

The obvious next step in this program of work is to consider whether such features exist in real turbulent flows.

References

1. Fessler, J.R., Kulick, J.D. and Eaton, J.K., 1994, Preferential concentration of heavy particles in a turbulent channel flow, *Phys. Fluids* **6**(11), 3742–3749.

2. Marchioli, C. and Soldati, A., 2002, Mechanisms for particle transfer and segregation in turbulent boundary layers, *J. Fluid Mech.* **468**, 283–315.
3. Pinsky, M.B. and Khain, A.P., 1997, Turbulence effects on droplet growth and size distribution in clouds – A review, *J. Aerosol Sci.* **28**, 1117–1214.
4. Ahmed, A.M. and Elghobashi, S., 2000, On the mechanisms of modifying the structure of turbulent homogeneous shear flows by dispersed particles, *Phys. Fluids* **12**(11), 2906–2930.
5. Reeks, M.W., 1993, On the continuum equations for dispersed particles in nonuniform flows, *Phys. Fluids A* **5**(3), 750–761.
6. Maxey, M.R., 1987, The gravitational settling of aerosol particles in homogeneous turbulence and random flow fields, *J. Fluid Mech.* **174**, 441–465.
7. Reeks, M.W., 2001, Particle drift in turbulent flows, the influence of local structure and inhomogeneity, Paper 187, *Proc. of the ICMF-2001, International Conference on Multiphase Flows*, New Orleans, Louisiana, May 27–June 1, 2001.
8. Reeks, M.W., 2002, Particle interaction with turbulent structures, the application of random walk models, in *10th Workshop on Two-Phase Flow Predictions*, M. Sommerfeld (ed.), Merseberg, April 9–12, 2002.
9. Février, P., Simonin, O. and Legendre, D., 2004, Particle dispersion and preferential concentration dependence on turbulent, in *International Conference on Multiphase Flows*, Yokohama, Japan, June 2004, Paper No. 797, pp. 1–12.
10. Davila, J. and Hunt, J.C.R., 2001, Settling of particles near vortices and in turbulence, *J. Fluid Mech.* **440**, 117–145.
11. Reeks, M.W., 1977, On the dispersion of small particles in an isotropic turbulent flow, *J. Fluid Mech.* **83**, 529–546.
12. Squires, K.D. and Eaton, J.K., 1992, Preferential concentration of particles by turbulence, *Phys. Fluids A* **3**, 1169–1178.

Use of a Stochastic Method to Describe Dispersion and Deposition in an Idealized Annular Flow

Thomas J. Hanratty and Yoichi Mito

University of Illinois, Urbana, USA; e-mail: hanratty@scs.uiuc.edu, ymito@uiuc.edu

Abstract. A stochastic representation of fluid turbulence has been developed to study the behavior of very dilute suspensions of solid spheres in a turbulent flow. Particular emphasis is given to the understanding of deposition in an idealized annular pattern. The accuracy of the stochastic method was checked by comparing with calculations done in a DNS at $Re_\tau = 150$. The striking aspect of the study is that calculations of the dimensionless deposition constant in a horizontal channel are presented for $V_T^+ = 0$ to 3.0, $Re_\tau = 590$ and $\tau_p^+ = 0$ to 20,000. For some runs it was necessary to use computation times of $t^+ = 2 \times 10^8$ in order to insure that a fully-developed condition was realized. Such an extensive study would not be possible if the turbulence was represented by a DNS.

1 Description of stochastic method

Considerable progress has been made in understanding particle dispersion and deposition by studying the behavior of spherical particles in a DNS of turbulent flow. However, this approach has limitations in that it is impractical to use in studying a wide range of variables, large Reynolds numbers and large times. This prompted our examination of a stochastic method which uses a modified Langevin equation to represent the fluid turbulence seen by a particle. This paper evaluates this approach and demonstrates its usefulness by considering an idealized model of the dispersed flow that exists in an annular pattern. Large density ratios, ρ_p/ρ_f, are assumed. Lift forces are neglected. The suspension is considered to be dilute enough that particle-particle interactions and feedback can be ignored.

A point source of solid spheres is described by the equations

$$\frac{dx_i}{dt} = V_i, \tag{1}$$

$$\frac{dV_i}{dt} = -\frac{3\rho_f C_D}{4d_p\rho_p}|V - U|(V_i - U_i) + g_i, \tag{2}$$

where U_i is a component of the fluid velocity and V_i is a component of the particle velocity and C_D is the drag coefficient. In all of the calculations a fully-developed

S. Balachandar and A. Prosperetti (eds), Proceedings of the IUTAM Symposium on Computational Multiphase Flow, 31–38.

flow in a channel is assumed so that $\overline{U}_1(x_2)$ is the mean velocity profile of the fluid, where x_2 is the coordinate perpendicular to the wall, x_3 is in the spanwise direction and x_1 is in the flow direction. A modified Langevin equation is used to represent the fluid velocity fluctuations

$$d\left(\frac{u_i}{\sigma_i}\right) = -\frac{u_i}{\sigma_i \tau_i}dt + \overline{d\mu_i} + d\mu_i',$$ (3)

where $\sigma_i(x_2)$ is the Eulerian root-mean-square of the velocity fluctuations and $\tau_i(x_2)$ is a Lagrangian time constant. The forcing function is assumed to be Gaussian [5]. The mean drift and variances are given as

$$\overline{d\mu_i} = \frac{\partial\left(\dfrac{\overline{u_2 u_i}}{\sigma_i}\right)}{\partial x_2}dt,$$ (4)

$$\overline{d\mu_i' d\mu_j'} = \frac{\overline{u_i u_j}}{\sigma_i \sigma_j}\left(\frac{1}{\tau_i} + \frac{1}{\tau_j}\right)dt.$$ (5)

The mean velocity profile of the fluid, $\overline{U}_1(x_2)$, and all of the parameters in Equations (1–4) except τ_i are given from Eulerian measurements in a DNS.

For the case of a point source of fluid particles

$$\frac{dx_i}{dt} = U_i$$ (6)

with u_i given by Equation (3). By considering a uniform distribution of instantaneous point sources of fluid particles, it can be shown that Equations (3–5) satisfy the well-mixedness condition defined by Thomson [9]. Calculations [5] for the Lagrangian correlation coefficients of point sources of fluid particles originating at different x_2 in a DNS at $Re_\tau = 150, 300$ were used to specify $\tau_i(x_2)$. Equations (3–5) do not capture the small scale turbulence in that they give Lagrangian correlation coefficients that vary as $\exp(-t/\tau_i)$. Therefore, the model is accurate in representing the correlation only in an integral sense. Comparisons of the model with DNS studies of the dispersion of fluid particles originating from point sources in the fluid yielded good, but not exact, agreement. The model was tested further by carrying out calculations of the dispersion of heat markers from wall sources [6]. This was done by adding random displacements to Equation (6) in order to capture the effects of molecular diffusion. Temperature profiles and heat transfer coefficients agree with Eulerian DNS calculations and with calculations in a DNS which uses Lagrangian methods.

Original plans about using Equations (3–5) to represent fluid turbulence seen by a dispersing particle was to adjust the time constants in the Langevin equation to take account of the inability of the particles to follow the fluid turbulence exactly. This was explored by comparing calculations with studies of the behavior of point sources of solid spheres in a channel flow [4, 7]. It was found that the time scales determined from studies of dispersion of fluid particles could be used for τ_2 and τ_3.

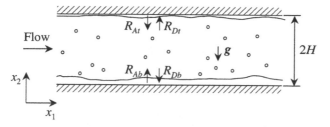

Fig. 1. Idealized annular flow.

The time scale τ_1 for fluid particles was too large by a factor of about 1.5. However, calculations with the stochastic model were found to be insensitive to the choice of τ_1 so no systematic attempts were made to adjust it.

2 Idealized "annular flow"

In the annular regime, observed for gas-liquid flow in a pipe, part of the liquid flow along the wall as an annular film and part as drops entrained in gas flows. Drops are injected into the gas flow by unstable waves on the film. Under fully-developed conditions, atomization is balanced by the deposition of drops entrained in the turbulent gas [10]. The idealized model of an annular flow which we have explored considers a horizontal channel for which the bottom and top walls are arrays of sources (see Figure 1). Drops are represented by solid spherical particles. The atomization process is modeled by injecting the particles from $x_2 = d_p/2$ with a velocity of $(15, 1, 0)$ v^* and a rate per unit area of R_{Ab} and from $x_2 = 2H - d_p/2$ with a velocity of $(15, -1, 0)$ v^* and a rate per unit area of R_{At}, where v^* is the friction velocity and H is the half-height of the channel. The particles are removed from the field when they hit a wall. The ratio of the strengths of the sources at the two boundaries is adjusted so that at long times a fully-developed condition is reached for which the rates of atomization and deposition are equal and the net flux in the x_2-direction is zero [1, 7]. In the calculation of a fully-developed concentration field the sources are assumed to be uniformly distributed on the wall.

The theoretical problem is to represent the behavior of a single instantaneous source. Particles from these sources eventually deposit on the two walls, so that at large enough times none remain in the field. Under fully-developed conditions the concentration at a given x_2 has contributions from sources that have been in the field for different lengths of time. The main contribution of this work was to reveal physical mechanisms which control deposition and mixing.

3 Concentration profiles

Figure 2 presents concentration profiles resulting from an instantaneous source located on the bottom wall. The + superscript indicates that a quantity has been made di-

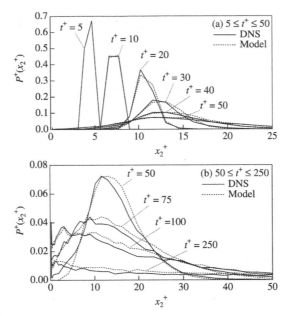

Fig. 2. Distribution of particles from a source located on the bottom wall ($\tau_p^+ = 20$, $V_T^+ = 0.11$, $d_p^+ = 0.368$, $Re_\tau = 150$).

mensionless using the friction velocity and the kinematic viscosity. Particles injected into the flow at $t^+ = 0$ move away from the wall. Eventually they mix with the fluid turbulence and are dispersed in both the plus and minus x_2-directions. Both gravity and turbulence bring particles back to the bottom wall where they are removed from the field. The solid curves and dashed curves, respectively, represent calculations in which the fluid turbulence is represented by a DNS and by a stochastic model. Good agreement is noted. The calculations in Figure 2 were done for $\tau_p^+ = 20$, a dimensionless free-fall velocity of $V_T^+ = 0.11$, a particle size of $d_p^+ = 0.368$ and a Reynolds number of $Re_\tau = 150$.

Fully-developed concentration profiles, for which the rate of injection equals the rate of deposition, such as shown in Figure 3, were calculated with the stochastic method for $Re_\tau = 590$ by adding the contributions from a number of sources. The influence of gravity on the symmetry is clearly seen.

Particles deposit on the bottom wall with a range of velocities, V_d^+. The distribution functions calculated for the case of $V_T^+ = 0$ are shown in Figure 4. The solid line represents the average of the fluid velocity fluctuations at a location of one particle radius from the wall. The curves for $\tau_p^+ \leq 3$ represent situations for which the particles accumulate close to the wall and eventually deposit in response to local fluid velocity fluctuations. For $\tau_p^+ \geq 10$ deposition is governed by particle velocities characteristic of particle (or fluid) velocity fluctuations outside the viscous wall

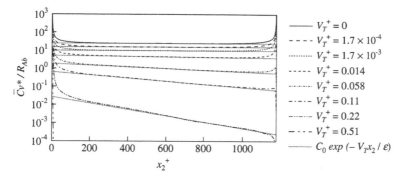

Fig. 3. Concentration profiles for $\tau_p^+ = 10$.

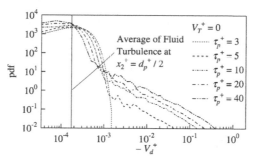

Fig. 4. Probability density functions of the depositing particles for $V_T^+ = 0$.

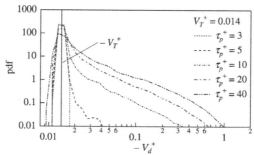

Fig. 5. Probability density functions of the depositing particles for $V_T^+ = 0.014$.

layer. The picture emerges that particles become detached from the fluid turbulence and move in free-flight to the wall.

Figure 5 shows the distribution functions for $V_T^+ = 0.014$. A maximum is noted at $V_d^+ \cong V_T^+$. This suggests that particles accumulating near the wall are carried to the wall by gravitational settling. Thus the rate of deposition can be enhanced for very small gravitational effects.

A deposition constant can be defined as

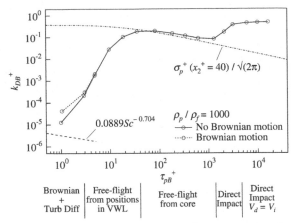

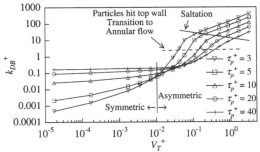

Fig. 6. Deposition constants defined with the bulk concentrations for $V_T^+ = 0$.

Fig. 7. Deposition constants defined with the bulk concentrations for $V_T^+ \neq 0$.

$$R_D = \frac{R_{Db} + R_{Dt}}{2} = k_{DB}C_B, \qquad (7)$$

where C_B is the average concentration over the channel cross section. Calculations of k_{DB} for $g_i = 0$ are given in Figure 6. These are for $V_T^+ = 0$, $\rho_p/\rho_f = 1000$ and are representative of what would be found for a vertical flow. The calculations shown in Figure 6 represent a range of $\tau_p^+ = 1$–20,000. They were done for diffusion times as large as $t^+ = 2 \times 10^8$. The calculations demonstrate the transition amongst four mechanisms for deposition: Brownian motion, turbulent diffusion, free-flight, and unidirectional trajectories.

Figure 7 shows the influence of the g_2 component of gravity. A range of terminal velocities of $V_T^+ = 0$–3.2 are covered for $\tau_p^+ = 3, 5, 10, 20, 40$. The enhancement of deposition with increasing V_T^+ is clearly seen. This occurs both because of an increase in the deposition velocity at the bottom wall and because of the development of asymmetries in the concentration profiles. Transitions to an asymmetric configuration ($V_T^+ = 0.01$), to an annular pattern for which particles reach the top wall ($g^+ = 0.012$) and to a saltation regime are clearly defined.

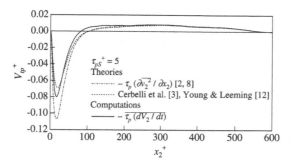

Fig. 8. Test of theories of turbophoretic velocity ($V_T^+ = 0$).

4 Calculation of concentration profiles

The fully-developed concentration fields considered in these calculations provide a simple system with which to evaluate the Boussinesq model, that is commonly used. The flux of particles in the x_2-direction is zero at all x_2 so that

$$\overline{C}\,\overline{V}_2 + \overline{cv_2} = 0, \tag{8}$$

$$\overline{V}_2 = V_{tp} - \overline{\tau}_p g. \tag{9}$$

This simply represents a balance amongst gravitational settling, turbophoresis and Reynolds transport, $-\overline{v_2 c}$.

Young and Hanratty [11] and Young and Leeming [12] have shown that the turbophoretic velocity can be given as

$$V_{tp} = -\tau_p \overline{\frac{dV_2}{dt}}. \tag{10}$$

This was evaluated directly in the calculations. An example is given in Figure 8 for $\tau_{pS}^+ = 5$ and $V_T^+ = 0$. Thus all of the terms in Equations (8, 9), with the exception of $-\overline{v_2 c}$, could be evaluated.

The Boussinesq approximation represents the turbulent mixing as

$$-\overline{v_2 c} = \varepsilon \frac{d\overline{C}}{dx_2}, \tag{11}$$

where ε is a turbulent diffusivity. The integration of Equation (8) gives

$$\frac{d \ln \overline{C}}{dx_2} = \frac{\overline{V}_2}{\varepsilon}. \tag{12}$$

This relation works in the center of the channel for large V_T. Then

$$\overline{V}_2 \cong -V_T \tag{13}$$

and

$$\varepsilon \sim H v^*. \tag{14}$$

In general, calculated concentration profiles can be used to evaluate the variation of ε with x_2. The results are not reasonable. For example in the case of $V_T = 0$ negative values of ε are obtained.

Clearly, turbulent mixing is controlled by large scale motions which are not captured by a linear diffusion mechanism.

5 Concluding remarks

Particles accumulate in the viscous-sublayer due to turbophoresis. We show that the turbophoretic velocity can be obtained by evaluating $\overline{dV_2/dt}$.

Several regimes for k_D are defined: (1) Turbulent diffusion of particles that accumulate close to the wall. The presence of gravity greatly affects the importance of this mechanism. (2) Free flight deposition. This mechanism is understood but a mathematical theory is not available. For annular flows free-flights start from outside the viscous wall layer and $k_D \sim (\overline{v_2^2})^{1/2}$ at $x_2^+ = 40$. (3) An annular flow regime can be defined if particles ejected from the bottom wall can reach the top wall. (4) A saltation regime is defined when the effect of fluid turbulence can be neglected. Two behaviors can be defined. In one of these particles do not reach the top wall. When particles reach the top wall a direct impaction mechanism is defined for annular flow. Under these circumstances the deposition velocity can be approximately equal to the injection velocity.

Acknowledgments

This work is supported by DOE under grant DEFG0286ER 13556. Computer resources have been provided by the National Center for Supercomputing Applications located at the University of Illinois.

References

1. Binder, J.L. and Hanratty, T.J., 1992, *Int. J. Multiphase Flow* **18**, 803–820.
2. Caporaloni, M., Tampieri, F., Trombetti, F. and Vittori, O., 1975, *J. Atmos. Sci.* **32**, 565–568.
3. Cerbelli, S., Giusti, A. and Soldati, A., 2001, *Int. J. Multiphase Flow* **27**, 1861–1879.
4. Iliopoulos, I., Mito, Y. and Hanratty, T.J., 2003, *J. Multiphase Flow* **29**, 375–394.
5. Mito, Y. and Hanratty, T.J., 2002, *Flow Turbul. Combust.* **68**, 1–26.
6. Mito, Y. and Hanratty, T.J., 2003, *Int. J. Heat Mass Tran.* **46**, 1063–1073.
7. Mito, Y. and Hanratty, T.J., 2003, *Int. J. Multiphase Flow* **29**, 1373–1394.
8. Reeks, M.W., 1983, *J. Aerosol Sci.* **14**, 729–739.
9. Thomson, D.J., 1987, *J. Fluid Mech.* **180**, 529–556.
10. Woodmansee, D.E. and Hanratty, T.J., 1969, *Chem. Engng. Sci.* **24**, 299–307.
11. Young, J.B. and Hanratty, T.J., 1991, *J. Fluid Mech.* **231**, 665–688.
12. Young, J. and Leeming, A., 1997, *J. Fluid Mech.* **340**, 129–159.

On Momentum Coupling Methods for Calculation of Turbulence Attenuation in Dilute Particle-Laden Gas Flows

John K. Eaton and Judith C. Segura

Department of Mechanical Engineering, Stanford University, Building 500, 488 Panama Mall, Stanford, CA 94305, USA: e-mail: eaton@vk.stanford.edu

Many experiments have shown that small particles uniformly dispersed in a turbulent gas flow can cause substantial attenuation of the turbulence at volume loadings as small as 10^{-4}. Very small particles closely follow the turbulent motions and therefore have no effect on the turbulence. The superposed wakes of randomly moving large particles create substantial turbulence augmentation. In an intermediate particle size range, the particles respond sluggishly to the turbulent motions thereby damping energy containing eddies. Extensive review of the experimental literature showed that the largest observed turbulence attenuation occurs for cases where the particle diameter is of the same order as the turbulence Kolmogorov scale, and the Stokes number based on the Kolmogorov time scale is around 50. Since this parameter range includes important natural and technological flow fields, it is important to develop models capable of predicting turbulence attenuation over a range of flow and particle parameters.

Model-free simulation requires resolving the Navier–Stokes equations over the entire flow domain, including accounting for the boundary conditions on the surface of each of thousands to millions of particles. Such simulations of particle/turbulence interactions have been done, but only for simplified single-particle systems [1, 3]. Burton and Eaton [3] showed that accurate (<1% error) prediction of the force applied by the fluid onto the particle requires around 1 million grid points in a local spherical grid of diameter $D_{\mathrm{grid}} = 25d_p$ where d_p is the particle diameter. Full resolution of a many particle system is well beyond current computer resources, so a model is needed that will accurately reflect the effect of an individual particle on the turbulence using a much coarser grid resolution.

A commonly applied model is the point-force coupling model first used by Squires and Eaton [9]. Improvements to this model have been proposed by Lonholt et al. [5] and Sundaram and Collins [10], however all such models implicitly assume that the particles are much smaller than both the Kolmogorov scale and the grid spacing [2]. A realistic example is instructive at this point. We consider

S. Balachandar and A. Prosperetti (eds), Proceedings of the IUTAM Symposium on Computational Multiphase Flow, 39–42.
© 2006 *Springer. Printed in the Netherlands.*

the particle-laden channel flow studied by Paris and Eaton [7]. The air channel was 40 mm wide, operated at a Reynolds number $U_{bulk}h/\nu = 13,800$, and had a center-line Kolmogorov scale of 170 microns. A 20% mass loading ratio ($\phi = 0.2$) of 150 micron diameter glass beads (0.00008 volume fraction) produced strong turbulence attenuation. The average interparticle spacing was approximately 2.9 mm, comparable to the 3 mm Taylor microscale. The particle Stokes number based on the center-line Kolmogorov time scale was about 50 and the average particle Reynolds number was approximately 19. While this is just a single case, the parameters are typical of flows producing strong turbulence attenuation.

The point-force model is routinely added into direct numerical simulation codes adapted from single-phase applications. Direct simulation of turbulence requires a grid resolution on the order of the Kolmogorov scale, comparable to the particle diameter in the above example. Implementation of the point force model on such a grid cannot resolve the local flow distortions around the particle and therefore cannot capture the extra viscous dissipation associated with the particle motion. Nevertheless, many believe that such a model will still capture the effects of particles on the energy-containing eddies. Segura (2004) performed a well-resolved large-eddy simulation of the Paris and Eaton experiment using the point-force coupling model. The simulation used second order staggered grid discretization and second-order time advancement. A stretched mesh normal to the channel walls was used with 16 nodes below $y^+ = 30$ and 1 node below $y^+ = 1$. The baseline single-phase flow and the particle motions for light loading cases were accurately predicted by the code. Turbulence statistics calculated using only the resolved motions agreed closely with both the experimental data of Paris and Eaton and with DNS results of Moser et al. [6]. This indicates that the simulation had near-DNS resolution. However, the turbulence attenuation was grossly underpredicted as shown in Figure 1. Segura increased the particle-loading by a factor of ten producing good agreement with the data across the full profile.

It is important to understand why point-force coupled simulations underpredict turbulence attenuation. One issue is that the particle drag and lift are underpredicted by typical Lagrangian tracking models because the models do not account for the effects of small scale turbulence. Secondly, most models relate the particle drag to the undisturbed fluid velocity. However, the undisturbed velocity is not available in a two-way coupled simulation. The failure to capture the high levels of local dissipation around the particles is probably the biggest failing of the point-force method. Burton and Eaton [4] compared the viscous dissipation from a fully resolved simulation to that calculated using the point-force approximation with a grid resolution equal to the particle diameter. The viscous dissipation in a local region around the particle was underpredicted by as much as 50%. Rather than dissipating the energy associated with the particle motions, the point force scheme transfers much of that energy to different scales.

New modeling approaches are needed that will reproduce the correct statistical distribution of individual particle effects on turbulence and the overall changes to the turbulence. One possibility is to represent the extra dissipation that occurs at sub-grid scales using an effective "particle-field sub-grid scale viscosity". It is likely that this

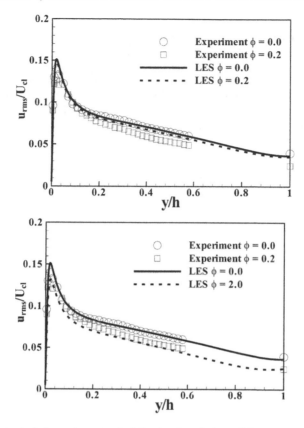

Fig. 1. Streamwise turbulence intensity in fully developed channel flow. Experiment Paris and Eaton [7], LES Segura [8].

would be a function of the local particle Reynolds number, the small-scale turbulence properties, and the grid resolution. The values can be estimated by developing a library of well-resolved simulations of particles in turbulent flows. The strong viscous dissipation is confined to a region within about 5 particle diameters of the particle center. Since the interparticle separation is often much larger than 5 diameters, we expect that the sub-grid scale viscosity field will vary in both time and space depending on the particle motions. Such a model potentially could mimic both the small and large-scale effects of particles on turbulence.

References

1. Bagchi, P. and Balachandar, S., 2003, *Phys. Fluids* **15**, 3496–3513.
2. Boivin, M., Simonin, O. and Squires, K.D., 1998, *J. Fluid Mech.* **375**, 235–263.
3. Burton, T.M. and Eaton, J.K., 2003, Report TSD-151, ME Department, Stanford University.

4. Burton, T.M. and Eaton, J.K., 2004, *Proc. Int. Conf. Multiphase Flow*, Yokohama.
5. Lonholt, S., Stenum, B. and Maxey, M.R., 2002, *Int. J. Multiphase Flow* **28**, 225–246.
6. Moser, R., Kim, J. and Mansour, N., 1999, *Phys. Fluids* **11**, 943–945.
7. Paris, T. and Eaton, J.K., 2001, Turbulence attenuation in a particle-laden channel flow, Report TSD-137, Department of Mechanical Engineering, Stanford University.
8. Segura, J.C., 2004, Ph.D. Thesis, ME Department, Stanford University.
9. Squires, K.D. and Eaton, J.K., 1990, *Phys. Fluids* **A2**, 1191–1203.
10. Sundaram, S. and Collins, L.R., 1999, *J. Fluid Mech.* **379**, 105–143.

Multifractal Concentrations of Heavy Particles in Random Flows

Jérémie Bec

CNRS, Laboratoire Cassiopée, Observatoire de la Côte d'Azur, BP 4229, 06304 Nice Cedex 4, France; e-mail: bec@obs-nice.fr

1 Introduction

Small-size impurities such as dust or droplets suspended in turbulent incompressible flows typically have a finite size and a mass density larger than the carrier fluid. They cannot be described as simple passive tracers, that is point-like particles with negligible mass advected by the fluid; an accurate model for their motion must take into account inertia effects. These *inertial particles* generally interact with the fluid through a viscous Stokes drag and thus their motion typically lags behind that of passive tracers. The dynamics of the latter is governed by a conservative dynamical system when the carrier flow is incompressible (because volume is conserved), but inertial particles have *dissipative* dynamics. While an initially uniform distribution of tracers remains uniform at any later time, the spatial distribution of inertial particles develops strong inhomogeneities.

Such a phenomenon of *preferential concentration* refers to the presence of regions with either extremely high or low concentrations. Their characterization plays an essential role in natural and industrial phenomena. Instances are optimization of combustion processes in the design of Diesel engines [1], the growth of rain drops in sub-tropical clouds [2], the formation of the planets in the Solar system [3], co-existence between several species of plankton [4], etc. For such applications it is recognized that a key problem is the prediction of the collision or reaction rates and their associated typical time scales. The time scales obtained using diffusion theory exceed by one or several orders of magnitude those observed in experiments or numerical simulations. A full understanding of particle clustering and, in particular, of the fine structures appearing in the mass distribution is crucial for identifying and quantifying the mechanisms responsible for this drastic reduction in time scales.

We propose here an original approach leading to a systematic description of inertial particles clustering. This approach is in part inspired by recent breakthroughs in the study of passive scalar advection by turbulent flows, using Lagrangian techniques [5]. Preferential concentrations can be interpreted as the convergence of particle trajectories onto certain dynamically evolving sets in the position–velocity phase space called *attractors*. Use of dissipative dynamical systems tools and, in particular, of

43

S. Balachandar and A. Prosperetti (eds) Proceedings of the IUTAM Symposium on Computational Multiphase Flow, 43–52.

methods borrowed from the study of random dynamics, allows a rather complete characterization of particle distribution.

We focus on very diluted suspensions where collisions, particle-to-particle hydrodynamical interactions and retroaction of the particles on the fluid can be ignored. When the particle radius a is much smaller than the Kolmogorov dissipation scale η and when the Reynolds number based on the particle size and its relative velocity to the fluid is sufficiently small, the dynamics of the particles is described by the standard model of Maxey and Riley [6]. The forces exerted on the particles are then buoyancy, the Stokes viscous drag, the added mass effect and the Basset–Boussinesq history force. The goal here is to study a simple model for the dynamics able to capture qualitatively most aspects of inertial particle dynamics. We hence consider very heavy particles whose trajectories are solutions of the Newton equation

$$\ddot{\mathbf{X}} = -\frac{1}{\tau_p}[\dot{\mathbf{X}} - \mathbf{u}(\mathbf{X}, t)], \tag{1}$$

where $\mathbf{u}$ is a prescribed fluid velocity field and where we have neglected the effects of gravity. The response time τ_p, frequently referred to as the Stokes time, is defined as $\tau_p = (2\rho_p a^2)/(9\rho_f \nu)$, where the particle-fluid mass density ratio ρ_p/ρ_f is assumed to be very large; ν is the kinematic viscosity of the fluid. One usually introduces the Stokes number $St = \tau_p/\tau_\eta$, defined by non-dimensionalizing the response time by the smallest characteristic time of the turbulent fluid flow, i.e. the eddy turnover time associated to the Kolmogorov scale $\tau_\eta = \varepsilon^{-1/2}\nu^{1/2}$.

2 Local dynamics and Lyapunov exponents

The temporal evolution of the separation $\delta\mathbf{R}(t)$ between two infinitesimally close trajectories is given by the linearized (tangent) system

$$\delta\ddot{\mathbf{R}} = \frac{1}{\tau_p}\,\sigma(t)\,\delta\mathbf{R} - \frac{1}{\tau_p}\,\delta\dot{\mathbf{R}}\,, \tag{2}$$

where $\sigma(t)$ denotes the strain matrix of the carrier flow along a reference trajectory: $\sigma_{ij}(t) \equiv \partial_j u_i(\mathbf{X}(t), t)$. This second-order equation needs to be studied in the full position-velocity phase space of dimension $2 \times d$ where d is the dimension of the physical space. Infinitesimal distances, surfaces, volumes... of the phase space are expanded or contracted exponentially in time by the linearized dynamics (2). One usually introduces the *stretching rates* $\mu_1(t) \geq \cdots \geq \mu_{2d}(t)$ as the instantaneous exponential rates, μ_1 measuring the growth of distances between two neighboring trajectories, $\mu_1 + \mu_2$ that of areas defined by three trajectories, etc. The sum of the $2d$ stretching rates controls the time evolution of $2d$-dimensional phase-space volumes. When the fluid flow is incompressible (i.e. $\nabla \cdot \mathbf{u} = 0$), it is easily shown that $\mu_1 + \cdots + \mu_{2d} = -d/\tau_p < 0$, meaning that all phase-space volumes are subject to a uniform exponential contraction by the dynamics.

The long-time behavior of the local dynamics is dominated by the almost-sure convergence of the stretching rates to the classical *Lyapunov exponents* $\lambda_j =$

$\lim_{t \to \infty} \mu_j(t)$. Under some ergodicity hypothesis on the dynamics, the Lyapunov exponents are independent of both the realization of the random carrier flow and of the peculiar trajectory $\mathbf{X}(t)$ around which the linearized dynamics is considered. The Lyapunov exponents are linked to many fundamental features of the dynamics. For instance, when the largest Lyapunov exponent λ_1 is negative, the stability of the linearized system is ensured and all the trajectories are converging together, thereby leading to a somewhat degenerate statistical steady state in which all the mass is concentrated in discrete time-dependent points in phase space.

3 Fractal clustering

When λ_1 is positive, the dynamics is said to be chaotic and the long-time evolution has richer features characterized by the attraction of all particle trajectories to complex dynamical structures. Clearly convergence to such *attractors* cannot be detected by considering a single trajectory but requires to have a global approach of the dynamics and to study the phase-space flow defined by (1). For this we consider the phase-space density of particles $f(\mathbf{x}, \mathbf{v}, t)$ which evolves according to

$$\partial_t f + \nabla_\mathbf{x} \cdot (f \, \mathbf{v}) + \frac{1}{\tau_p} \nabla_v \cdot [f \, (\mathbf{u}(\mathbf{x}, t)) - \mathbf{v})] = 0, \tag{3}$$

with $f(\mathbf{x}, \mathbf{v}, 0)$ being the position-velocity joint distribution of particles at the initial time. Note that f is not averaged with respect to the fluid flow realizations but it can be interpreted as the probability (or mass) density to have at time t a particle at position $\mathbf{x}$ with velocity $\mathbf{v}$ for a given realization of $\mathbf{u}$. When the fluid flow is statistically stationary, f converges at large times to a singular density with support on a (dynamically evolving) fractal set, the attractor.

Focusing on positions while ignoring velocities leads to consider the physical-space density

$$\rho(\mathbf{x}, t) = \int f(\mathbf{x}, \mathbf{v}, t) \, d^d v. \tag{4}$$

In the statistical equilibrium reached at large times, this integration over velocities amounts to project the attractor from the $2d$-dimensional phase space onto the d-dimensional space of particle positions. A standard result on fractal sets ensures that, when the fractal dimension $\mathcal{D}$ of the attractor is less than the dimension of the projection space (here, d), the projection of the fractal is itself a fractal set with dimension $\mathcal{D}$. If however $\mathcal{D} > d$, the projection has dimension d. As a consequence, according to the dimension of the phase-space attractor, the physical space density ρ will have or not a fractal support.

This is illustrated numerically by considering inertial particles suspended in a two-dimensional random incompressible flow. The velocity field $\mathbf{u}$ is generated as the superposition of few Fourier modes (here eight) which are independent Gaussian random processes with finite correlation time τ_f and with variances such as to ensure statistical isotropy. The Stokes number is here defined as $St = \tau_p/\tau_f$. This flow was

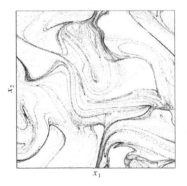 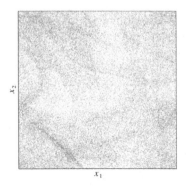

Fig. 1. Snapshot of the position of 10^5 heavy particles in a two-dimensional synthetic flow. Right: $St = 10^{-2}$, value below the threshold for which fractal clustering is observed. Left: $St = 1$ above the threshold; here particles fill the whole domain.

chosen mainly to mimic dissipative range dynamics. The particle trajectories are then integrated using a fourth-order Runge–Kutta scheme. Figure 1 shows snapshots of the position of $N = 10^5$ particles in the statistical steady state reached at large times for two different values of the Stokes number illustrating the two possible regimes mentioned above. When the Stokes number is sufficiently small, the particles concentrate on a dynamically evolving fractal set. At large Stokes numbers, the particles distribute also inhomogeneously but they fill the whole domain. Similar simulations were made in three dimensions leading to the same qualitative observation when considering two-dimensional cuts.

The presence of a threshold in Stokes number can be observed in a more quantitative fashion. The convergence to an attractor can be seen as the result of a competition between stretching and folding effects that occur during the chaotic motion of the particles. The properties of the attractor, and in particular its fractal dimension, thus depend on the stretching rates of the dynamics. The positive μ_j's are responsible for stretching in their associated eigendirections, while the negative rates give contraction and hence folding. This picture lead Kaplan and Yorke [7] to propose an estimate of the dimension of the attractor in terms of the Lyapunov dimension

$$\mathcal{D}_{KY} \equiv J - (\lambda_1 + \cdots + \lambda_J)/\lambda_{J+1}, \tag{5}$$

where J is such that $\lambda_1 + \cdots + \lambda_J \geq 0$ and $\lambda_1 + \cdots + \lambda_{J+1} < 0$. This non-random number can be interpreted heuristically as the dimension of phase-space objects that keep a constant volume during time evolution. It was actually shown in [8] that the Lyapunov dimension is equal to the information dimension associated to the steady-state phase-space density of particles and, as we shall see in Section 4, is thus related to the small-scale properties of the mass distribution of inertial particles.

The Lyapunov dimension is particularly convenient for estimating numerically fractal dimensions since it does not require box-counting. The Lyapunov exponents can indeed be computed using very efficient and fast-converging methods [10]. This was exploited for particles suspended in the random flow described above to confirm

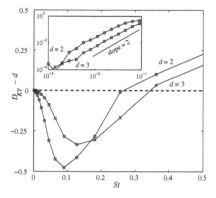 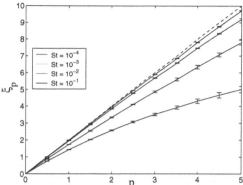

Fig. 2a. Lyapunov dimension as a function of St for $d = 2$ (circles) and $d = 3$ (squares) in a synthetic flow (see text). Inset: same in log-log coordinates. Note the quadratic behavior as $St \to 0$ predicted in [9].

Fig. 2b. Scaling exponents ξ_p of the moments of mass as a function of p for four different values of St. The dashed line represents the exponents of a uniform distribution.

the presence of a threshold in Stokes number for the concentration of particle on fractal sets [11]. Figure 2a represents the Lyapunov dimension $\mathcal{D}_{KY}$ as a function of the Stokes number in two and three dimensions. The particle positions form fractal clusters when $\mathcal{D}_{KY} < d$. The threshold corresponds to the value of St such that $\mathcal{D}_{KY} = d$. For the random flow considered here, it occurs at $St \approx 0.2$ for $d = 2$ and $St \approx 0.3$ for $d = 3$.

Note finally that there is a maximum of clustering (minimum of the dimension) for $St \approx 0.09$ when $d = 2$ and $St \approx 0.13$ when $d = 3$. These values of the Stokes number are clearly flow-dependent and can hardly be compared to observations stemming from direct numerical simulations of Navier–Stokes turbulence where the maximal clustering is observed for St order unity [12].

4 Scaling properties of the mass distribution

Once we know that particles form fractal clusters, the next step is to understand the statistical properties of the mass distribution, what is generally referred to as the *multifractal properties* of the attractor. The relevant quantity is the so-called *quasi-Lagrangian* coarse-grained mass

$$m(r, t) = \int_{|\mathbf{y}| < r} \rho(\mathbf{y} - \mathbf{X}(t), t) \, d^d y \,, \tag{6}$$

that is the mass of particles in a ball of radius r centered on a given particle trajectory $\mathbf{X}(t)$. In the large-time statistical equilibrium, the moments of $m(r, t)$ are expected to behave algebraically at small ball radii, i.e.

$$\langle m^p(r, t) \rangle \sim r^{\xi_p} \quad \text{when } r \to 0. \tag{7}$$

The scaling exponents ξ_p's are related to the multifractal spectrum of dimensions [13, 14] (see also [15]) by $\mathcal{D}_p = \xi_{p-1}/(p-1)$; $\mathcal{D}_0$ denotes the Hausdorff dimension of the attractor, $\mathcal{D}_1$ the information dimension, $\mathcal{D}_2$ the correlation dimension, etc. Note that for situations close to that of inertial particles as for instance tracers advected by compressible flows, the scaling exponents ξ_p can be expressed in terms of the joint probability density function of the stretching rates μ_j's introduced in the previous section [16].

The exponents ξ_p were computed numerically in [17] for the synthetic two-dimensional flow. They are shown in Figure 2b as a function of their order p and for various Stokes numbers. Already when $St = 10^{-4}$ deviations from a uniform distribution are observed, a quantitative sign of preferential concentration. The non-linear behavior of ξ_p as a function of p is a signature of multifractality of the particle spatial distribution.

It was checked numerically that the information dimension $\mathcal{D}_1$, obtained as the slope of $p \mapsto \xi_p$ at $p = 0$, is equal to the Lyapunov dimension $\mathcal{D}_{KY}$ as shown in [8]. This implies that for almost every realization of the fluid velocity field and for almost every time, the coarse-grained mass around a particle trajectory has the asymptotic scaling

$$\frac{\ln m(r, t)}{\ln r} \to \mathcal{D}_{KY} \quad \text{as } r \to 0. \tag{8}$$

The scaling behavior (7) of the moments of mass which was observed numerically implies that for small but finite r the mass deviates from this limiting form. More precisely, the stationary distribution of the fluctuating finite scaling exponents $h \equiv (\ln m(r, t))/(\ln r)$ takes the large deviation form

$$\mathcal{P}(h, r) \propto r^{d-D(h)}, \tag{9}$$

where $D(h)$ is a convex rate function with a maximum equal to d attained for $h = \mathcal{D}_1 = \mathcal{D}_{KY}$. This function can be seen as the dimension of the set on which the mass m scales as r^h and is frequently referred to as the *multifractal spectrum*. $D(h)$ is related to the scaling exponents ξ_p by a simple Legendre transform: $\xi_p = \inf_h(ph + d - D(h))$. The small-radii behavior (9) was confirmed by numerical experiments in random flows (see [17]).

5 Ghost collisions

We now focus on the enhancement of collision rates induced by particle inertia. The minimal model (1) for particle dynamics allows for a systematic study which will help us to identify the main physical mechanisms leading to this effect. A step in this direction has been taken in [9] for very strong viscous drag, i.e. $St \ll 1$. The dynamics can then be approximated by that of simple tracers advected by an effective flow with a small compressible component [18]. Extending this approach to

(a) 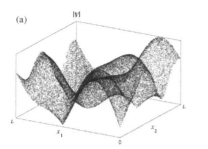 (b)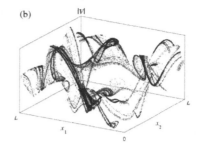

Fig. 3. Snapshot of the position of particles in phase space (a) for $St = 3.8 \, 10^{-4}$ and (b) for $St = 0.38$. To represent positions in a four-dimensional space, the modulus of particle velocities is showed on the z axis.

larger values of St requires to take into account the full position-velocity phase-space dynamics. To tackle collisions we make use of a ghost-particle approximation [19], which amounts to record collision events but to let particles overlap. This approach allows to consider statistical stationarity and to use a Lagrangian approach.

To estimate the binary collision rate between particles with the same Stokes number St, the relevant quantity is the *approaching rate* $\kappa(r, t)$, defined as the flux of particles at a distance r from a reference particle that are approaching it. This rate can be interpreted as an average velocity difference weighted by the probability that the two particles are at a distance r. Trivial correlations between these two quantities simplify the estimation of κ. Clearly, the probability density that the two particles are at a small distance r has a power-law behavior with exponent $\mathcal{D}_2 - 1 = \xi_1 - 1$. In the asymptotics $St \ll 1$, the velocity difference between particles trivially depends on their separation since the Stokes drag is very strong and the particle velocity is close to that of the fluid. We hence have $\kappa(r, t) \sim r^{\mathcal{D}_2}$. When $St \gg 1$, the particle inertia is so large that the motion is almost ballistic. The velocity difference is then essentially independent of the separation. The approaching rate thus behaves as $\kappa(r, t) \sim r^{\mathcal{D}_2 - 1}$.

For intermediate Stokes numbers, κ behaves as a power-law with an exponent γ that cannot be trivially related to $\mathcal{D}_2$. This is due to a balance between two competing effects in phase space: folding of the attractor in the $\mathbf{v}$-direction and its relaxation toward the surface defined by the instantaneous fluid velocity at a rate given by the Stokes time. For $St = 0$ this happens infinitely fast, preventing folding and thus the presence of positions where the particle velocity field is multi-valued (see Figure 3a). As the Stokes number increases the probability of finding particles at the same position with different velocities becomes larger (see Figure 3b). Such points correspond to self-intersections of the attractor once projected in position space; this is a phase-space interpretation of what is known as the *sling effect* [9]. The dependence on St of the two exponents $\mathcal{D}_2$ and γ is illustrated in Figure 4a for suspensions in a two-dimensional synthetic flow. The two asymptotics of small and large Stokes numbers are confirmed.

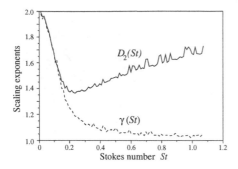 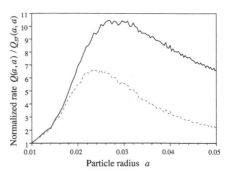

Fig. 4a. Scaling exponents $\mathcal{D}_2$ and γ of the separation probability and of the approaching rate respectively, as a function of the Stokes number.

Fig. 4b. Solid curve: ratio between the collision rate $Q(a)$ for equal-size particles and that obtained when neglecting inertia. Dashed curve: same for that obtained when neglecting correlations between the velocity difference and the density.

The collision rate $Q(a)$ between particles with the same size a is given by the approaching rate computed at a distance $r = 2a$. For small particle radii, this leads to $Q(a) = \kappa(2a) \propto (2a)^{\gamma(St)}$ where the Stokes number itself depends on the radius: $St \propto a^2$. This approach leads to an estimate of $Q(a)$ (details are given in [21]). Results of numerical simulations in a synthetic flow are shown in Figure 4b. In order to quantify the importance of particle inertia, we represented the ratio between the measured kernel and that obtained using the Saffman–Turner approach [20] where particles are assumed to be simple tracers. To disentangle the effects of clustering and of density-velocity correlations, the kernel obtained when the velocity difference is assumed to behave linearly with the separation is also shown. The two curves coincide at very small radii a (i.e. small Stokes numbers): in this regime, the enhancement of collision rates is mainly due to clustering effects and there are no large velocity differences between particles. Discrepancies between the two curves appear rather soon and tend to a constant as the Stokes number increase. It is clear from Figure 4b that preferential concentration alone is responsible of an increase of roughly one order of magnitude in the collision rate of inertial particles compared to that of tracers. However, the measured values of the kernel differ markedly from those obtained when only clustering effects are considered. Therefore, away from the two asymptotics of small and large particle inertia, it is crucial not to consider as independent the effects of clustering and enhanced relative velocity.

This approach can be extended to binary collisions between particles with different Stokes numbers. The pair dynamics is then characterized by the presence of a critical separation $r_\star \propto |St_1 - St_2|$. Below it, the two motions are essentially uncorrelated. Correlations due to the fact that particles are suspended in the same flow show up for length scales above $r_\star$. The origin of this characteristic length is understood in terms of the pair-separation dynamics, which is dominated by the Stokes difference for $r < r_\star$ (the accelerative mechanism), and by the fluid velocity (the shear mechanism) when $r > r_\star$. This crossover length separates in two distinct regimes the scale dependence of both the probability distribution of particle separations and their rate of approach. This has consequences on estimations of the collision rates that are detailed in [21].

6 Concluding remarks

We have shown that the dynamics of inertial particles in smooth flows is character-ized at small scales by the convergence of their trajectories to dynamically-evolving attractors. Such sets are generally fractal leading to a particle mass distribution with multifractal properties. This behavior is illustrated by numerical experiments in ran-dom time-correlated flows that mimic the dynamics of turbulent flows at scales be-low the Kolmogorov length scale. Recent numerical experiments in fully developed turbulent flows (see [22]) show that this picture is rather robust and that the ran-dom flows considered here catches the qualitative features and the main physical mechanisms present in particles dynamics. An important consequence of the fractal clustering of inertial particles is that there exists no scale for particle concentrations within the dissipative range. This may in particular lead to some difficulties in the design of sub-grid models for the numerical integration of particle suspensions.

An important open issue concerns clustering at inertial-range scales, where the velocity field is not differentiable. Experiments indeed show that preferential con-centration appears also at those scales [23]. Non-trivial clustering properties have been observed numerically also in the inverse-cascade range of two-dimensional tur-bulent flows, namely the formation of holes in the distribution of particles [24]. It is important to remark that clustering at the inertial scales may influence the probability for two particles to arrive below the dissipative scale and thus the collision rates. In the inertial range the dynamics of the fluid is close to Kolmogorov 1941 theory and, as a consequence, tracers separate according to the celebrated Richardson's $t^{3/2}$ law. For inertial particles, one needs to understand the competition between this algebraic separation and clustering due to dissipative dynamics. In this direction, it may be useful to further extend to inertial particles recent models and techniques developed in the framework of passive scalars (for a recent review, see [5]).

References

1. Post, S. and Abraham, J., 2002, *Int. J. Multiphase Flow* **28**, 997–1019.
2. Pinsky, M. and Khain, A., 1997, *J. Aerosol Sci.* **28**, 1177–1214.
3. Weidenschilling, S., 1995, *Icarus* **116**, 433–435.
4. Squires, K. and Yamazaki, H., 1995, *Deep Sea Res.* Part I, **42**, 1989–2004.
5. Falkovich, G., Gawędzki, K. and Vergassola, M., 2001, *Rev. Mod. Phys.* **73**, 913–976.
6. Maxey, M. and Riley, J., 1983, *Phys. Fluids* **26**, 883–889.
7. Kaplan, J. and Yorke, J., 1979, Functional differential equations and the approximation of fixed points, in *Lecture Notes in Math.* Vol. 730, H. Peitgen and H. Walther (eds), Springer, Berlin.

8. Ledrappier, F. and Young, L.-S., 1988, *Comm. Math. Phys.* **117**, 529–548.
9. Falkovich, G., Fouxon, A. and Stepanov, M., 2002, *Nature* **419**, 151–154.
10. Benettin, G., Galgani, L., Giorgilli, A. and Strelcyn, J., 1980, *Meccanica* **15**, 9–30.
11. Bec, J., 2003, *Phys. Fluids* **15**, L81–L84.
12. Sundaram, S. and Collins, L., 1997, *J. Fluid Mech.* **335**, 75–109q.
13. Grassberger, P., 1983, *Phys. Lett. A* **97**, 227–230.
14. Hentschel, H. and Procaccia, I., 1983, *Physica D* **8**, 435–444.
15. Paladin, G. and Vulpiani, A., 1987, *Phys. Rep.* **156**, 147–225.
16. Bec, J., Gawędzki, K. and Horvai, P., 2004, *Phys. Rev. Lett.* **92**, 224501.
17. Bec, J., 2005, *J. Fluid Mech.* **528**, 255–277.
18. Maxey, M., 1987, *J. Fluid Mech.* **174**, 441–465.
19. Wang, L.-P., Wexler, A. and Zhou, Y., 1998, *Phys. Fluids* **10**, 266–276.
20. Saffman, P. and Turner, J., 1956, *J. Fluid Mech.* **1**, 16–30.
21. Bec, J., Celani, A., Cencini, M. and Musacchio, S., 2005, *Phys. Fluids* **17**, 073301.
22. Bec, J., Biferale, L., Boffetta, G., Celani, A., Cencini, M., Lanotte, A., Musacchio, M. and Toschi, F. (2005) preprint, http://arxiv.org/nlin.CD/0508012.
23. Eaton, J. and Fessler, J., 1994, *Int. J. Multiphase Flow* **20**, 169–209.
24. Boffetta, G., De Lillo, F. and Gamba, A., 2004, *Phys. Fluids* **16**, L20–L23.

Turbulence Modulation by Micro-Particles in Boundary Layers

Maurizio Picciotto, Andrea Giusti, Cristian Marchioli and Alfredo Soldati

Centro Interdipartimentale di Fluidodinamica e Idraulica and Dipartimento di Energetica e Macchine, Università di Udine, 33100 Udine, Italy; E-mail: soldati@uniud.it

Abstract. Turbulent dispersed flows over boundary layers are crucial in a number of industrial and environmental applications. In most applications, the key information is the spatial distribution of inertial particles, which is known to be highly non-homogeneous and may exhibit a complex pattern driven by the structures of the turbulent flow field. Theoretical and experimental evidence shows that fluid motions in turbulent boundary layers are intermittent and have a strongly organized and coherent nature represented by the large scale structures. These structures control the transport of the dispersed species in such a way that the overall distribution will resemble not at all those given by methods in which these motions are ignored.

In this paper, we study from a statistical viewpoint turbulence modulation produced by different-size dispersed particles and we examine how near-wall particle concentration is modified due to the action of particles themselves in modulating turbulence. The physical mechanisms and the statistics proposed are based on Direct Numerical Simulation (DNS) of turbulence and Lagrangian particle tracking, considering a two-way coupling between particles and fluid.

1 Introduction

In a number of environmental and industrial problems involving turbulent dispersed flows, the information on particle distribution is a crucial issue. In particular, the relevant information sought is the local concentration of particles which controls all relevant exchange mechanisms (e.g. momentum exchange, reaction and deposition rates, mass transfer, evaporation and so on). Accurate three-dimensional, time-dependent simulations together with precise experiments are required to gain physical insights on the effect of the flow on particles distribution and of particles on the flow field. The simplest computational approach to investigate on dispersed flows is to consider particles as passive species under the one-way coupling assumption, which is valid for dilute flows characterized by volume fraction $\Phi_V < 10^{-3}$ and mass fraction $\Phi_M < 10^{-3}$ [1, 2].

Simulations performed under dilute flow conditions have shown that turbulent flow fields in general are of a strongly organized and coherent nature represented

S. Balachandar and A. Prosperetti (eds), Proceedings of the IUTAM Symposium on Computational Multiphase Flow, 53–62.

by large scale structures. These structures, because of their coherence and persistence, have a significant influence on the transport of dispersed particles. Specifically, coherent structures generate preferentially directed, non-random motion of particles leading to non-uniform concentration and to long-term accumulation. The local effect of coherent flow structures on particles is related to their mutual interaction which, in turn, is modulated by inertia [3–5] and their action is not captured by engineering models [6–8].

Preferential accumulation of particles induced by turbulent coherent structures has been examined previously in a number of theoretical and experimental works [3–5, 9–11]. In the case of homogeneous turbulence [3–5, 11], the particle concentration field will be characterized by local particle accumulation in low-vorticity, high-strain regions. In the case of non-homogeneous turbulence [9, 10], the local interaction between particles and turbulence structures produces a remarkably macroscopic behavior leading to long-term particle accumulation in specific flow regions within the viscous sublayer [12–14]. When particles segregate in specific flow regions, the dilute flow assumption is no longer valid locally. In particular, if particles are heavy (solid/liquid in gas), their overall volume may be negligible, yet the momentum coupling with the fluid may be such to induce significant modifications in the flow field [2, 15–17]. These effects will modify flow transport properties which eventually will change particle distribution. This may be of fundamental significance in applications as particle abatement, flow reactors and control of momentum, heat and mass fluxes at a wall.

In this paper, we examine from a statistical viewpoint the two-way interaction between particles and fluid in non-homogeneous turbulence. In particular, we aim at studying turbulence modifications due to particles having different inertia when gravity is neglected.

2 Methodology

The balance equations governing the turbulent channel flow are (in dimensionless form):

$$\frac{\partial u_i}{\partial x_i} = 0, \tag{1}$$

$$\frac{\partial u_i}{\partial t} = -u_j \frac{\partial u_i}{\partial x_j} + \frac{1}{Re_\tau} \frac{\partial^2 u_i}{\partial x_j^2} - \frac{\partial p}{\partial x_i} + \delta_{1,i} + \tilde{\mathbf{f}}_{2w}, \tag{2}$$

where u_i is the ith component of the velocity vector, p is the fluctuating kinematic pressure, $\delta_{1,i}$ is the mean pressure gradient driving the flow, Re_τ is the shear Reynolds number, while $\tilde{\mathbf{f}}_{2w}$ is an equivalent body force accounting for the action of the dispersed particles onto the fluid ($\tilde{\mathbf{f}}_{2w} = 0$ for simulations run under the one-way coupling assumption). For a generic volume of fluid Ω_p containing a particle, the action-reaction law imposes that:

$$\int_{\Omega_p} \tilde{\mathbf{f}}_{2w}(\mathbf{x}) \, d\Omega = -\mathbf{f}_{fl}, \tag{3}$$

where $\mathbf{f}_{fl}$ is the force exerted on the particles by the fluid. The term $\tilde{\mathbf{f}}_{2w}$ can be obtained by adding the contributions of each particle:

$$\tilde{\mathbf{f}}_{2w} = \sum_{p=1}^{n_p} (\mathbf{f}_{2w}^k)_p, \qquad (4)$$

where n_p is the number of particles. With the point-source approximation [15, 18], $\mathbf{f}_{2w}(\mathbf{x}) = -\mathbf{f}_{fl}\,\delta(\mathbf{x} - \mathbf{x}_p)$, where $\delta(\mathbf{x})$ is the Dirac's delta function.

Equations (1) and (2) are solved using pseudo-spectral Direct Numerical Simulation (DNS); details of the numerical method can be found elsewhere [19].

Particle motion is described by a set of ordinary differential equations for particle velocity and position. For particles much heavier than the fluid ($\rho_p/\rho \gg 1$, where ρ_p is particle density and ρ is fluid density), the only significant forces are Stokes drag and buoyancy, whereas Basset force can be neglected being an order of magnitude smaller [20]. Since the present contribution represents the first step taken at our laboratory towards the study of turbulence modulation by particles in boundary layers, several other simplifications have been made. The idea is to improve fundamental understanding on this topic starting with the most simplified simulation setting and dealing with a manageable set of parameters. To this aim, the effects of gravity and shear-induced lift in the equations of particle motion have been neglected. Also, inter-particle collisions are not taken into account at the current stage of our simulations. Following [21], this simplification is reasonable when the mean inter-particle spacing, L/D, in the near-wall region is $O(10)$. For the range of volume fractions considered in this work, L/D is still in the range 6 : 7 even when concentration levels become 50 times larger than their initial value. Of course, the mean spacing between neighbouring particles will decrease as simulations continue and concentration levels become higher. When this spacing will become smaller than a given threshold, than the effect of inter-particle collisions will be no longer negligible.

With the above simplifications the following Lagrangian equation for the particle velocity is obtained [22]:

$$\frac{d\mathbf{v}}{dt} = -\frac{3}{4}\frac{C_D}{d_p}\left(\frac{\rho}{\rho_p}\right)|\mathbf{v} - \mathbf{u}|(\mathbf{v} - \mathbf{u}), \qquad (5)$$

where $\mathbf{v}$ and $\mathbf{u}$ are the particle and fluid velocity vectors, d_p is particle diameter. The drag coefficient C_D is given by:

$$C_D = \frac{24}{Re_p}(1 + 0.15 Re_p^{0.687}), \qquad (6)$$

where the particle Reynolds number is equal to $Re_p = d_p|\mathbf{v} - \mathbf{u}|/\nu$, ν being fluid kinematic viscosity. Correction for C_D is necessary since Re_p does not necessarily remain small, in particular for depositing particles.

3 Numerical simulations

The flow into which particles are introduced is a turbulent Poiseuille channel flow of air assumed incompressible and Newtonian. The reference geometry consists of two infinite flat parallel walls: the origin of the coordinate system is located at the center of the channel and the x-, y- and z-axes point in the streamwise, spanwise and wall-normal directions respectively. Periodic boundary conditions are imposed on the fluid velocity field in x and y, no-slip boundary conditions are imposed at the walls. All variables are normalized by the wall shear velocity u_τ, the fluid kinematic viscosity ν and the half channel height h. The shear velocity is defined as $u_\tau = (\tau_w/\rho)^{1/2}$, where τ_w is the mean shear stress at the wall. Calculations are performed on a computational domain of $1885 \times 942 \times 300$ wall units in x, y and z discretized with $128 \times 128 \times 129$ nodes. The shear Reynolds number is $Re_\tau = u_\tau h/\nu = 150$. The time step used is $\Delta t^+ = 0.045$ in wall time units.

A Lagrangian particle tracking code coupled with the DNS code was developed to calculate particles paths in the flow field. The code interpolates fluid velocities at Eulerian grid nodes onto the particle position by means of 6th order Lagrangian polynomials, and integrates the equations of particle motion forward in time by means of a 4th order Runge–Kutta scheme. Four sets of 10^5 particles were considered, characterized by different values of the relaxation time, defined as $\tau_p = \rho_p d_p^2/18\mu$, where μ is the fluid dynamic viscosity. Particle relaxation time is made dimensionless using wall variables and the Stokes number for each particle set is obtained. In this work, we considered $\tau_p^+ = St = 1, 5$ and 25, as shown in Table 1 which summarizes all relevant simulation parameters.

At the beginning of the simulation, particles are distributed homogeneously over the computational domain and their initial velocity is set equal to that of the fluid at particle position. Also, particles are assumed to be pointwise, rigid and spherical. Periodic boundary conditions are imposed on particles in both streamwise and spanwise directions, elastic reflection is applied when the particle centre is at a distance less than $d_p/2$ from the wall. Elastic reflection was chosen since it is the most conservative assumption when studying the particle prefential concentration in a turbulent boundary layer. Interparticle collisions are neglected.

Table 1. Parameters relative to the simulation of particle dispersion. The superscript + identifies dimensionless variables: particle relaxation time τ_p^+ (equivalent to particle Stokes number St), particle density ρ_p^+, particle diameter d_p^+ and particle settling velocity v_{sett}^+. Φ_V and Φ_M represent the average volume fraction and the average mass fraction of the particles, respectively.

$\tau_p^+ (=St)$	d_p^+	ρ_p^+	v_{sett}^+	Φ_V	Φ_M	n_p	ΔT_p^+
1.0	0.153	769.23	0.0942	$3.52 \cdot 10^{-7}$	$2.71 \cdot 10^{-4}$	10^5	1080
5.0	0.342	769.23	0.4710	$3.93 \cdot 10^{-6}$	$3.02 \cdot 10^{-3}$	10^5	1080
25.0	0.765	769.23	2.3350	$4.40 \cdot 10^{-5}$	$3.38 \cdot 10^{-2}$	10^5	1080

4 Results

4.1 Flow field modification by particles

Object of this paper is to study the modification of turbulence due to the two-way interaction between fluid and particles having different inertia in the absence of gravity. In this section, we will compare results obtained from two-way coupling simulations with available results from previous one-way coupling simulations [23], in which particles are not allowed to influence the fluid motion ($\tilde{\mathbf{f}}_{2w} = 0$ in Equation (2)). Similar studies have been performed previously for the case of homogeneous isotropic turbulence [20, 24].

The effect of particles with different inertia on the streamwise component $\langle u_x^+ \rangle$ of the mean fluid velocity is shown in Figure 1a, where lines refer to benchmark one-way coupling simulations and symbols refer to two-way coupling simulations accounting for particle feedback on turbulence. We do not show the spanwise and the wall-normal components of the mean fluid velocity since they exhibit the expected behavior and do not add to the discussion. Velocity profiles, averaged in both space (over the streamwise and spanwise directions) and time (over a time span of 1080 t^+) and normalized by the shear velocity u_τ of the particle-free flow, deviate only slightly, if not negligibly, from each other. Deviations correspond to reductions of the channel flowrate no larger than 0.4% with respect to one-way coupling simulations. A careful examination of Figure 1a indicates that velocity profiles computed under two-way coupling conditions are slightly shifted towards higher values in the buffer region ($5 < z^+ < 30$) and towards smaller values in the outer region ($z^+ > 30$).

Though small, more noticeable differences are observed for turbulence intensities (RMS of fluid velocity fluctuations). Streamwise, spanwise and wall-normal turbulence intensities are shown in Figures 1b, 1c and 1d, respectively. As in Figure 1a, lines refer to one-way coupling simulations whereas symbols refer to two-way coupling simulations. It appears that particles do not affect much turbulence intensities in the outer flow. However, for both the spanwise component ($\langle u'_{y,\mathrm{rms}}^+ \rangle$ in Figure 1c), and the wall-normal component ($\langle u'_{z,\mathrm{rms}}^+ \rangle$ in Figure 1d) and regardless of particle size, particles do substantially increase turbulence intensities at the wall, particularly in the region where profiles develop a peak. Conversely a slight decrease in the RMS along the streamwise direction ($\langle u'_{x,\mathrm{rms}}^+ \rangle$, Figure 1b) is observed in correspondence of the maximum values.

The modifications in the RMS is likely to cause a modification in heat and mass transfer since the wall-normal velocity fluctuations are responsible for transport processes at the wall.

The Reynolds stress profiles, shown in Figure 2 for one-way coupling (line) and two-way coupling simulations (symbols), do show modifications due to particles outside the viscous wall region. The effect of particles is noticeable in the buffer layer, where the Reynolds stress increases. The Reynolds stress in the very-near-wall region ($z^+ < 5$ roughly) does not exhibit significant changes.

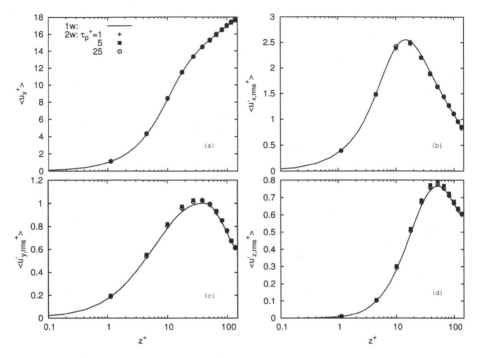

Fig. 1. Mean and RMS fluid velocity profiles for one-way coupling (lines) and two-way coupling (symbols).

4.2 Influence of flow field modification on particle statistics

The issue addressed in this section is: how turbulence modulation by particles influences the distribution of the particles? We will try to answer this question by comparing results on particle statistics obtained from simulations with and without particle feedback on turbulence.

Figure 3 shows the streamwise ($\langle v_x^+ \rangle$) and the wall-normal ($\langle v_z^+ \rangle$) components of the mean particle velocity for one-way coupling (solid line with empty circles) and two-way coupling (black circles) simulations. Figures 3a and 3b are relative to $\tau_p^+ = 1$ particles, Figures 3c and 3d are relative to $\tau_p^+ = 5$ particles, Figures 3e and 3f are relative to $\tau_p^+ = 25$ particles. Modifications to the mean streamwise velocity are pretty small: profiles shown in Figures 3a, 3c and 3e overlap almost perfectly regardless of particle size, and only slight deviations can be observed for the larger particles two-way coupled with the fluid.

More noticeable (and meaningful) differences are observed for the wall-normal velocity, shown in Figures 3b, 3d and 3f. Under the one-way coupling assumption, profiles of particle wall-normal velocity develop a peak in the buffer layer, which increases monotonically with particle inertia. Correspondingly, particle wall-normal turbophoretic accumulation increases with particle inertia. A two-way coupling between particles and fluid appears to modify the shape of the profiles from

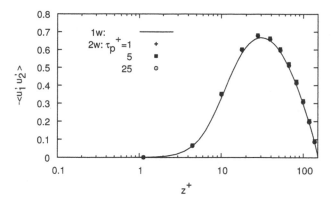

Fig. 2. Reynolds stress for one-way coupling (lines) and two-way coupling (symbols).

a quantitative (though not qualitative) viewpoint by shifting them towards smaller values for $\tau_p^+ = 1$ and 25 and towards larger values for $\tau_p^+ = 5$.

Results on second-order moments for the particle velocity field (not shown) provide evidence that RMS velocity fluctuations are not much affected by the two-way coupling in the outer region and slightly increase near the wall. The behavior qualitatively resembles that of the fluid flow field (see Figure 1).

Figure 4 shows the time evolution of particle concentration profiles along the wall-normal direction, for $\tau_p^+ = 1$ particles (Figure 4a), $\tau_p^+ = 5$ particles (Figure 4b) and $\tau_p^+ = 25$ particles (Figure 4c), respectively. Lines with empty symbols refer to one-way coupling simulations, whereas black symbols refer to two-way coupling simulations. Profiles are averaged in space (along the streamwise and spanwise directions), smoothed by time-averaging over spans of 360 time units and normalized with respect to the initial uniform concentration. It is apparent that particle interactions with turbulence act to decrease the near-wall peak of concentration. This behavior is in agreement with the decrease of particle drift velocity in the wall-normal direction previously observed in Figures 3b and 3f for $\tau_p^+ = 1$ and 25, respectively. Surprisingly, this is not the case for $\tau_p^+ = 5$ for which, despite of a larger wall-ward wall-normal velocity, the peaks of accumulation are also reduced with two-way coupling. This effect is more evident for the smaller particles ($\tau_p^+ = 1$, Figure 4a) and increases monotonically with particle inertia.

5 Concluding remarks

This paper addresses the issue of particle concentration in a fully developed turbulent boundary layer with specific reference to the influence of particle inertial response to the underlying flow field under one-way and two-way coupling assumptions.

Statistical analysis of particle and fluid velocity fields computed from numerical simulations run under dilute flow conditions provides evidence of the crucial effect of inertia in determining particle drift toward the wall and particle sampling of specific

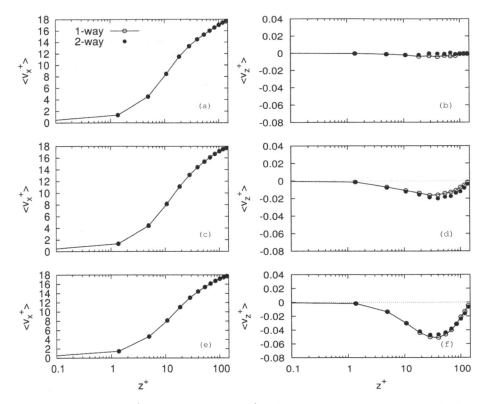

Fig. 3. Streamwise ($\langle v_x^+ \rangle$) and wall-normal ($\langle v_z^+ \rangle$) components of mean particle velocity for one-way coupling (solid line with empty circles) and two-way coupling (black circles). (a) and (b) $\tau_p^+ = 1$, (c) and (d) $\tau_p^+ = 5$, (e) and (f) $\tau_p^+ = 25$.

flow regions: as a consequence, particles accumulate in the near-wall region, this trend being enhanced by increasing particle inertia.

When particles segregate in specific flow regions, the effect of the dispersed phase on turbulence is no longer negligible and the dilute flow assumption is not valid locally. Simulations with a two-way coupling between particles and fluid were performed to investigate on turbulence modifications due to dispersion and segregation of particles with different inertia in the flow. For the particle sizes investigated in this work, turbulence modulation by particles appears rather small. This may be due to the small volume fraction occupied by the particles and to the fact that only the effect of the drag force was considered in the balance equation of particle motion. However, it was possible to observe that particle accumulation in the near-wall region is overestimated when the feedback of the dispersed phase onto the flow field is neglected. More detailed studies, focusing also on particle wall fluxes, are currently underway and will be addressed in forthcoming papers.

Further development of this work will be the analysis of additional effects on the mechanisms by which particles modulate turbulence. As mentioned in the paper,

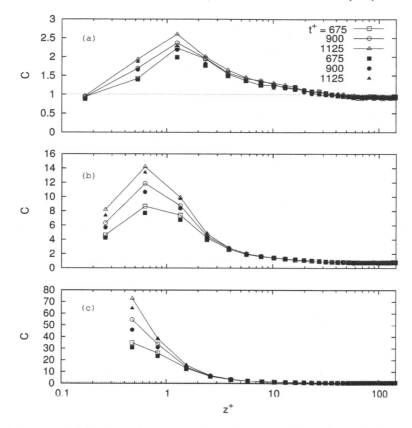

Fig. 4. Time evolution of particle concentration profiles along the wall-normal direction for one-way coupling (lines with empty symbols) and two-way coupling (black symbols) simulations. (a) $\tau_p^+ = 1$, (b) $\tau_p^+ = 5$, (c) $\tau_p^+ = 25$.

several potentially important effects (for instance lift, gravity, particle rotation, inter-particle collisions) have been neglected for the time being. Yet, we are aware of the importance of such effects in real two-phase systems and our future efforts will be devoted to their inclusion in the simulation setting. Another important issue is the effect of the particle time-scale on turbulence modulation. This effect can be easily singled out by varying particle size for fixed mass and volume fractions.

References

1. W.S. Uijttewaal and R.V.A. Oliemans, 1996, Particle dispersion and deposition in direct numerical and large eddy simulations of vertical pipe flows, *Phys. Fluids* **8**, 2590–2604.
2. K.T. Kiger and C. Pan, 2002, Suspension and turbulence modification effects of solid particulates on a horizontal turbulent channel flow, *J. Turbulence* **3**, 1–21.
3. L.P. Wang and M.R. Maxey, 1993, Settling velocity and concentration distribution of heavy particles in homogeneous isotropic turbulence, *J. Fluid Mech.* **256**, 27–68.

4. M.W. Reeks, 1977, On the dispersion of small particles suspended in an isotropic turbulent fluid, *J. Fluid Mech.* **83**, 529–546.
5. M.R. Maxey, 1987, The gravitational settling of aerosol particles in homogeneous turbulence and random flow fields, *J. Fluid Mech.* **174**, 441–465.
6. S.E. Elghobashi and T.W. Abou Arab, 1983, A two equation turbulence model for two-phase flows, *Phys. Fluids* **26**, 931–938.
7. J. Young and A. Leeming, 1997, A theory of particle deposition in turbulent pipe flow, *J. Fluid Mech.* **340**, 129–159.
8. S. Cerbelli, A. Giusti and A. Soldati, 2001, ADE approach to predicting dispersion of heavy particles in wall bounded turbulence, *Int. J. Multiphase Flow* **27**, 1861–1879.
9. M. Caporaloni, F. Tampieri, F. Trombetti and O. Vittori, 1975, Transfer of particles in nonisotropic air turbulence, *J. Atmos. Sci.* **32**, 565–568.
10. M.W. Reeks, 1983, The transport of discrete particles in inhomogeneous turbulence, *J. Aerosol Sci.* **310**, 729–739.
11. J.K. Eaton and J.R. Fessler, 1994, Preferential concentration of particles by turbulence, *Int. J. Multiphase Flow* **20**, 169–209.
12. C. Marchioli and A. Soldati, 2002, Mechanisms for particle transfer and segregation in turbulent boundary layer, *J. Fluid Mech.* **468**, 283–315.
13. C. Marchioli, A. Giusti, M.V. Salvetti and A. Soldati, 2003, Direct numerical simulation of particle wall transfer and deposition in upward turbulent pipe flow, *Int. J. Multiphase Flow* **29**, 1017–1038.
14. M. Picciotto, C. Marchioli and A. Soldati, 2005, Characterization of near-wall accumulation regions for inertial particles in turbulent boundary layers, *Phys. Fluids* **17**, 098101, DOI 10.1063/1.2033573.
15. M. Boivin, O. Simonin and K.D. Squires, 1998, Direct numerical simulation of turbulence modulation by particles in isotropic turbulence, *J. Fluid Mech.* **375**, 235–263.
16. A.A. Mostafa and H.C. Mongia, 1988, On the interaction of particles and turbulent fluid flow, *Int. J. Heat Mass Transfer* **31**, 2063–2075.
17. S. Dasgupta, R. Jackson and S. Sundaresan, 1998, Gas-particle flow in vertical pipes with high mass loading of particles, *Powd. Tech.* **96**, 6–23.
18. S. Sundaram and L.R. Collins, 1999, A numerical study of the modulation of isotropic turbulence by suspended particles, *J. Fluid Mech.* **379**, 105–143.
19. K. Lam and S. Banerjee, 1992, On the condition of streak formation in bounded flows, *Phys. Fluids A* **4**, 306–320.
20. S. Elghobashi and G.C. Truesdell, 1992, Direct simulation of particle dispersion in a decaying isotropic turbulence, *J. Fluid Mech.* **242**, 655–700.
21. C. Crowe, M. Sommerfeld and Y. Tsuji, 1998, *Multiphase Flows with Droplets and Particles*, CRC Press, New York.
22. M.R. Maxey and J.K. Riley, 1983, Equation of motion for a small rigid sphere in a nonuniform flow, *Phys. Fluids A* **26**, 883–889.
23. M. Picciotto, C. Marchioli, M. Reeks and A. Soldati, 2005, Statistics of velocity and preferential accumulation of micro-particles in boundary layer turbulence, *Nucl. Eng. Des.* **235**, 1239–1249.
24. O.A. Druzhinin, 2001, The influence of particle inertia on the two-way coupling and modification of isotropic turbulence by microparticles, *Phys. Fluids* **13**, 3738–3755.

Stochastic Diffusion of Finite Inertia Particles in Non-Homogenous Turbulence

Eric Loth[1] and Todd L. Bocksell[2]

[1] Department of Aerospace Engineering, University of Illinois at Urbana-Champaign, USA
[2] Pratt & Witney, East Hartford, CT, USA

Abstract. Several Continuous Random Walk (CRW) models were constructed to predict turbulent particle diffusion based only on mean Eulerian fluid statistics. The particles were injected near the wall ($y^+ = 4$) of a turbulent boundary layer that is strongly anisotropic and inhomogeneous near the wall. To assess the performance of the models for wide range of particle inertias (Stokes numbers), the CRW results were compared to particle diffusion statistics gathered from a Direct Numerical Simulation (DNS). The results showed that accurate simulation required a modified (non-dimensionalized) Markov chain for the large gradients in turbulence based on fluid-tracer simulations. For finite-inertia particles, a modified drift correction for the Markov chain (developed herein to account for Stokes number effects) was critical to avoiding non-physical particle collection in low-turbulence regions. In both cases, inclusion of anisotropy in the turbulent kinetic energy was found to be important, but the influence of off-diagonal terms was found to be weak.

1 Introduction

Simulating particle diffusion due to turbulence is important to many engineering systems. A common approach is to utilize the time-averaged velocity ($\overline{u_f}$) and turbulence properties (k, ε) from a Reynolds–Averaged Navier–Stokes (RANS) solution along with a Continuous Random Walk (CRW)model to simulate the instant aneous fluid fluctuation velocities (u'_f) seen by the particles in a Lagrangian frame. By tracking a large number of particles, mean particle statistical information is then obtained. This approach can also be used to model the sub-grid stress fluctuations for Large Eddy Simulations [1]. A key issue in using a Markov chain is that inhomogeneous turbulent flow can lead to a non-physical numerical diffusion of particles if the inhomogeneity is not included in the stochastic model. Several CRW studies have sought to take into account the inhomogeneous drift correction based on tracer (zero-inertia) particles. MacInnes and Bracco [2] investigated the performance of a CRW model, similar to the one of Legg and Raupach [3], in 2-D inhomogeneous turbulent flows of a turbulent mixing layer and an axisymmetric jet, and determined that a drift correction of

S. Balachandar and A. Prosperetti (eds.) Proceedings of the IUTAM Symposium on Computational Multiphase Flow, 63–74.
© 2006 Springer. Printed in the Netherlands.

$$\delta \overline{u'_{f_1}} = \tau_\Lambda \left\{ 1 - \exp\left(\frac{-\Delta t}{\tau_\Lambda}\right) \right\} \frac{\partial}{\partial x_j} (\overline{u'_{f_k} u'_{f_1}}) \delta_{ijkl} \qquad (1)$$

should be included in the Markov chain, where τ_Λ is the integral turbulent time-sale, $\overline{u'_{f_k} u'_{f_1}}$ is the Reynolds stress tensor, δ_{ijkl} is the Kronecker delta tensor, u'_{f_1} is the instantaneous fluid velocity fluctuation, and Δt is the time step. This can also be approximated to first-order as [4]

$$\delta \overline{u'_{f_i}} = \Delta t \left(\overline{\frac{Du'_{f_i}}{Dt}} \right) = \Delta t \left(\overline{u'_{f_j} \frac{\partial u'_{f_i}}{\partial x_j}} \right), \qquad (2)$$

Without this drift correction, errors of up to 500% for the *tracer* particle number concentration were found (while inclusion reduced the error to around 10%). Iliopoulos and Hanratty [5] utilized a normalized Langevin equation as the basis for a Markov chain in their analysis of near-wall fluid-tracer diffusion to avoid errors arising due to large turbulence gradients. For a single, uncorrelated velocity perturbation this gives

$$u'_f(t + \Delta t) = u'_f(t) \exp\left(\frac{-\Delta t}{\tau_L}\right) \frac{\sigma(t + \Delta t)}{\sigma(t)}$$

$$+ \left\{ 1 - \exp\left(\frac{-2\Delta t}{\tau_L}\right) \right\}^{1/2} \sigma(t + \Delta t)\xi(t). \qquad (3)$$

This was shown to work well for low inertia particles in a turbulent channel flow, where particle response times were small compared to the fluid time-scale. However, no previous studies (to the authors' knowledge) have derived the drift correction for finite-inertia particles, and it is often assumed that the drift correction is independent of Stokes number ($= \tau_p/\tau_\Lambda$, where τ_p is the particle response time and τ_Λ is the fluid time-scale). In this study, we aim to fulfill this need in the simulation technology by evaluating this new CRW method with DNS-derived statistics of particle diffusion data (thereby eliminating any issues associated with turbulence modeling).

2 Methodology

2.1 DNS solution and RANS-like statistics

The continuous phase solution for the turbulent boundary layer was obtained from DNS of the incompressible Navier–Stokes equations, assuming the particle concentration is dilute (does not effect the carrier phase) and negligible particle-particle interactions (a one-way coupled multiphase flow). The DNS code was developed by Spalart and Watmuff [6] to simulate a three-dimensional, spatially developing turbulent boundary layer with zero-streamwise pressure gradient. The Reynolds number for the present study is 4500 ($\mathrm{Re}_\delta = \rho_f U_\infty \delta / \mu_f$, where δ is the boundary layer thickness U_∞ is the free-stream velocity, and μ_f and ρ_f are the fluid viscosity and density). Grid-independent results were obtained for a domain discretized by 256

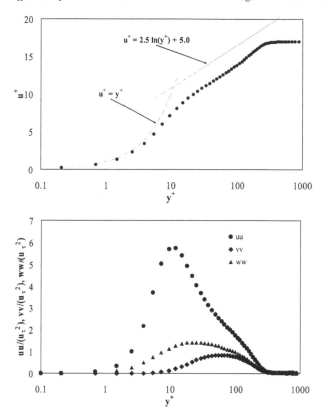

Fig. 1. DNS statistics at the particle injection plane: (a) mean velocity profile, (b) turbulent auto-correlation profiles.

nodes in the stream direction, 96 in the span direction, and 55 in the transverse direction for a total of 1,351,680 nodes in the three-dimensional mesh. Spatial evolution aspects and time integration details are given by Bocksell [7] and Dorgan [8].

From the DNS flow, Eulerian time and spanwise averaged statistics of the fluid properties near the injection location are shown in Figure 1. The transverse profiles of the mean velocity (in wall units) are shown in Figure 1a. For the mean velocity profile, there is evidence of the viscous sublayer below $y^+ \approx 20$, transition to a logarithmic curve is seen by $y^+ \approx 50$, and a boundary layer edge is located at roughly $y^+ \approx 270$ (i.e. $Re_\tau = 270$, where Re_δ is the Reynolds number based on δ and u_τ). The conventional "law of the wall" curves are included for the purpose of comparing the data to high Reynolds number boundary layers. The turbulent velocity fluctuations (normalized by u_τ^2) are shown in Figure 1b where the peak values for $v'_{f,\mathrm{rms}}$ are similar in magnitude to experimental results of Klebanov [9] at $Re_\tau = 2800$, though at somewhat larger y^+ locations. These Eulerian statistics, along with the full Reynolds-stress tensor and the turbulent dissipation (ε), were used to construct a typical RANS-like turbulent boundary layer flow solution [7]. Also noted by Bock-

Table 1. Particle conditions for the DNS particle simulations.

St_δ	$\langle St_\Lambda \rangle$	St^+	$\Delta t/\tau_p$	$\Delta t/\tau^+$
10^{-4}	7.8×10^{-4}	0.027	6.6×10^{-2}	0.57
10^{-3}	7.8×10^{-3}	0.27	6.6×10^{-3}	0.57
10^{-2}	8.3×10^{-2}	2.7	6.6×10^{-4}	0.57
0.1	1.2	27	6.6×10^{-5}	0.57
1	13.9	270	6.6×10^{-6}	0.57

sell [7], the Lagrangian turbulent time and length scales (τ_Λ and Λ) in a boundary layer are not approximated well by the free-shear flow assumption and instead depend on distance from the wall (y^+). Therefore, the DNS was also utilized to obtain the Lagrangian values of τ_Λ and Λ using fluid tracer statistics as functions of distance from the wall (which compared well with the estimates of Kallio and Reeks [10]).

3 Particle equation of motion

The particle equation of motion for both the DNS and CRW simulations is

$$m_p \frac{du_{p_i}}{dt} = 3\pi \mu_f d_p (u_{f_i} - u_{p_i}) + m_p g_i, \qquad (4)$$

where m_p is the particle mass, d_p is the particle diameter, u_{p_i} is the particle velocity vector, and g_i is the gravity vector. This equation assumes spherical solid particles with a Stokesian drag and a particle density that is much greater than the fluid density so that other forces (lift, stress gradient, and Basset history) are negligible. The equation of motion was integrated using a modified version of the exponential-Lagrangian method first described by Barton [11] and later generalized and improved by Bocksell [7]. This method is an Adams–Bashforth multistep integration scheme, implemented in a predictor-corrector fashion that is second-order accurate in time.

For each set of test conditions, roughly 100,000 particles (for each test condition) are injected over a range of times and spanwise positions at $y^+ = 4$ with an elastic reflection imposed at $y^+ = 1$ for all downstream wall interactions. The time step used for the DNS (both fluid and particle simulations) was constant for all cases and appears in Table 1 in various non-dimensional forms. The following definitions are used for the test conditions: the particle integral-scale Stokes number ($St_\Lambda = \tau_p/\tau_\Lambda$); the wall-based Stokes number is the ratio of the particle relaxation time to the wall based time scale ($St^+ = \tau_p \rho_f u_\tau^2/\mu_f$); the outer Stokes number is the ratio of particle relaxation time to the outer time scale ($St = \tau_p U_\infty/\delta$; the drift parameter is the ratio of particle terminal velocity ($V_{\text{term}} = g\tau_p$) to the root-mean-square of the fluid fluctuation velocities ($\gamma = V_{\text{term}}/u'_{f,\text{rms}}$); and the particle Reynolds number based on the particle terminal velocity ($Re_{p,\text{term}} = \rho_f d_p V_{\text{term}}/\mu_f$). In order to understand how inertia influences particle diffusion in a boundary layer, five different particle inertias (Table 1) were selected with outer Stokes number varying from $St_\delta = 10^{-4}$ to $St_\delta = 1$, all at a constant γ of 10^{-2} (with V_{term} directed away from the wall). The low value of γ ensures that the terminal velocity is small compared to the fluid velocity fluctuations so that inertia and turbulent diffusion effects dominate the particle dispersion (as opposed to gravity). By recording the observed integral fluid Lagrangian time scale (τ_Λ) along the particle path, an average local integral-scale Stokes number was obtained, $\langle St_\Lambda \rangle = \tau_p/\langle \tau_\Lambda \rangle$, for each class of particles and these values also appear in Table 1. Details of the particle dispersion physics are given in Dorgan [8].

4 Investigated CRW methods

One of the main goals of this research is to evaluate the performance of the CRW model with regard to the amount of turbulence information available. Conventional and normalized Markov chains are described here for three different types of simulations: (1) isotropic turbulence, (2) anisotropic turbulence but no Reynolds stresses, and (3) anisotropic turbulence and including the Reynolds stresses. For consistency, the time-scale treatment, the velocity fluctuation treatment, and the incremental drift correction treatment were all identical in terms of the level of turbulence information resulting in three types of CRW simulations to evaluate the importance of the anisotropy. These three simulation types are summarized in Table 2 and defined in detail by Bocksell [7].

For a boundary layer, the only non-zero cross-correlation is the u–v cross correlation so the conventional Markov chain for the "full" simulations is

$$
\begin{bmatrix} u(t+\Delta t) \\ v(t+\Delta t) \\ w(t+\Delta t) \end{bmatrix} = \begin{bmatrix} k_u & 0 & 0 \\ 0 & k_v & 0 \\ 0 & 0 & k_w \end{bmatrix} \begin{bmatrix} u(t) \\ v(t) \\ w(t) \end{bmatrix}
$$

$$
+ \begin{bmatrix} \sigma_u\sqrt{1-k_u^2} & 0 & 0 \\ 0 & \sigma_v\sqrt{1-k_v^2} & 0 \\ 0 & 0 & \sigma_w\sqrt{1-k_w^2} \end{bmatrix} \begin{bmatrix} \sqrt{1-b^2} & b & 0 \\ 0 & 1 & 0 \\ 0 & 0 & 1 \end{bmatrix} \begin{bmatrix} \xi_u \\ \xi_v \\ \xi_w \end{bmatrix}, \quad (5)
$$

$$
k_u = \exp\left(\frac{-\Delta t}{\tau_{L_u}}\right), \quad k_v = \exp\left(\frac{-\Delta t}{\tau_{L_v}}\right), \quad k_w = \exp\left(\frac{-\Delta t}{\tau_{L_w}}\right)
$$

Table 2. Summary of types of CRW simulations.

Name	Turbulence Type	Time-scale Type
Isotropic	$\overline{uu} = \overline{vv} = \overline{ww} = 2k/3, \overline{uv} = 0$	$\tau_{L_u} = \tau_{L_v} = \tau_{L_w}$
Diagonal	$\overline{uu} \neq \overline{vv} \neq \overline{ww}, \overline{uv} = 0$	$\tau_{L_u} \neq \tau_{L_v} \neq \tau_{L_w}$
Full	$\overline{uu} \neq \overline{vv} \neq \overline{ww}, \overline{uv} \neq 0$	$\tau_{L_u} \neq \tau_{L_v} \neq \tau_{L_w}$

$$b = \frac{R_{uv}\left[1 - \exp\left(\frac{-\Delta t}{\tau_{Lu}}\right)\exp\left(\frac{-\Delta t}{\tau_{Lv}}\right)\right]}{\left[1 - \exp\left(\frac{-2\Delta t}{\tau_{Lu}}\right)\right]^{1/2}\left[1 - \exp\left(\frac{-2\Delta t}{\tau_{Lv}}\right)\right]^{1/2}}, \quad \text{and} \quad R_{uv} = \frac{\overline{uv}}{\sigma_u \sigma_v}. \quad (6)$$

The normalized Markov chain for the "full" simulations is

$$\begin{bmatrix} u(t + \Delta t) \\ v(t + \Delta t) \\ w(t + \Delta t) \end{bmatrix} = \begin{bmatrix} k_u & 0 & 0 \\ 0 & k_v & 0 \\ 0 & 0 & k_w \end{bmatrix} \begin{bmatrix} \frac{\sigma_u^*(t+\Delta t)}{\sigma_u(t)} & 0 & 0 \\ 0 & \frac{\sigma_v^*(t+\Delta t)}{\sigma_v(t)} & 0 \\ 0 & 0 & \frac{\sigma_v^*(t+\Delta t)}{\sigma_w(t)} \end{bmatrix} \begin{bmatrix} u(t) \\ v(t) \\ w(t) \end{bmatrix}$$

$$+ \begin{bmatrix} \sigma_u^*(t + \Delta t)\sqrt{1 - k_u^2} & 0 & 0 \\ 0 & \sigma_v^*(t + \Delta t)\sqrt{1 - k_v^2} & 0 \\ 0 & 0 & \sigma_w^*(t + \Delta t)\sqrt{1 - k_w^2} \end{bmatrix}$$

$$\times \begin{bmatrix} \sqrt{1 - b^2} & b & 0 \\ 0 & 1 & 0 \\ 0 & 0 & 1 \end{bmatrix} \begin{bmatrix} \xi_u \\ \xi_v \\ \xi_w \end{bmatrix}, \quad (7)$$

The main difference between the conventional Markov chain of (5) and the normalized Markov chain of (7) is the ratio of the root-mean-square of the velocity fluctuations from the previous time step and the next time step. Essentially this decorrelates (in time) the velocity fluctuations along a particle path in regions where gradients in the mean turbulence quantities are large (near the wall).

5 Particle drift correction for finite-inertia particles

As noted in the introduction, a drift correction for a finite mass particle for both the conventional and normalized Markov chains has been developed in this study. This is different than previous fluid-tracer drift corrections [2, 3, 5] since the total differential of the fluid velocity fluctuation along a particle trajectory includes both the fluid and particle velocities:

$$\frac{du'_{f_i}}{dt} = \frac{\partial u'_{f_i}}{\partial t} + u_{p_j}\frac{\partial u'_{f_i}}{\partial x_j}. \quad (8)$$

Taking the Eulerian time-average results in

$$\overline{\frac{du'_{f_i}}{dt}} = u_{p_j}\overline{\frac{\partial u'_{f_i}}{\partial x_j}}. \quad (9)$$

The goal is to replace the right-hand-side correlation between the particle velocity and fluid velocity fluctuation gradient with particle characteristics and Eulerian fluid correlations. Starting from the particle equation of motion (4), introducing Reynolds averaging, utilizing Laplace transforms, and then taking the limit as (as discussed by Bocksell [7]) results in the "finite-inertia incremental drift correction" as

$$\overline{u_{p_j} \frac{\partial u'_{f_i}}{\partial x_j}} = \overline{u'_{f_j} \frac{\partial u'_{f_i}}{\partial x_j}} \left(\frac{1}{1 + \text{St}_\Lambda} \right). \tag{10}$$

This finite-inertia drift correction tends to the proper fluid-tracer correction (2) as the particle inertia becomes negligible ($\text{St}_\Lambda \to 0$) and it tends to zero as the particle inertia becomes high ($\text{St}_\Lambda \ll 0$). This latter limit is consistent with the eventual elimination of the correlation between fluid and particle velocity fluctuations for very large particles. Note that these limits would be observed even if a non-linear drag coefficient were used such that (10) is expected to be at least qualitatively reasonable at high particle Reynolds numbers.

The finite-inertia drift correction for the normalized Markov chain is similarly obtained for the normalized Langevin equation as

$$\overline{u_{p_j} \frac{\partial}{\partial x_j} \left(\frac{u'_{f_k}}{\sigma_{u_1}} \right)} \delta_{ikl} = \overline{u'_{f_j} \frac{\partial}{\partial x_j} \left(\frac{u'_{f_k}}{\sigma_{u_1}} \right)} \delta_{ikl} \left(\frac{1}{1 + \text{St}_\Lambda} \right). \tag{11}$$

Thus, for both the conventional and normalized Markov chains, the factor used to transform the particle-fluid correlation to fluid-fluid correlations, $1/(1 + \text{St}_\Lambda)$, is identical and the same limits occur.

When implementing the incremental drift correction for the CRW simulations, the turbulence correlations for the drift correction are treated consistently for the total Markov chain. For example, if the time and length scales for the CRW simulation are assumed isotropic, then the turbulence correlations in the incremental drift correction are also assumed isotropic (various forms of the tested CRW models are given in Table 3).

6 CRW results

6.1 Transverse concentration profiles

To test the drift corrections, fluid-tracer particles ($m_p \approx 0$, $\text{St}_\delta \approx 0$) were injected uniformly (with respect to mass flux) throughout the boundary layer (from $y = 0$ to $y > \delta$) and by conservation of mass, the concentration profile should remain uniform as they move downstream (in an averaged sense). The results of the CRW simulations of this type of tracer particle injection with isotropic turbulence and isotropic time scale using the conventional Markov chain of (5) gave very poor results (as expected) since no drift correction was applied [7]. Including the incremental drift correction for the conventional, isotropic Markov chain substantially reduced but did not eliminate the non-physical peaks of particle concentration in the near-wall of the boundary layer as shown in Figure 2a. Note that the correct result is a uniform

Table 3. Summary of the types of incremental drift velocities utilized for the CRW model.

Drift Type	Markov Chain	Increment Drift Formula
No Drift	Conventional & Normalized	$\overline{\delta u'_{f_i}} = 0$
Fluid-Tracer	Conventional	$\overline{\delta u'_{f_i}} = \Delta t\, \overline{u'_{f_j} \dfrac{\partial u'_{f_i}}{\partial x_j}}$
Finite-inertia	Conventional	$\overline{\delta u'_{f_i}} = \Delta t\, \overline{u'_{f_j} \dfrac{\partial u'_{f_i}}{\partial x_j}} \left(\dfrac{1}{1+\mathrm{St}}\right)$
Fluid-Tracer	Normalized	$\overline{\delta u'_{f_i}} = \Delta t\, \overline{u'_{f_j} \dfrac{\partial}{\partial x_j} \left(\dfrac{u'_{f_k}}{\sigma_{u_1}}\right)} \delta_{ikl}$
Finite-inertia	Normalized	$\overline{\delta u'_{f_i}} = \Delta t\, \overline{u'_{f_j} \dfrac{\partial}{\partial x_j} \left(\dfrac{u'_{f_k}}{\sigma_{u_1}}\right)} \delta_{ikl} \left(\dfrac{1}{1+\mathrm{St}}\right)$

concentration as shown by the solid line, but there is a significant amount of wall-peaking ($C/C_0 > 2.0$). This case also yielded a high number of non-physical wall collisions for the fluid-tracer particles (about 50 collisions for every 1000 particle injected). Thus, there is an incorrect description of the CRW velocity perturbations for tracer trajectories as particles approach the wall such that the fluctuation velocity seen by the tracer particle does not de-correlate as quickly as the real system (a true fluid particle should *never* bounce). This situation can occur frequently with the conventional Markov chain since a fluid particle approaching the wall can have a negative (wall-ward) transverse velocity fluctuation whereby $|v'_f(t)| > \tau_{\mathrm{int}}/y_p$ for $10 < y^+ < 20$. This problem is rectified by using the normalized Markov chain, which is simply a transformation from inhomogeneous turbulence to homogenous turbulence. Application of this normalized Markov chain yielded an order of magnitude reduction in wall collisions. This improvement is also reflected in Figure 2b which shows the concentration profiles and it can be seen that the CRW results are close to the exact solution throughout the boundary layer. Because of this, all the CRW simulations shown hereafter employ the normalized Markov chain (other results with the conventional Markov chain are given in [7]).

Figure 3 contains the results for particle simulations at the furthest downstream collection plane, $x/\delta = 15$, such that $t \gg \tau_\Lambda$ for the two extremes in Stokes number conditions ($\mathrm{St}_\delta = 10^{-4}$ and $\mathrm{St}_\delta = 1$). In Figure 3a, the concentration profiles from the diagonal and full Reynolds-stress CRW simulations for the near tracer particle case ($\mathrm{St}_\delta = 10^{-4}$) are quite close to the results from the DNS (the full Reynolds-stress results are slightly better). However, the simulations utilizing the isotropic, normalized Markov chain significantly under predict the particle diffusion in the $10 < y^+ < 100$ region. This is reasonable since the isotropic definition of the transverse velocity fluctuations from the kinetic energy results in an over-estimate of the actual $v'_{f,\mathrm{rms}}$ values (Figure 1b) and thus causes the particles to diffuse faster away

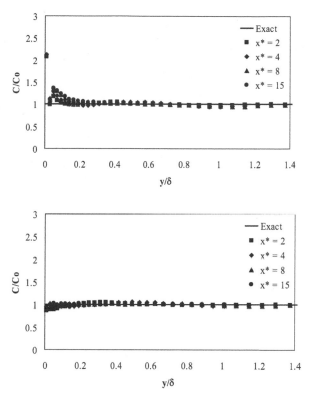

Fig. 2. Fluid-tracer particle concentration profiles for the turbulent boundary layer (injected uniformly with respect to mass flux) at four downstream locations for the CRW model with: (a) conventional fluid-tracer drift correction, (b) normalized fluid-tracer drift correction.

from the wall and reduce the near-wall concentration. These results indicate that the anisotropy in a boundary layer should be included in the Markov chain to obtain accurate near-wall results. Similarly, the $St_\delta = 10^{-4}$ and $St_\delta = 10^{-1}$ results show [7] substantial improvement with an anisotropic turbulence model but the improvement with adding the off-diagonal terms is slight. The results for particles with the largest Stokes number ($St_\delta = 1$) appear in Figure 3b and show the same trends. Results for other streamwise locations for the full range of particle Stokes numbers exhibited the same features [7] and also showed that the neglecting to use the finite inertia drift correction developed here gave poor results for $St_\delta > 10^{-2}$ (this will be demonstrated in the next paragraph).

In order to assess the mean particle trajectory movement normal to the wall, Figure 4 shows a comparison of the particle velocity averaged along the particle trajectory, $\langle v_p \rangle$, normalized by V_{term}, and plotted as a function of particle Stokes number. Since $\langle v_p \rangle$ is always greater than V_{term}, the net movement away from the wall is generally dominated by turbulent diffusion rather than gravitational settling. This is especially true for the tracer-like particles ($St_\delta = 10^{-4}$) for which the velocity

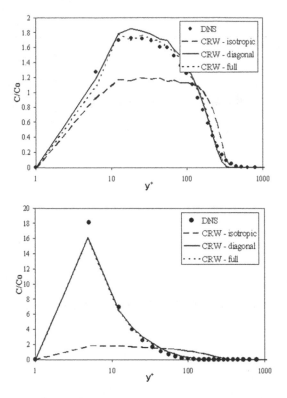

Fig. 3. Particle concentration profiles from DNS and CRW simulations for finite-inertia incremental drift correction at $x/\delta = 15$ with (a) $St_\delta = 10^{-4}$ and (b) $St_\delta = 1$.

ratio is nearly thirty. However, for the largest particles ($St = 1$) the mean transverse velocity approaches V_{term}. Movement away/toward the wall of V_{term} would occur for very large particles that are not immune to the effects of the fluid turbulence. Figure 4a contains the results from the CRW simulations with the tracer-particle drift correction and Figure 4b contains the CRW simulations with the finite-inertia drift correction; both cases also compare the isotropic and anisotropic diagonal CRW models. In general, the anisotropic effect is important at all Stokes numbers while the isotropic model consistently over-predicts the mean transverse velocity. For the anisotropic CRW simulations, the fluid-tracer drift correction is reasonable up to $St_\delta = 10^{-2}(St_\Lambda = 0.083)$. However, for particles with $St_\delta = 10^{-1}(St_\Lambda = 1)$ and larger, the finite-inertia drift correction is needed and gives good results.

Acknowledgments

The authors would like to acknowledge Dr. P. K. Yeung for help with obtaining the DNS continuous-phase results, Mr. A. Dorgan help with obtaining the DNS particle-

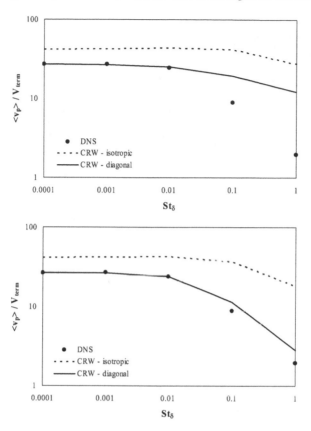

Fig. 4. Plot of the averaged vertical (wall-normal) particle velocity normalized by the terminal velocity *versus* particle Stokes numbers for (a) fluid-tracer particle drift correction and (b) finite-inertia drift correction.

phase results, the Defense Advanced Research Projects Agency (DARPA) for funding support, and MHPCC for computer allocation time.

References

1. Loth, E., 2000, Numerical approaches for motion of dispersed particles, droplets, and bubbles, *Progress in Energy and Combustion Science* **26**, 161–223.
2. MacInnes, J.M. and Bracco, F.V., 1992, Stochastic particle dispersion modeling and the tracer-particle limit, *Physics of Fluids A* **12**, 2809–2824.
3. Legg, B.J. and Raupach, M.R., 1982, Markov-chain simulation of particle dispersion in inhomogeneous flows: The mean drift velocity induced by a gradient in the Eulerian velocity variance, *Boundary-Layer Meteorology* **24**, 3–13.
4. Bocksell, T.L. and Loth, E., 2001, Random walk models for particle diffusion in free-shear flows, *AIAA Journal* **39**(6), 1086–1096.

5. Iliopoulos, I. and Hanratty, T.J., 1999, Turbulent dispersion in a non-homogeneous field, *Journal of Fluid Mechanics* **392**, 45–71.
6. Spalart, P.R. and Watmuff, J.H., 1993, Experimental and numerical study of a turbulent boundary layer with pressure gradients, *Journal of Fluid Mechanics* **249**, 337–371.
7. Bocksell, T.L., 2004, Numerical simulation of turbulent particle diffusion, Ph.D. Thesis, University of Illinois at Urbana-Champaign, Urbana, IL.
8. Dorgan, A.J., 2003, Boundary layer dispersion of near-wall injected particles of various inertias, M.S. Thesis, University of Illinois at Urbana-Champaign, Urbana, IL.
9. Hinze, J.O., 1975, *Turbulence*, McGraw-Hill, New York.
10. Kallio, G.A. and Reeks, M.W., 1989, A numerical simulation of particle deposition in turbulent boundary layers, *International Journal of Multiphase Flow* **15**(3), 433–446.
11. Barton, I.E., 1996, Exponential-Lagrangian tracking schemes applied to Stokes law, *Journal of Fluids Engineering* **118**, 85–89.

Accumulation of Heavy Particles in Bounded Vortex Flows

Rutger H.A. IJzermans and R. Hagmeijer

University of Twente, P.O. Box 217, 7500 AE Enschede, The Netherlands;
e-mail: r.h.a.ijzermans@ctw.utwente.nl, r.hagmeijer@ctw.utwente.nl

Abstract. Much research has been done on the motion of heavy particles in simple vortex flows. In most of this work, particle motion is investigated under the influence of fixed vortices. In the context of astrophysics, the motion of heavy particles in rotating two-dimensional flows has been investigated; the rotation follows from the laws of Kepler. In the present paper, the motion of heavy particles in potential vortex flow in a circular domain is investigated. The vortex describes a circular trajectory due to the presence of the boundary, so that a steadily rotating flow is obtained. In order to isolate the effect of particle inertia, only Stokes drag is taken into account in the equation of motion. The numerical simulations are based on a one-way coupling. They show that small heavy particles accumulate in an ellitpic region of the flow, counterrotating with respect to the vortex. When the particle Stokes number exceeds a threshold, depending on the vortex configuration, particles are expelled from the circular domain. A stability criterion for this particle accumulation is derived analytically. These results are qualitatively comparable to those obtained by others in astrophysics.

1 Introduction

Gas-particle separators are used in some industrial processes. Their purpose is to separate liquid droplets or small heavy particles from gas flows. In general the separators consist of a cylindrical tube containing a region of high vorticity. In some applications this region of high vorticity has a helical shape. The goal of the present research is to determine the influence of this coherent structure of vorticity on the properties of heavy particle separation.

The configuration of a steady helical vortex filament in a cylindrical tube is sketched in Figure 1. The three-dimensional (potential) velocity field for this situation was first derived by Alekseenko et al. [1]. The calculation of this velocity field is far from trivial due to the torsion of the helical vortex filament.

If, however, the pitch of the helix is sufficiently large compared to the tube radius, the contribution due to the three-dimensionality of the helical vortex filament can be neglected. In this limit, the velocity field reduces to a superposition of a constant axial velocity and a time-dependent two-dimensional flow in the cross-sectional

S. Balachandar and A. Prosperetti (eds), Proceedings of the IUTAM Symposium on Computational Multiphase Flow, 75–85.

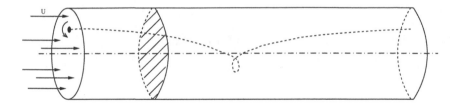

Fig. 1. Typical configuration of gas-liquid separator.

plane, moving with velocity U (see Figure 1). Here, we use this two-dimensional approximation. The two-dimensional flow is characterized by an eccentrically placed point vortex in a circular domain. The vortex rotates at constant angular velocity due to its self-induced motion.

The motion of heavy particles in dilute suspensions has received much attention in the past two decades. Investigations (e.g. [5–7]), have reported the motion of small heavy particles in elementary vortex flows. Most of them focussed on the motion of particles near fixed vortices. The general conclusion is that heavy particles are expelled from regions of high vorticity and tend to accumulate in regions of high strain. The particle segregation was shown to be highest for particles whose relaxation time corresponds to a typical time scale of the flow [4]. This causes also the effect of preferential concentration observed in turbulent flows [11].

The motion of heavy particles in two-dimensional rotating flows has been investigated in the context of planet formation from the solar nebula [3, 10]. The solar nebula is a collection of gas particles situated on a large disk, which rotates following the laws of Kepler. The turbulent flow in the solar nebula is approximately two-dimensional, so that large coherent vortex structures are likely to occur. Provenzale [10] gives a good overview of the motion of heavy particles in a two-dimensional flow field with a finite vorticity distribution. Chavanis [3] makes an analytical estimate of the time it takes to capture a heavy particle in an anticyclonic vortex, by assuming the flow to be a superposition of a prescribed elliptic patch of uniform vorticity and a steadily rotating Keplerian disk.

In this paper we investigate the motion of heavy particles in closed circular domains containing a point vortex. The presence of the boundary gives rise naturally to a steadily rotating flow field [9]. The focus in this paper will be on the accumulation of particles in certain flow regions due to their inertia. In order to isolate the effect of the particle inertia, the simulations are based on a one-way coupling. Gravity is neglected, since it is typically a minor effect in industrial gas-liquid separators. A stability criterion for particle accumulation is derived for any steadily rotating flow field which can be expressed in terms of a stream function. It is shown that the general results correspond to those obtained by Chavanis [3], Provenzale [10] and others.

The paper is organized as follows. In Section 2 we present the dynamical equations governing the motion of a point vortex on a unit disk. Besides, we give the equation of motion of passive tracers in such flow, and the equation of motion of heavy particles. In Section 3 we present the numerical results of motion of heavy particles in a circular domain containing one vortex; analysis is used to explain the results for the trajectories of heavy particles in such flows. Finally, a summary and conclusions are given in Section 4.

2 Dynamical equations

The goal of the present research is to investigate the motion of heavy particles in a flow of one vortex on a disk. The governing equations are related to the motion of the point vortex under its self-induced velocity, to the motion of passive tracers in the flow and to the motion of small heavy particles in such flows. The equations governing these three types of motion are presented in this section.

2.1 Point vortex motion on a unit disk

Flows with N point vortices are singular solutions of the 2D Euler equations and can be seen as a Hamiltonian system. If the velocity field is divergence-free ($\nabla \cdot \boldsymbol{u} = 0$), the motion of passive tracers is governed by a stream function Ψ which plays the role of a Hamiltonian. It is well-known [9] that the motion of point vortices is Hamiltonian, too.

We consider the example of one point vortex on a disk. All variables are made dimensionless by the vortex strength and the cylinder radius. The distance from the vortex to the disk center is denoted by r_v. In order to satisfy the boundary condition (zero normal velocity on the circular boundary), a counter-rotating image vortex is placed outside the domain, on a distance $1/r_v$ ([9]).

The Hamiltonian, governing the motion of the vortex, becomes:

$$H = \frac{1}{4\pi} \ln \left[1 - x_v^2 - y_v^2 \right],$$ (1)

so the motion of the vortex is:

$$\dot{x}_v = \frac{\partial H}{\partial y_v} = \frac{1}{2\pi} \left(\frac{-y_v}{1 - x_v^2 - y_v^2} \right), \qquad \dot{y}_v = -\frac{\partial H}{\partial x_v} = \frac{1}{2\pi} \left(\frac{x_v}{1 - x_v^2 - y_v^2} \right). \quad (2)$$

This shows that the vortex moves on a circle of constant radius $\sqrt{x_v^2 + y_v^2} = r_v$ with constant angular velocity. This angular velocity is here called $\dot{\theta}_v$ and is given by:

$$\dot{\theta}_v = \frac{1}{2\pi} \left(\frac{1}{1 - r_v^2} \right).$$ (3)

2.2 Passive tracers in bounded vortex flow

The time-dependent stream function governing the motion of passive tracers reads:

$$\Psi(x, y, t) = -\frac{1}{4\pi} \ln \frac{(x - x_v)^2 + (y - y_v)^2}{\left(x - (x_v/r_v^2)\right)^2 + \left(y - (y_v/r_v^2)\right)^2}.$$ (4)

Then, the velocity of passive tracers follows from:

$$U = \frac{\partial \Psi}{\partial y}, \qquad V = -\frac{\partial \Psi}{\partial x}. \tag{5}$$

The stream function can be simplified by applying the following coordinate transform:

$$\xi(x, y, t) \equiv x \cos\theta_v + y \sin\theta_v,$$
$$\eta(x, y, t) \equiv -x \sin\theta_v + y \cos\theta_v.$$

This means that a reference frame is chosen that rotates with the vortex. In this frame, we define:

$$\overline{\Psi}(\xi(x, y, t), \eta(x, y, t)) \equiv \Psi(\xi(x, y, t), \eta(x, y, t), 0) = \Psi(x, y, t). \tag{6}$$

Substituting the expression for $\overline{\Psi}$ and the coordinate transform into Equation (5) yields:

$$U = \sin\theta_v \frac{\partial \overline{\Psi}}{\partial \xi} + \cos\theta_v \frac{\partial \overline{\Psi}}{\partial \eta}, \tag{7}$$

$$V = -\cos\theta_v \frac{\partial \overline{\Psi}}{\partial \xi} + \sin\theta_v \frac{\partial \overline{\Psi}}{\partial \eta}. \tag{8}$$

Besides, it is easily derived that the velocity in the co-rotating frame, denoted by (υ, v) satisfies:

$$\upsilon = U \cos\theta_v + V \sin\theta_v + \dot{\theta}_v \eta, \tag{9}$$
$$v = -U \sin\theta_v + V \cos\theta_v - \dot{\theta}_v \xi. \tag{10}$$

In order to obtain a stream function $\hat{\Psi}$ in the co-rotating frame such that:

$$\upsilon = \frac{\partial \hat{\Psi}}{\partial \eta}, \qquad v = -\frac{\partial \hat{\Psi}}{\partial \xi}, \tag{11}$$

we define:

$$\hat{\Psi}(\xi, \eta) \equiv \overline{\Psi}(\xi, \eta) + \frac{1}{2}\dot{\theta}_v(\xi^2 + \eta^2). \tag{12}$$

The total stream function $\hat{\Psi}$ then reads:

$$\hat{\Psi}(\xi, \eta) = \frac{1}{2}\dot{\theta}_v(\xi^2 + \eta^2)^2 - \frac{1}{4\pi} \ln \frac{(\xi - r_v)^2 + \eta^2}{(r_v\xi - 1)^2 + r_v^2\eta^2}, \tag{13}$$

where, for convenience, the vortex is placed on the positive ξ-axis. Contour lines of the stream function are plotted in Figure 2 (see also [9], p. 135). The boundary of the circular domain is a streamline of the flow, as it should be in order to guarantee zero wall-normal velocity on the boundary.

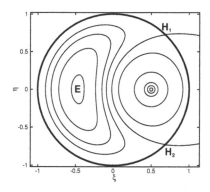

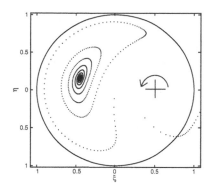

Fig. 2. Contour lines of stream function describing the motion of passive tracers in a one-vortex system, plotted in the frame rotating with the vortex; $r_v = 0.5$. H_1 and H_2 are hyperbolic stagnation points, E is an elliptic stagnation point.

Fig. 3. Poincaré sections of 2 slipping particles in one-vortex system. $r_v = 0.5$; $St = 0.5$.

Stagnation points correspond to critical points of the stream function, i.e. points where the flow velocity is zero. The Hessian, which is defined as:

$$\mathcal{H} \equiv \hat{\Psi}_{\xi\xi}\hat{\Psi}_{\eta\eta} - \hat{\Psi}_{\xi\eta}^2, \tag{14}$$

is used to determine the character of the stagnation point (the subscripts indicate differentiation). With the Hessian in the stagnation point denoted by $\mathcal{H}_0$, the following classification can be made:

$$\begin{aligned} \mathcal{H}_0 < 0 &\Leftrightarrow \text{saddle point (hyperbolic point)}, \\ \mathcal{H}_0 > 0 &\Leftrightarrow \text{extremum (elliptic point)}. \end{aligned} \tag{15}$$

With help of Equation (12), the Hessian can also be rewritten in terms of the stream function $\overline{\Psi}(\xi, \eta)$. Since $\nabla^2\overline{\Psi}(\xi, \eta) = 0$ (irrotational flow), it follows that:

$$\mathcal{H} = -\overline{\Psi}_{\xi\xi}^2 - \overline{\Psi}_{\xi\eta}^2 + \dot{\theta}_v^2. \tag{16}$$

From this it follows that if $\dot{\theta}_v = 0$, which corresponds to the instantaneous flow field in the quiescent frame, only hyperbolic stagnation points exist. If, on the other hand, $\dot{\theta}_v > 0$, then also an elliptic stagnation point may arise. This elliptic stagnation point is always counter-rotating (anticyclonic) with respect to $\dot{\theta}_v$.

An example of a rotating point vortex flow field with both hyperbolic and elliptic stagnation points is shown in Figure 2. This is the flow field induced by one single point vortex in a circular boundary, plotted in the frame rotating with the vortex. In this frame, the streamlines are independent of time.

2.3 Motion of heavy particles

Using the one-vortex flow as the background flow field, we now consider the motion of heavy particles in such a flow. The particles in relevant applications (such as small iced droplets in gas-liquid separators) are small and to good approximation spherical. In most relevant applications of gas-liquid separators, the influence of gravity can be neglected. For the sake of simplicity, effects of inter-particle collisions are not taken into account. The particles are assumed not to influence the gas flow, so the approach presented here is based on a one-way coupling.

At the beginning of the simulation, the particles are assumed to have the same velocity as the local gas flow. The particles are allowed to cross the circular boundary, but this does not have a significant effect on the results: a particle that has left the domain does not enter it again.

The dynamical equations for small spherical particles have been established by Maxey and Riley [8]. Under the assumptions above they reduce to the following equation, which reads in a quiescent frame and in dimensionless form:

$$\frac{dx_p}{dt} = u_p,$$ (17)

$$\frac{du_p}{dt} = \frac{1}{St}(u_g - u_p).$$ (18)

where x_p and u_p are the position and the velocity of the particle respectively, u_g is the velocity of the gas. The parameter St is the Stokes number. This is the particle relaxation time made dimensionless with respect to the vortex strength and the cylinder radius:

$$St \equiv \frac{\tau_p \Gamma}{R^2}.$$ (19)

Particles with $St = 0$ will react instantaneously to changes in the flow and will thus behave as passive tracers, whereas particles with $St \to \infty$ will be insensitive to the flow field.

In the rest of this paper, it turns out to be practical to rewrite the equations of motions in a rotating reference frame:

$$\frac{d\xi_p}{dt} = v_p,$$ (20)

$$\frac{dv_p}{dt} = \frac{1}{St}(v_g - v_p) + 2\dot{\theta}_v \wedge v_p + \dot{\theta}_v^2 \xi_p,$$ (21)

where ξ and v denote the position and the velocity in the rotating frame. The two additional terms on the RHS, which depend on the rotation rate $\dot{\theta}_v$, are the Coriolis force and the centrifugal force.

Consider the trajectories of two particles, which are initially very close. The initial differences in position and velocity are small and therefore denoted by $\delta\xi_p$ and δv_p, respectively. Now, the 4-dimensional separation vector $R \equiv [\delta\xi_p, \delta v_p]^T$ is introduced (see also [2]). If the separation between the two trajectories is small, the time development of the separation vector can be expressed in the following form:

$$\frac{d}{dt}R(t) = MR(t),$$ (22)

where the matrix M reads:

$$M = \begin{pmatrix} 0 & 0 & 1 & 0 \\ 0 & 0 & 0 & 1 \\ \frac{1}{St}\frac{\partial v_g}{\partial \xi} + \dot{\theta}_v^2 & \frac{1}{St}\frac{\partial v_g}{\partial \eta} & -\frac{1}{St} & 2\dot{\theta}_v \\ \frac{1}{St}\frac{\partial v_g}{\partial \xi} & \frac{1}{St}\frac{\partial v_g}{\partial \eta} + \dot{\theta}_v^2 & -2\dot{\theta}_v & \frac{1}{St} \end{pmatrix}. \tag{23}$$

Clearly, the separation vector can only be used in smooth flows for which the gradient of the velocity field exists. This will be no problem in our test cases. When all eigenvalues of the matrix M have a real part smaller than zero, the separation vector goes to 0 for $t \to \infty$. This means that the two particles converge towards each other.

3 Results: heavy particle motion in a bounded one-vortex flow

Now we investigate the motion of heavy particles in bounded vortex flows. Each particle is traced individually by using a fourth-order Runge–Kutta method. First, the equations of motion (Equation 18) are integrated using a fixed time step. Subsequently, the same integration is done with half of the time step. This procedure is repeated until the difference between two subsequent solutions is below a certain preset level.

In Figure 3, two different particle trajectories are plotted for the case $r_v = 0.5$. One particle, released on $(\xi, \eta) = (0.25, -0.2)$, is quickly expelled from the circular boundary and moves increasingly far away from the origin. The other particle, released on $(\xi, \eta) = (0, 0)$, is trapped in one particular attraction point within the circular domain.

This behavior is better perceptible when the positions of a group of heavy particles in the course of time are considered. In this case, we have taken 7495 particles which are uniformly distributed over the circular domain at the start of the simulation ($t = 0$). The particle positions are plotted in the frame rotating with the vortex in Figure 4. Clearly, many particles accumulate in the same point. This means that in physical space the particles approach to a circular trajectory periodic with the vortex motion.

The particle accumulation within the circular boundary occurs for a wide variety of initial conditions for the particle position. As an illustration, the particle trapping efficiency P, defined as:

$$P \equiv \frac{(\text{number of particles with } r < 1 \text{ for } t \to \infty)}{(\text{total number of initially uniformly distributed particles})} \times 100\%, \tag{24}$$

is calculated for three different configurations of a bounded one-vortex flow: r_v is taken 0.3, 0.5 and 0.7, respectively. The results are plotted in Figure 5.

For the particle accumulation to occur, two conditions must be met: firstly, a fixed point of the dynamical equations (20) and (21) can be found, and secondly, the fixed point has to be stable, thus attracting particles. Both conditions will be treated in the remainder of this section.

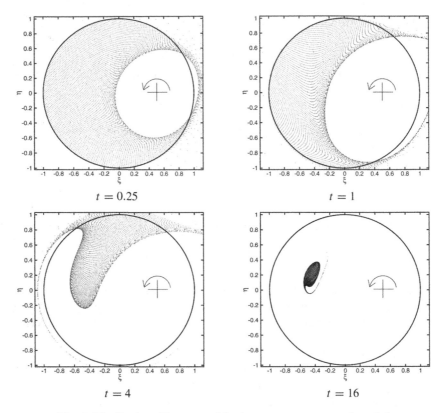

Fig. 4. Distribution of heavy particles in one-vortex system; $St = 0.6$.

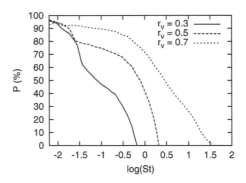

Fig. 5. Percentage of particle trapping as a function of St, for three different vortex configurations.

3.1 Location of fixed points in co-rotating frame

A trapped particle, rotating with the same speed as the vortex, has zero velocity in the co-rotating frame. Hence, the RHS of Equations (20) and (21) goes to 0 for such a particle in the fixed point, say $\boldsymbol{\xi}^*$. In order for this to happen, the Stokes drag has to balance the centrifugal acceleration:

$$(v_g(\boldsymbol{\xi}^*)) + St\dot{\theta}_v^2 \boldsymbol{\xi}^* = 0. \tag{25}$$

From this equation, it follows immediately that for small Stokes numbers also v_g has to be small. Consequently, in the limit of $St \downarrow 0$ the fixed point is situated near a stagnation point of the gas velocity in the co-rotating frame. The only reasonable candidate for this is the elliptic stagnation point situated on the negative ξ-axis, since the hyperbolic stagnation points are unstable by definition.

3.2 Stability of fixed points in co-rotating frame

Now a linear stability analysis is made of the particle approaching the fixed point $\boldsymbol{\xi}^*$. If the particle is close enough to the attraction point, its equation of motion can be approximated by:

$$\frac{d}{dt}R^* = M R^*, \tag{26}$$

where R^* is a vector denoting the separation between the attracted particle and the fixed point:

$$R^* \equiv [\boldsymbol{\xi}_p - \boldsymbol{\xi}^*, v_p]^T. \tag{27}$$

The matrix M is given in Equation (23). In this case, the matrix can be evaluated in the fixed point.

If the real parts of all eigenvalues $\lambda_1, \ldots, \lambda_4$ of M are negative, the fixed point $\boldsymbol{\xi}^*$ is called stable. The eigenvalues read:

$$\lambda_{1,2,3,4} = \frac{-1 \pm \sqrt{1 - 4\dot{\theta}_v^2 St^2 \pm 4St\sqrt{-\mathcal{H}^*}}}{2St}, \tag{28}$$

where $\mathcal{H}^*$ denotes the Hessian, defined in Equation (14), evaluated in the fixed point. For small Stokes numbers, the fixed point is situated close to the elliptic stagnation point, so that $\mathcal{H}^* > 0$. Then, the eigenvalues can be approximated by:

$$\lambda_{1,2,3,4} \simeq \frac{-1 \pm 1}{2St} + St\left(\mathcal{H}^* - \dot{\theta}_v^2\right) \pm i\sqrt{\mathcal{H}^*}. \tag{29}$$

Using the property of the total Hessian in a steadily rotating reference frame, given in Equation (16), we obtain:

$$\lambda_{1,2,3,4} \simeq \frac{-1 \pm 1}{2St} - St\left\{\overline{\Psi}_{\xi\xi}^2 + \overline{\Psi}_{\xi\eta}^2\right\} \pm i\sqrt{\mathcal{H}^*}. \tag{30}$$

Hence, for small Stokes numbers, the real part is always smaller than 0, indicating that the fixed point is stable and does attract particles. So, if a counter-rotating elliptic stagnation point exists in some steadily rotating reference frame, small heavy particles are attracted to it.

When the Stokes number becomes larger, the fixed point will be situated further away from the center of the elliptic island. Then, particles have too much inertia and will be expelled from the domain. Hence, the number of particles trapped inside the domain decreases with increasing Stokes number. This behavior is visible in Figure 5.

Please note that the stability analysis above is not only restricted to the flow induced by a point vortex in a circular domain, but can as well be applied to an other incompressible inviscid flow, as long as it is steady in some steadily rotating reference frame. Examples of this comprise the motion of vortices on a regular polygon on an infinite plane or on a disk (whose origin coincides with the barycenter) or an approximation of the flow field on a Keplerian disk as given by Chavanis [3]. Chavanis prescribes an anticyclonic vortex region a priori; in our case, the elliptic island is formed naturally just by the presence of a cyclonic vortex. Still, the results found here are qualitatively in correspondence with those obtained by Chavanis: small heavy particles are attracted towards a fixed point in a steady anticyclonic island.

4 Conclusions

In this paper, the trajectories of heavy particles in a bounded point vortex flow have been calculated numerically. The simulations are based on a one-way coupling. The results reveal that heavy particles may accumulate in certain regions where the centrifugal and the drag forces acting on the particles balance each other, thus causing an equilibrium trajectory.

A linear stability analysis shows that particles are always attracted to a fixed point, as long as the Stokes number is below a critical value, depending on the particular flow properties. The analysis is shown to be valid not only for point vortex flows but also for any steadily rotating flow field which can be expressed in terms of a stream function.

These results can also be relevant for the swirling pipe flow discussed in Section 1. Small inertial particles tend to accumulate in regions far away from the helical vortex filament, but inside the pipe. Although many other effects play a role in the particle motion on small scales, the inertia is believed to be a dominant effect in macro-scale motion of particles in this situation.

References

1. Alekseenko, S.V., Kuibin, P.A., Okulov, V.L. and Shtork, S.I., 1999, *J. Fluid Mech.* **382**, 195–243.
2. Bec, J., 2003, *Phys. Fluids* **15**(11), L81–L84.
3. Chavanis, P.H., 2000, *Astron. Astrophys.* **356**, 1089–1111.
4. Crowe, C.T., Gore, R. and Troutt, T.R., 1985, *Particulate Science and Technology*, **3**, 149–158.

5. Druzhinin, O.A., 1995, *Phys. Fluids* **7**(9), 2132–2142.
6. Marcu, B., Meiburg, E. and Newton, P.K., 1995, *Phys. Fluids* **7**(2), 400–410.
7. Maxey, M.R., 1990, *Phil. Trans. R. Soc. London A* **333**, 289–307.
8. Maxey, M.R. and Riley, J.J., 1983, *Phys. Fluids* **26**(4), 883–889.
9. Newton, P.K., 2001, *The N-Vortex Problem – Analytical Techniques*, Springer Verlag, Berlin.
10. Provenzale, A., 1999, *Ann. Rev. Fluid Mech.* **31**, 55–93.
11. Squires, K.D. and Eaton, J.K., 1991, *Phys. Fluids* **3**(5), 1169–1178.

PART II

LATTICE-BOLTZMANN AND MOLECULAR DYNAMIC SIMULATIONS

A Numerical Study of Planar Wave Instabilities in Liquid-Fluidized Beds

Jos Derksen[1,2] and Sankaran Sundaresan[2]

[1]*Permanent address: Kramers Laboratorium, Delft University of Technology, Prins Bernhardlaan 6, 2628 BW Delft, The Netherlands; e-mail: jos@klft.tn.tudelft.nl*
[2]*Department of Chemical Engineering, Princeton University, Princeton, NJ 08544, USA; e-mail: sundar@princeton.edu*

Abstract. We present direct simulations with interface resolution of dense, fluidized solid-liquid suspensions. The flow of interstitial fluid is solved by the lattice-Boltzmann method (LBM). The monodisperse, spherical particles move under the influence of gravity, hydrodynamic forces stemming from the LBM, subgrid-scale lubrication forces, and hard-sphere collisions. The cases we study have been derived from the experimental work by Duru et al. [1]. We first show that the experimentally observed waves are well represented by the simulations. Subsequently we use the detailed information contained in the simulation results to assess two-fluid closures, with a focus on the role of compaction and dilation of the particle phase.

1 Introduction

Dense fluidized beds exhibit a rich variety of complex, inhomogeneous flow structures, ranging from one-dimensional traveling waves to bubble-like voids. The hierarchy of these structures has been a subject of many theoretical and experimental studies [1–3]. An Eulerian two-phase flow model, which treats the fluid and particle phases as interpenetrating continua, coupled with simple phenomenological closures for the effective stresses and the fluid-particle interaction force, seems to capture the experimentally observed structures in a qualitatively correct manner; however, quantitative predictions remain elusive [3].

Recently Duru et al. [1] measured the particle volume fraction profiles in fully developed one-dimensional traveling waves in liquid-fluidized beds. Their wave data are particularly valuable, as they are made up of regions where the particle assemblies undergo dilation and regions where they compact. As compaction and dilation of particle assemblies are ubiquitous in granular and fluid-particle flows, it is important to test and validate closure models through clean model problems where both compaction and dilation occur. One-dimensional waves in fluidized beds serve as excellent model problems for this purpose.

Critical assessment of the closure relations requires detailed data on the spatial variation of particle and fluid velocity fields, collision statistics, etc. in these travel-

S. Balachandar and A. Prosperetti (eds.) Proceedings of the IUTAM Symposium on Computational Multiphase Flow, 89–98.

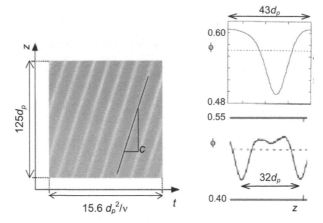

Fig. 1. Left: experimental space-time plot of the solids volume fraction at $\bar{\phi} = 0.540$. The wave speed (c) can be derived from the slope of the light lines representing the void regions. Right: solids volume fraction wave profiles for $\bar{\phi} = 0.57$ (top) and $\bar{\phi} = 0.49$ (bottom). Reprinted from [1].

ing waves, in addition to the particle volume fraction profiles. These are not easily measured in dense suspensions; to date, such measurements have not been made. However, one can use computer simulations to obtain the missing data. By performing detailed simulation of the flow of the fluid and the particles corresponding to these waves, all the detailed data required for critical evaluation of the closures can be extracted. The experimental data [1] can be used to validate the computer simulations.

2 Flow system

Duru et al. [1] carried out an extensive experimental program in which they studied the onset and characteristics of planar waves in relatively narrow, vertically oriented liquid fluidized beds. Their experimental variables were the solids volume fraction, the (solid over fluid) density ratio, the particle size, the fluid viscosity, and the size ratio (particle diameter divided by tube diameter). Even though planar waves form spontaneously, the authors excited specific wave frequencies so that clean, high-quality data could be obtained. Therefore the excitation frequency and amplitude are also inputs in the experiments. Figure 1 shows a typical experimental result: a space-time plot of the solids volume fraction ϕ. Clearly visible are regions of low particle volume fraction ("voids") that travel with a well-defined speed in the vertical (z) direction. Figure 1 also shows two of the many waveforms measured in the experiments. The top and bottom panels show traveling waves with a single hump and two humps, respectively.

In the simulations we represent the experimental system by a set of spherical particles all having the same size (diameter d_p) immersed in a fluid. The three-dimensional domain has fully periodic boundary conditions. The flow is induced by a force in the negative z-direction on the particles (gravity), and a body force acting on the fluid that balances the gravity force on the particles. The body force on the fluid mimics the overall pressure gradient that in real life drives the flow. If we write the gravitational acceleration as $-g\mathbf{e}_z$, then the net gravity force acting on each sphere is

$$\mathbf{F}_G = -(\rho_s - \bar{\rho})\frac{\pi}{6}d_p^3 g\mathbf{e}_z,$$

and the force per unit volume acting on the fluid is

$$\mathbf{f}_B = (\bar{\rho} - \rho_f)g\mathbf{e}_z, \tag{1}$$

with $\bar{\rho} = \bar{\phi}\rho_s + (1 - \bar{\phi})\rho_f$ the density of the fluid-solid mixture, and $\bar{\phi}$ the overall (spatially averaged) solids volume fraction.

We need to translate the physical parameters of the experiments into LB-parameters. The spatial resolution of the simulations can be expressed in terms of the number of lattice spacings Δ spanning a particle diameter d_p. This number was set to 16 (after comparing preliminary results obtained with $d_p = 10\Delta$, 16Δ and 24Δ). The gravitational acceleration g and the fluid viscosity ν are now chosen such that the dimensionless group $(gd_p^3)/\nu^2$ is the same in experiment and simulation, and the terminal settling velocity of a single particle in unbounded fluid is of the order of 0.02 in LB units (distance traveled in lattice spacings per time step). The latter condition assures that fluid velocities stay well below the speed of sound of the numerical scheme so that incompressible flow is simulated. The density ratio and the solids volume fraction are dimensionless numbers that can be directly represented in the simulations. The specific experiments that we selected from [1] had particles with $d_p = 685 \pm 30$ μm, a density ratio $\rho_s/\rho_f = 4.1$, and a fluid viscosity of $\nu = 0.90 \cdot 10^{-6}$ m^2/s (these settings being denoted "Combination 7" in [1]).

Simulations were performed in three-dimensional periodic domains ($6d_p \cdot 6d_p \cdot 20d_p$), for three different average particle volume fractions ($\bar{\phi} = 0.580, 0.505$, and 0.488). After initializing a nearly homogeneous bed, gravity and body force were turned on. As the lateral dimensions of the box are small ($6d_p$), there is very little opportunity for any persistent lateral structure to evolve, but one can readily see non-uniform structures that travel in the direction of the mean fluid flow.

3 Numerical setup

In our simulations, we consider a three-dimensional (Cartesian) domain which is discretized into a number of lattice nodes residing on a uniform, cubic grid. In the LBM, fluid particles move from each node to its neighbors according to pre-scribed rules. It can be proven that (with the proper grid topology and collision rules) in the low Mach number limit this system obeys the incompressible Navier–Stokes equations (see e.g. [4]). The specific implementation used in our simulations has been described by [5], which is a variant of the widely used Lattice BGK scheme to handle the collision integral (e.g. [6]).

Pioneering work on the application of lattice-Boltzmann methods for suspension simulations was done by Ladd [7, 8]. In our code, the no-slip condition at the solid-fluid boundaries is introduced through a forcing scheme [9, 10]. In this scheme, body forces acting on the fluid are determined such that at the surface of the sphere the fluid velocity matches the local velocity of the solid surface (that is the sum of the linear velocity $\mathbf{v}_p$ and $\mathbf{\Omega}_p \times (\mathbf{r} - \mathbf{r}_p)$ with $\mathbf{\Omega}_p$ the angular velocity of the particle). The collection of forces acting on the fluid at the sphere's surface is subsequently used to determine the hydrodynamic force and torque acting on the sphere. We follow the calibration procedure of Ladd [7] to find the hydrodynamic diameter of the particles.

An artifact of the forcing scheme is that there is fluid inside the spherical particles. As long as the density of the solid is higher than the density of the fluid, the effects of the internal fluid can be effectively corrected for: The force (and torque) acting on the fluid determined by the LB/forcing method is the sum of the force (torque) needed to accelerate the internal fluid and the force (torque) of the particle acting on the external fluid. Since the internal fluid largely behaves as a solid body (i.e. it approximately has the linear and angular velocity of the sphere), the force (torque) on the fluid (internal *and* external) due to the solid particle is

$$\mathbf{F}_{\text{LB}} = \mathbf{F}_{\text{ext}} + \rho_f \frac{\pi}{6} d_p^3 \frac{d\mathbf{v}_p}{dt}, \quad \mathbf{T}_{\text{LB}} = \mathbf{T}_{\text{ext}} + \rho_f \frac{\pi}{60} d_p^5 \frac{d\mathbf{\Omega}_p}{dt}. \tag{2}$$

The force (torque) that the external fluid exerts on the particle is $-\mathbf{F}_{\text{ext}}$ ($-\mathbf{T}_{\text{ext}}$).

A second effect of the internal fluid that needs to be corrected for in the equation of linear motion of the particles is the body force acting on the fluid (Equation (1)) that not only acts on the external fluid, but also on the internal fluid. This (non-physical) force $(\pi/6)d_p^3(\bar{\rho} - \rho_f)g\mathbf{e}_z$ acting on the internal fluid needs to be compensated by an equal and opposite force on the particle.

If the equation of linear motion of a spherical particle without internal fluid on which the external fluid exerts a force $\mathbf{F}_{\text{ext}}$ is

$$\rho_s \frac{\pi}{6} d_p^3 \frac{d\mathbf{v}_p}{dt} = -\mathbf{F}_{\text{ext}} - (\rho_s - \bar{\rho}) \frac{\pi}{6} d_p^3 g\mathbf{e}_z, \tag{3}$$

the corrections described above lead to the following equation for a particle with internal fluid:

$$(\rho_s - \rho_f) \frac{\pi}{6} d_p^3 \frac{d\mathbf{v}_p}{dt} = -\mathbf{F}_{\text{LB}} - (\rho_s - \rho_f) \frac{\pi}{6} d_p^3 g\mathbf{e}_z. \tag{4}$$

The equation of angular motion (for a particle with internal fluid) is

$$(\rho_s - \rho_f) \frac{\pi}{60} d_p^5 \frac{d\mathbf{\Omega}_p}{dt} = -\mathbf{T}_{\text{LB}}. \tag{5}$$

In order to test if the above procedure represents the dynamics of spheres immersed in liquid properly, we considered the transient motion of a single sphere that is accelerated starting from rest under the influence of gravity. In the limit of zero Reynolds number in an unbounded fluid the equation of motion of the sphere has been derived by Maxey and Riley [11]. Lattice-Boltzmann results are in excellent agreement with those obtained by integrating the Maxey and Riley equation even at a density ratio as low as 1.1.

The spheres mutually interact by means of binary, hard-sphere collisions and lubrication forces. For the former, we apply an event-driven collision algorithm: we move the collection of particles until two particles get into contact. At that moment we carry out the collision (i.e. update the velocities of the two particles taking part in the collision). Subsequently, the motion of all particles is continued until the next collision. The collision model that we apply (described in detail in [12]) has two parameters: a restitution coefficient e and a friction coefficient μ. As the default situation we consider fully elastic, frictionless collisions ($e = 1$, $\mu = 0$).

If particles are in close proximity (their distance being of the order or even less than the lattice-spacing), the hydrodynamic interaction between the particles cannot be properly accounted for anymore by the LBM. We then explicitly impose lubrication forces on the particles, in addition to the hydrodynamic forces stemming from the LBM [13]. Both radial and tangential lubrication have been considered. Lubrication is smoothly switched on once the spacing between two particles gets less than 1.6 times the lattice spacing ($0.1d_p$); it saturates at a distance of $10^{-4}d_p$. The latter we use for numerical reasons but also with the surface roughness of the particles and/or the mean-free-path of the fluid in mind. Further details of the implementation of the lubrication forces are discussed in [14].

4 Results

4.1 Waves and wave speeds

We start from a random distribution of (non-overlapping) spheres at rest in stagnant liquid. At $t = 0$ the gravity and the body force are switched on. The spheres start falling, and the fluid starts flowing. The system develops a wave instability in a time span of typically $1d_p^2/\nu$. The associated void travels in the direction opposite to gravity (i.e. the positive z direction), see Figure 2. Outside the void, the solids volume fraction is significantly higher than the average volume fraction. At the upper side of the void, particles detach from the dense region, "rain" through the void, and fall on the dense region below the void.

The simulated wave can be represented in a space-time plot similar to the experimental one. Examples of such plots are given in Figure 3. The wave amplitude and structure depend on the solids volume fraction: shallow waves at high $\bar{\phi}$, more complicated wave forms for lower $\bar{\phi}$. Duru et al. [1] measured wave speeds of $cd_p/\nu = 28$ and 29 (± 1.4) for their "Combination 7" system at $\bar{\phi} = 0.488$ and 0.496 respectively (with c the wave speed). The wave speeds that can be extracted from Figure 3 are 33 (± 2) (within the error margin c is independent of $\bar{\phi}$).

By z-shifting the set of instantaneous, one-dimensional solids volume fraction profiles that constitute the space-time plots by an amount ct and subsequently averaging over time produces smooth solids fraction profiles comparable to those measured in [1]. The resulting graphs are shown in Figure 4. The size of our domain in the

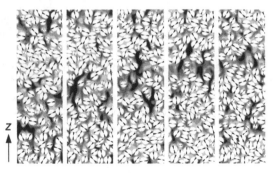

Fig. 2. Cross sections through the simulated solid-liquid field with $\bar{\phi} = 0.505$ at various moments in time. From left to right $tv/d_p^2 = 1.225, 1.277, 1.329, 1.381, 1.434$. The gray scale denotes the absolute value of the liquid velocity (dark is high).

Fig. 3. Simulated space-time plots of the solids volume fraction. From left to right: $\bar{\phi} = 0.580$, 0.505, and 0.488.

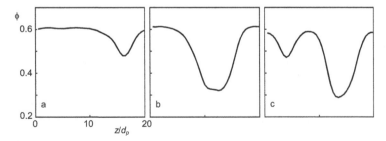

Fig. 4. Solids volume fraction wave profiles for (from left to right) $\bar{\phi} = 0.580$, 0.505, and 0.488.

z-direction ($20d_p$) is generally smaller than the measured wavelengths – the waves in Figure 1 have lengths of the order of $40d_p$ – which inhibits a quantitative comparison. Qualitatively there is good agreement: the asymmetric wave shape with smaller gradients at the compaction (= left) side of the wave; and double humped waves at lower solids volume fractions.

4.2 Quantitative analysis

During the simulations, data sets containing short-time averages (averaging time $t_a = 5.2 \times 10^{-4} d_p^2/v$) of volume fractions, velocities, forces, and stresses as a function of z were stored to disk. A series of 2000 of these sets (spanning a time $2000t_a$) are used to determine their profiles in a frame of reference moving with the (fully developed)

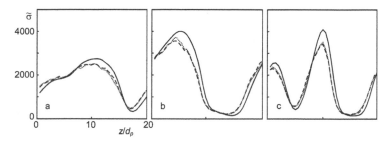

Fig. 5. Variation of the three collisional normal stresses (zz: drawn curve; xx: dashed, yy; dotted) along the wave. $\tilde{\sigma}$ is the dimensionless stress defined as $\tilde{\sigma} = \sigma \dfrac{d_p^2}{\rho_f v^2}$. From left to right: $\bar{\phi} = 0.580, 0.505$, and 0.488.

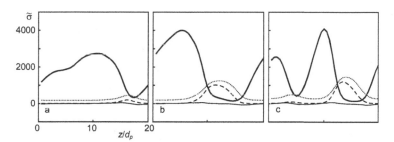

Fig. 6. Variation of zz-stresses along the wave ($\tilde{\sigma}$ has been defined in Figure 5). From left to right: $\bar{\phi} = 0.580, 0.505$, and 0.488. Thick, drawn line: collisional stress; thin drawn line: stress due to lubrication; dotted line: fluid streaming stress; dashed line: particle streaming stress.

wave. For this the same ct-shifting procedure that was applied to derive the solids volume fraction profiles (Figure 4) was used. In this section of the paper the focus will be on the momentum transfer mechanisms (i.e. stresses).

The wave clearly induces anisotropy. As an example we show in Figure 5 the three components of the normal collisional stress. As expected, the two lateral components (xx and yy) are approximately equal to one another, and the axial component (zz) differs appreciably from the other two. In the void-part of the wave the collisional stress is much lower than in the dense part.

The most important zz-stresses are presented in Figure 6. Collisions are largely responsible for the particle phase stress in these flows at high particle volume fractions. In the void, fluid and particle streaming stress are significant and of comparable magnitude. Lubrication plays only a modest role. The normal viscous stresses (not shown in Figure 6) are negligible.

In an Eulerian two-phase flow model, continuum equations of motion for the particle phase are based on the concepts of kinetic theory of dense gases and are referred to as kinetic theory of granular material (KTGM). This typically leads to a particle phase stress, σ_s, expressed in a compressive sense, of the form

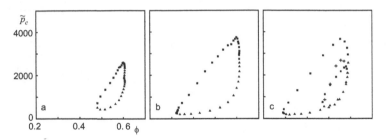

Fig. 7. Dimensionless collisional pressure $\tilde{p}_c$ as a function of solids volume fraction. From left to right: $\bar{\phi} = 0.580, 0.505$, and 0.488. The squares relate to the negative ϕ-slope (compaction), the triangles to the positive ϕ-slope (dilation). The plusses in the right graph are collisional pressures in the shallow void based on bulk viscosity estimates from the deep void.

$$\sigma_s = p_s \mathbf{I} - \kappa_s (\nabla \cdot \mathbf{v}_s)\mathbf{I} - \mu_s \left[(\nabla \mathbf{v}_s) + (\nabla \mathbf{v}_s)^T - \frac{2}{3}(\nabla \cdot \mathbf{v}_s)\mathbf{I} \right], \qquad (6)$$

where $\mathbf{v}_s$ is the particle phase velocity; and p_s, κ_s, and μ_s are the pressure, bulk viscosity, and shear viscosity of the particle phase respectively. Although the actual expressions for the shear and bulk viscosities differ slightly from one derivation to another, all derivations yield comparable values for them, with the shear viscosity being larger in magnitude than the bulk viscosity (e.g. [15]). It is also important to note that in all theories, the bulk and shear viscosities depend on local particle volume fraction and granular temperature, but not explicitly on the local rate of deformation. At prescribed particle volume fraction and granular temperature, the particle phase stress depends linearly on the rate of compaction or dilation of the particle phase (which is captured through the term); and the bulk and shear viscosities are independent of whether the assembly is undergoing compaction or dilation locally.

In Figure 7, we present the average of the three collisional normal stresses with the local particle volume fraction (taken from Figure 4). This average normal stress is the sum of the contributions of the particle phase pressure and the bulk viscosity term. Figures 7a and 7b take the form of a single lobe, as it corresponds to a single hump wave, and it shows unequivocally that the average normal stress is not a unique function of particle volume fraction and that it is dramatically higher in the compaction branch than in the dilation branch. Figure 7c shows two lobes as it corresponds to a double hump wave. The rates of compaction and dilation in the shallower void (Figure 4c) are much smaller than those in the deeper void; Figure 7c suggests that the average normal stress in the dilation branch is approximately independent of the rate of dilation. In contrast, the average normal stress in the compaction branch of the shallower void is appreciably lower than that of the deeper void, indicating a pronounced dependence on the rate of compaction.

In the kinetic theory, particle phase pressure depends on both volume fraction and the granular temperature, and the granular temperature is indeed higher in the compaction branch than in the dilation branch, but this difference is no more than 30%, and it cannot explain the factor of 4–6 difference seen in the average normal stresses

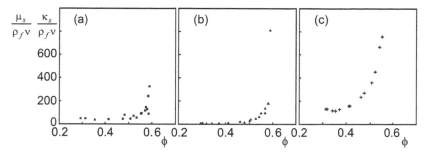

Fig. 8. Solids phase viscosity as a function of the solids volume fraction for $\bar{\phi} = 0.488$. (a) Shear viscosity under compaction, (b) shear viscosity under dilation, (c) the bulk viscosity estimates in the deep void.

in the two branches at intermediate concentrations. It is natural to begin by suspecting that the path dependence seen in Figure 7 is a consequence of the bulk viscosity, with the bulk viscosity being essentially zero under dilation and non-negligible upon compaction. In this line of thinking, the difference between the compaction and dilation branches in Figure 7 is exclusively attributed to the bulk viscosity effect. To test this further, we used the data in the outer lobe of Figure 7c and estimated bulk viscosity at different particle volume fractions. Using these bulk viscosity estimates and the compaction rate at different locations in the shallow hole, we calculated what the average normal stress at different locations in this wave must be; these results are shown in Figure 7c as plusses. Although not quantitative, these estimates are close to the actual average normal stress in the shallow hole, lending support to the argument that the bulk viscosity contribution to the particle phase stress is very significant. The bulk viscosity estimates corresponding to Figure 7c are shown in Figure 8. Also shown in these figures are the collisional shear viscosities extracted from this simulation. It is clear that the bulk viscosity estimate is appreciably larger than the shear viscosity.

5 Closure

Dense, fluidized solid-liquid suspensions have been simulated in great detail. Wave formation was found in qualitative agreement with experimental data from [1]. Subsequently we demonstrated the potential of the simulations for gaining a better insight in the physics of dense solid-liquid suspensions. At the upper and lower side of the wave the particle phase respectively dilates and compacts. It was shown that in order to capture the particle phase stresses by means of KTGM-based modeling, the particle-phase bulk viscosity, and the way it depends on the local variables involved (such as the solids volume fraction, and the granular temperature) needs further attention.

References

1. Duru, P., Nicolas, M., Hinch, J. and Guazelli, E., 2002, Constitutive laws in liquid-fluidized beds, *J. Fluid Mech.*, **452**, 371–404.
2. Anderson, K., Sundaresan, S. and Jackson, R., 1995, Instabilities and the formation of bubbles in fluidized-beds, *J. Fluid Mech.*, **303**, 327–366.
3. Sundaresan, S., 2003, Instabilities in fluidized beds, *Annu. Rev. Fluid Mech.*, **35**, 63–88.
4. Chen, S. and Doolen, G.D., 1998, Lattice-Boltzmann method for fluid flows, *Annu. Rev. Fluid Mech.*, **30**, 329–364.
5. Somers, J.A., 1993, Direct simulation of fluid flow with cellular automata and the lattice-Boltzmann equation, *Appl. Sci. Res.*, **51**, 127–133.
6. Qian, Y.H., d'Humieres, D. and Lallemand, P., 1992, Lattice BGK for the Navier–Stokes equations, *Europhys. Lett.*, **17**, 479–484.
7. Ladd, A.J.C., 1994, Numerical simulations of particle suspensions via a discretized Boltzmann equation. Part I: Theoretical foundation, *J. Fluid Mech.*, **271**, 285–309.
8. Ladd, A.J.C., 1994, Numerical simulations of particle suspensions via a discretized Boltzmann equation. Part II: Numerical results, *J. Fluid Mech.*, **271**, 311–339.
9. Goldstein, D., Handler, R. and Sirovich, L., 1993, Modeling a no-slip flow boundary with an external force field, *J. Comp. Phys.*, **105**, 354–366.
10. Ten Cate, A., Nieuwstad, C.H., Derksen, J.J. and Van den Akker, H.E.A., 2002, PIV experiments and lattice-Boltzmann simulations on a single sphere settling under gravity, *Phys. Fluids*, **14**, 4012–4025.
11. Maxey, M.R. and Riley, J.J., 1983, Equation of motion for a small rigid sphere in a nonuniform flow, *Phys Fluids*, **26**, 883–889.
12. Yamamoto, Y., Potthoff, M., Tanaka, T., Kajishima, T. and Tsuji, Y., 2001, Large-eddy simulation of turbulent gas-particle flow in a vertical channel: Effect of considering inter-particle collisions, *J. Fluid Mech.*, **442**, 303–334.
13. Kim, S. and Karrila, S.J., 1991, *Microhydrodynamics: Principles and Selected Applications*, Butterworth-Heinemann.
14. Derksen, J. and Sundaresan, S., 2005, DNS of dense suspensions: Planar wave instabilities in liquid-fluidized beds, in *Proceedings of the 11th Workshop on Two-Phase Flow Predictions*, Merseburg, Germany (CD Rom, ISBN 3-86010-767-4).
15. Gidaspow, D., 1994, *Multiphase Flow and Fluidization*, Academic Press, CA.

Lattice Boltzmann Simulations to Determine Forces Acting on Non-Spherical Particles

A. Hölzer and M. Sommerfeld

Institut für Verfahrenstechnik, Fachbereich Ingenieurwissenschaften,
Martin-Luther-Universität Halle-Wittenberg, D-06099 Halle (Saale), Germany

Abstract. The drag, lift and moment coefficient of differently shaped single particles with respect to the angle of incidence and to the particle Reynolds number under different conditions were determined. For this purpose simulations of the flow around these particles were performed using the three-dimensional Lattice Boltzmann method. The first case studied was a fixed particle in a plug flow, the second case a rotating particle in a plug flow to determine the Magnus lift force and the third case a fixed particle in a linear shear flow to determine the Saffman lift force. In the first case six particle shapes were considered, which are two spheroids, two cuboids and two cylinders with an axis ratio of 1 and 1.5, respectively. In the second and third case, only the sphere was considered. The particle Reynolds number was varied between 0.3 and 480.

Nomenclature

α	=	angle of incidence
$c_D = \dfrac{\|F_D\|}{\frac{1}{2}\rho \underline{u}^2 \frac{\pi}{4} d_V^2}$	=	drag coefficient
$c_L = \dfrac{\|F_L\|}{\frac{1}{2}\rho \underline{u}^2 \frac{\pi}{4} d_V^2}$	=	lift coefficient
$c_M = \dfrac{\|M\|}{\frac{1}{2}\rho \underline{u}^2 \frac{\pi}{4} d_V^2 d_V}$	=	moment coefficient
d_V	=	diameter of a volume-equivalent sphere
$\underline{F}_D$	=	drag force
$\underline{F}_L$	=	lift force
$\underline{M}$	=	torque
υ	=	kinematic viscosity
$\mathrm{Re}_{Pa} = \dfrac{\|u\|d_V}{\upsilon}$	=	particle Reynolds number
ρ	=	fluid density
$S = \dfrac{\|\omega\|\frac{d}{2}}{\|u\|}$	=	spin number
$\underline{u}$	=	fluid velocity
$\underline{\omega} = \underline{\omega}_{Pa} = \frac{1}{2}\underline{\nabla} \times \underline{u}$	=	particle angular velocity or half fluid vorticity

S. Balachandar and A. Prosperetti (eds), Proceedings of the IUTAM Symposium on Computational Multiphase Flow, 99–108.

1 Introduction

The motion of particles is very important for many technical processes. Examples are combustion of pulverised coal, pneumatic transport of solids, fluidised beds or fibre suspension flow in paper forming. But it also plays an important role in natural processes as well, e.g. in the pollutant transport in the atmosphere. The modelling of these processes relies mostly on the assumption of spherical particles. For describing the motion of non-spherical particles, detailed information on the fluid dynamic forces acting on such particles are necessary, but generally not available.

Analytical solutions exist only in the low Reynolds number limit for the drag of a sphere in a plug flow $\underline{F}_D = 3\pi d_V \rho v \underline{u}$ or $c_D = 24/\mathrm{Re}_{Pa}$, for the lift of a rotating sphere in a plug flow (Magnus force) [6] $\underline{F}_L = (\pi/8)d_V^3 \rho \underline{\omega} \times \underline{u}$ or, if $\underline{\omega}$ and $\underline{u}$ are perpendicular, $c_L = 2S$, for the lift of a fixed sphere in a linear shear flow (Saffman force) [7] $F_{Ly} = (6.46/4)d_V^2 \rho \sqrt{v} \sqrt{d|\underline{u}|/dy} |\underline{u}|$ or $c_L = 46.46/\pi \sqrt{S/\mathrm{Re}_{Pa}}$ and for the torque in the last two cases $\underline{M} = -\pi d_V^3 \rho v \underline{\omega}$ or $c_M = 16S/\mathrm{Re}_{Pa}$. Correlations for the drag coefficient of spheres exist in the whole range of particle Reynolds numbers [4]. Also averaged correlations for the drag coefficient of non-spherical particles are available which depend on the shape of the particles [3, 4]. Only very few three-dimensional numerical studies about the drag of non-spherical particle exist, e.g. [2]. This work is to my knowledge the first comprehensive three-dimensional study about the lift, drag or moment coefficient of non-spherical particles as function of the angle of incidence.

2 Numerical method

The fluid flow is simulated by the Lattice Boltzmann method which is an alternative approach to conventional methods. Whereas conventional models are based on the conservation laws formulated at the macroscopic level, the Boltzmann equation describes the behaviour of fluids at the molecular level. The BGK relaxation and the D3Q19 model is used for this work [5].

The curved no-slip boundary condition introduced in [1] is imposed on the particle surface. This boundary condition considers the exact position of the particle surface within a cell. The influence of the boundaries of the computational domain and of the particle resolution was asymptotically calculated and was used to correct the coefficients. The domain size is $170 \times 60 \times 74$ cells and the smallest dimension of every particle is 12 cells which allows a good resolution of the flow field around the particle. The coefficients converge to 99% of the terminal value at $\mathrm{Re}_{Pa} = 0.3$ after 10000 time steps and at $\mathrm{Re}_{Pa} = 240$ after 4000 times steps.

3 Coefficients for a fixed particle in a plug flow

For these studies a particle is fixed at a certain angle of incidence in the centre of a cuboid domain, see Figure 1. As inflow (at $x = 0$) a plug flow is assumed. The

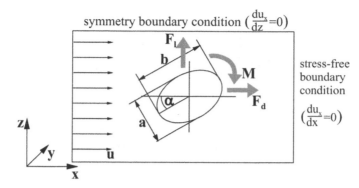

Fig. 1. Flow configuration.

other boundary conditions are symmetry boundary condition for the side walls and stress-free boundary condition for the outlet. Six different particles are considered, namely a sphere, a cube, a cylinder with an axis ratio of 1 and a spheroid, a cuboid and a cylinder with an axis ratio of 1.5, respectively. The drag force is positive in the x-direction, the lift force is positive in the z-direction and the torque is positive in the y-direction. Four different particle Reynolds numbers, i.e. 0.3, 30, 90 and 240 and for certain orientations also 480, are considered.

In Figures 2a–2d the drag coefficient is plotted as a function of the angle of incidence at the four different particle Reynolds numbers. At $Re_{Pa} = 0.3$ most of the non-spherical particles have a minimum in the drag coefficient around an angle of 45° since they have the best streamline shape for this orientation. At 0° and 90° of incidence the drag coefficient reaches maximum values. In contrast to that, at $Re_{Pa} = 90$ and $Re_{Pa} = 240$ the non-spherical particles have the largest drag at approximately this position, where the projected area reaches its greatest value, e.g. the cube at 45°. Thus the streamlining has more influence on the drag at low rather than at high particle Reynolds numbers and the projected area has more influence at high rather than at low particle Reynolds numbers. The lengthwise ($\alpha = 0°$) spheroid has the smallest drag and the crosswise ($\alpha = 90°$) or nearly crosswise cuboid the largest drag at every particle Reynolds number. The reason is the good streamline shape of the lengthwise spheroids and the bad streamline shape of cuboids because of the rough edges of cuboids. At $Re_{Pa} = 240$ all lengthwise particles with axis ratio 1.5 have a smaller drag than the sphere. In Figure 2e the drag coefficient of the lengthwise and crosswise particles is plotted versus the particle Reynolds number. It shows the increase of variation in drag with increasing particle Reynolds number.

Figures 3a–3d show the dependence of the lift coefficient on the orientation for the different particle Reynolds numbers. At $Re_{Pa} = 0.3$ the cube has practically no lift and the graph of the cuboid shows almost a parabolic shape with the minimum at about 45°. At higher particle Reynolds numbers the lift coefficient of the cube shows stronger variations. The shape of the lift coefficient curves for the cuboid becomes non-symmetric with increasing particle Reynolds number and the minimum

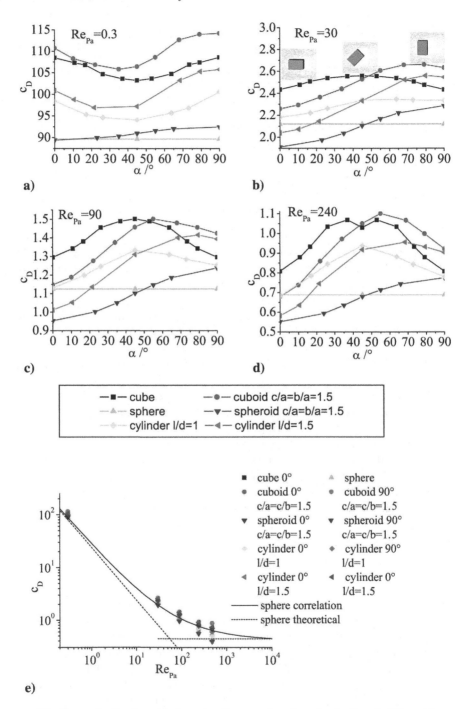

Fig. 2. c_D of a fixed particle in a plug flow as a function of α (a–d) and of Re_{Pa} (e).

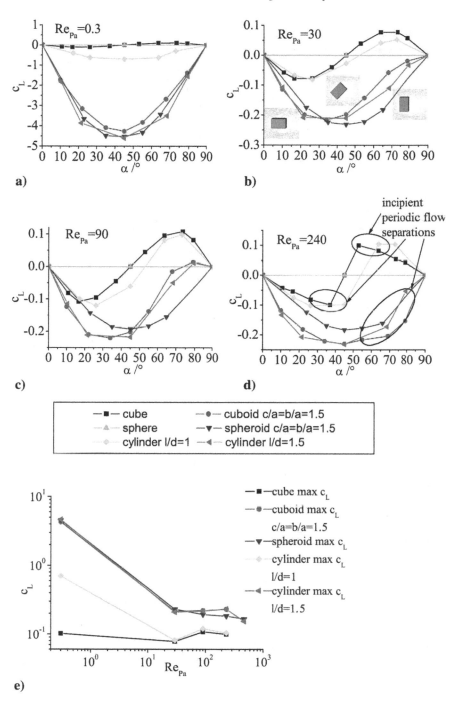

Fig. 3. c_L of a fixed particle in a plug flow as a function of α (a–d) and of Re_{Pa} (e).

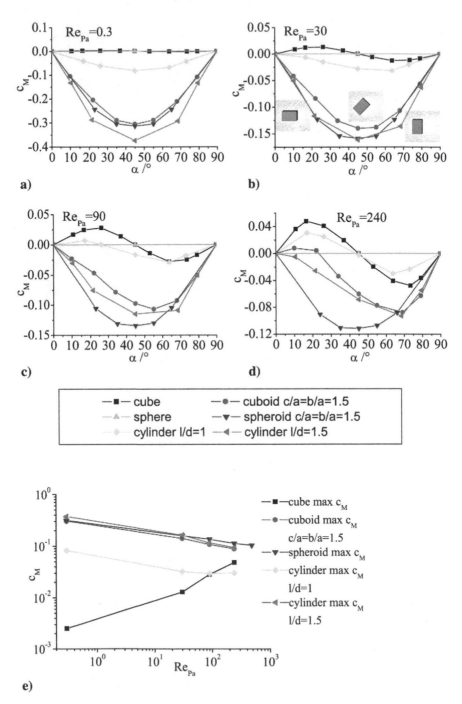

Fig. 4. c_M of a fixed particle in a plug flow as a function of α (a–d) and of Re_{Pa} (e).

is shifted towards smaller angles of incidence. For the non-spherical particles periodic flow separations appear at $Re_{Pa} = 240$ for angles of incidence where the projected area reaches its respective greatest values, yielding a increase of the lift coefficient. From Figure 3e, where the maximum lift coefficient is plotted versus the particle Reynolds number, it is obvious that between $Re_{Pa} = 0.3$ and $Re_{Pa} = 30$ the maximum lift coefficient decreases. The average decrease except for the cube is approximately proportional to $Re_{Pa}^{-0.5}$. At higher particle Reynolds numbers the lift remains almost constant.

The moment coefficient with respect to the angle of incidence plotted in Figures 4a–4d behaves similar to the lift coefficient, except that the periodic flow separations have no observable influence. A particle is in a stable position if the moment coefficient is zero and if the slope of the graph is positive. For the spheroid, cuboid and cylinder with axis ratio 1.5 a stable position is the crosswise position ($\alpha = 90°$) and for the cuboid at $Re_{Pa} = 240$ also the lengthwise position ($\alpha = 0°$). The cube has stable positions at $\alpha = 0°$ and $\alpha = 90°$. The cylinder with axis ratio 1 is in stable position always at $\alpha - 90°$ and at $Re_{Pa} = 90$ and $Re_{Pa} = 240$ also at $\alpha = 0°$. The maximum moment coefficients are plotted against the particle Reynolds number in Figure 4e. The average slope except for the cube is approximately proportional to $Re_{Pa}^{-0.18}$.

4 Coefficients for a rotating sphere in a plug flow

The flow conditions are the same as in the section before (see Figure 1) except that the particle is rotating with a constant angular velocity. Hence, the particle experience a lift which is the well known Magnus force. Four different angular velocities and therefore four different spin numbers (0.5, 1, 2 and 3) of a sphere are studied. The drag, lift and moment coefficient versus the particle Reynolds number are shown in Figures 5a–5c. Except for the sphere with $S = 3$ an increase of the spin number causes an increase of the drag. The difference becomes larger with increasing particle Reynolds number. The lift coefficient fits very well the theoretical solution at low particle Reynolds numbers, i.e. it remains almost constant. Beyond particle Reynolds numbers of about 1 the calculated lift becomes much smaller than the theoretical solution. Also the moment coefficient fits the theoretical prediction very well at low particle Reynolds numbers. At high particle Reynolds numbers the moment coefficient becomes slightly larger than the theoretical solution.

5 Coefficients for a fixed sphere in a linear shear flow

In contrast to Figure 1 no-slip boundary conditions with a moving wall are imposed for the bottom and upper wall, yielding a linear shear flow along the computational domain. Under these conditions a lift acts on a sphere in the direction toward larger velocities at low particle Reynolds numbers. This is the well-known Saffman force. The spin numbers, which are determined by the particle size and the ratio between the

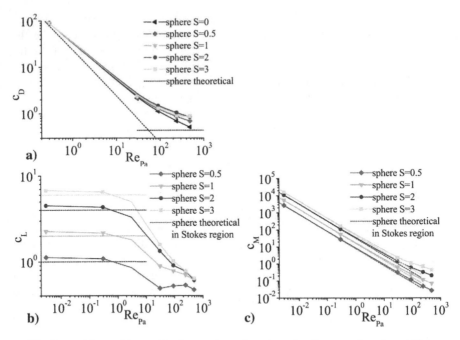

Fig. 5. c_W (a), c_A (b), c_M (c) of a rotating sphere in a plug flow as a function of Re_{Pa}.

velocity gradient and the velocity, are much lower than in the section before for the rotating sphere due to the numerical limitation of the maximum Reynolds number. The spin numbers studied are 0.04 and 0.08, see Figures 6a–6c. The drag coefficient is almost unaffected by the shear flow. The lift coefficient approaches the theoretical solution at low particle Reynolds numbers. At both spin numbers the lift changes its direction between $Re_{Pa} = 30$ and $Re_{Pa} = 90$. From the small diagram in Figure 6b, where the y-axis is linear, it can be observed that the change of direction occurs at nearly the same particle Reynolds number ($Re_{Pa} = 50$) for both spin numbers. The absolute value of the lift coefficient at the high spin number remains larger also after the change of direction. The moment coefficient fits the theoretical prediction again very well at low particle Reynolds numbers. Between $Re_{Pa} = 90$ and $Re_{Pa} = 240$ a drastic drop of the moment coefficient occurs and it remains almost constant beyond it.

6 Summary

The drag, lift and torque acting on particles depend, besides the particle Reynolds number, strongly on the particle shape and the angle of incidence. The particle streamlining has more influence on the drag at low rather than at high particle Reynolds numbers and the projected area has more influence at high rather than at low particle Reynolds numbers. A fixed cube in a plug flow shows practically no

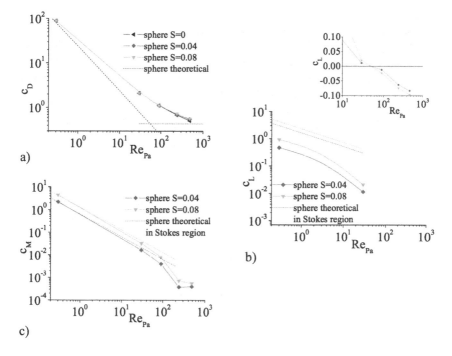

Fig. 6. c_W (a), c_A (b), c_M (c) of a fixed sphere in a linear shear flow as a function of Re_{Pa}.

lift or torque at low particle Reynolds numbers in contrast to high particle Reynolds numbers.

The drag of a sphere is only slightly influenced by rotation of the sphere or a shear flow. In addition to the drag a lift and torque act on a sphere if it is rotating or placed in a shear flow. The lift on a rotating sphere in a plug flow is the Magnus force and the lift on a sphere in a linear shear flow is the Saffman force. In both cases the lift and moment coefficients approach the theoretical solutions at low particle Reynolds numbers. At high particle Reynolds number the Magnus force becomes smaller and the corresponding torque slightly larger compared to the theoretical solution. The Saffman force becomes smaller with increasing particle Reynolds number until it changes its direction at approximately $Re_{Pa} = 50$.

References

1. Bouzidi, M., Firdaouss, M. and Lallemand, P., 2001, Momentum transfer of a Boltzmann-lattice fluid with boundaries, *Physics of Fluids* **13**(11), 3452–3459.
2. Dwyer, H.A. and Dandy, D.S., 1990, Some influences of particle shape on drag and heat transfer, *Physics of Fluids A* **2**(12), 2110–2119.
3. Ganser, G.H., 1993, A rational approach to drag prediction of spherical and nonspherical particles, *Powder Technology* **77**, 143–152.

4. Haider, A. and Levenssiel, O., 1989, Drag coefficient and terminal velocity of spherical and nonspherical particles, *Powder Technology* **58**, 63–70.
5. He, X. and Luo, L., 1997, Theory of the lattice Boltzmann method: From the Boltzmann equation to the lattice Boltzmann equation, *Physical Review E* **56**, 6811–6817.
6. Rubinow, S.I. and Keller, J.B., 1961, The transverse force on a spinning sphere moving in a viscous fluid, *Journal of Fluid Mechanics* **11**, 447–459.
7. Saffman, P.G., 1965, The lift on a small sphere in slow shear flow, *Journal of Fluid Mechanics* **22**, 385–400; Corrigendum, 1968, *Journal of Fluid Mechanics* **31**, 624.

Molecular Dynamics Simulations of Drop Motion on Uniform and Non-Uniform Solid Surfaces

J.B. McLaughlin, S.S. Saravanan, N. Moumen and R.S. Subramanian

Clarkson University, Potsdam, NY 13699-5705, USA

Abstract. Although much research has been performed on the motion of contact lines on solid surfaces, many questions remain. This paper presents results obtained with molecular dynamics ("MD") simulations that address some of these questions. Of specific interest is the nature of the frictional resistance to contact line motion.

Key words: contact angle, contact line motion, molecular dynamics.

1 Introduction

The motion of a liquid drop on a surface is opposed both by hydrodynamic forces and by resistance to displacement of the molecules of the liquid at the contact line. Brochard-Wyart and de Gennes [4] discussed the relative importance of these contributions. The resistance to the displacement of the molecules at the contact line is a major contributor in many experiments, and is dominant at the length scales that occur in MD simulations.

Blake and Haynes [2] developed a theoretical model of contact line motion. They used Eyring's kinetic theory of liquids [9] to describe the motion of molecules near the contact line. The motion of a contact line is, at a microscopic scale, related to the frequency of molecular jumps between sites on the solid substrate. Under suitable conditions, their result for the contact line velocity can be expressed in terms of a friction coefficient. The friction coefficient describes the molecular scale dissipation that occurs within a few molecular length scales of the contact line. Thus, it represents dissipation that is not described by macroscopic (viscous) dissipation. de Ruijter et al. [6] performed MD simulations of drops spreading on molecularly smooth, crystalline surfaces. They found that the Blake–Haynes theory described their results well.

The Blake–Haynes theory assumes that the dynamic contact angle differs from the equilibrium contact angle. On the other hand, Hocking [11], Hocking and Rivers [12], and Cox [5] assumed that the dynamic contact angle is equal to the equilibrium contact angle. One might assume that this difference in views is due to the fact that

S. Balachandar and A. Prosperetti (eds), Proceedings of the IUTAM Symposium on Computational Multiphase Flow, 109–118.
© 2006 *Springer. Printed in the Netherlands.*

the above studies considered only viscous dissipation, while Blake and Haynes considered molecular scale dissipation. Brochard [3], Brochard-Wyart and de Gennes [4] and others have, however, assumed that the dynamic contact angle may differ from the equilibrium contact angle even when only hydrodynamic dissipation is important. It is beyond the scope of the present paper to address this issue in general. For the drops that were studied in the present work, however, the dynamic contact angle differed significantly from the equilibrium contact angle even for uniform surfaces.

In what follows, the results of MD simulations of drops on crystalline surfaces will be presented. The results differ from those presented by de Ruijter et al. in several respects. First, results will be presented for the spreading of drops on a variety of crystalline surfaces that differ in their wettability for the drop phase. It will be seen that the friction coefficient increases rapidly with the wettability. A second difference is that results will be presented for receding contact lines as well as advancing contact lines. The Blake–Haynes theory did not distinguish between these cases, but it will be seen that the friction coefficient for a receding contact line is significantly larger than for an advancing contact line. The difference between the friction coefficients for advancing and receding contact lines increases with wettability. Finally, MD results for a drop migrating on a wettability gradient will be presented and the ability of the Blake–Haynes theory to describe this phenomenon will be discussed.

2 Blake–Haynes kinetic theory

Blake and Haynes [2] developed a kinetic theory model of contact line motion. Their model indicates that a non-hydrodynamic frictional resistance to contact line motion exists. The origin of the frictional resistance is an energy barrier that molecules must cross to move from the liquid drop to available sites on the solid substrate. The result for the velocity of the contact line obtained by Blake and Haynes is:

$$v = 2K_0 \lambda \sinh\left(\frac{w}{2nk_BT}\right), \tag{1}$$

where K_0 is the molecular jumping frequency when equilibrium conditions prevail, λ is the characteristic displacement distance, n is the density of surface adsorption sites for the fluid molecules, and w is the work per unit area done by the driving force. Blake and Haynes argued that w could be related to the surface tension and the dynamic and equilibrium contact angles through $w = \gamma(\cos\theta_e - \cos\theta)$. If one assumes that $w \ll 2nk_BT$, one obtains the following expression for the velocity of the contact line:

$$v = \frac{\gamma}{\varsigma_0}(\cos\theta_e - \cos\theta). \tag{2}$$

In Equation (2), the friction coefficient, ς_0, is given by $\varsigma_0 = nk_BT/K_0\lambda$. The Blake–Haynes theory makes no distinction between advancing and receding contact lines. Therefore, according to their theory, the above expression for the friction coefficient should apply to both advancing and receding contact lines.

In general, one should expect viscous dissipation as well as molecular scale dissipation at the contact line to play a role in determining the spreading or migration of a drop. To incorporate dissipation, the "wedge" solution presented by Cox [5] was used to obtain an estimate of the drag force and this was added to the resistance associated with the molecular scale dissipation at the contact line. For the case of radial spreading, the contact line velocity is given by

$$v = \frac{\gamma}{\zeta_0 + 2\mu E_A \ln(1/\varepsilon)}(\cos\theta_e - \cos\theta). \tag{3}$$

In Equation (3), μ is the dynamic viscosity of the drop phase, E_A is a function of the dynamic contact angle that was given by Cox [5], and ε is the ratio of the smallest hydrodynamic length scale to the largest length scale; the smallest length scale was taken to be the thickness of the drop-vapor interface, and the largest length scale was taken to be the planform radius of the drop. The viscosity was determined by performing a MD simulation of single phase Couette flow for the liquid used in the drop simulations to be discussed.

De Ruijter et al. [6] used MD simulations of spreading fluid drops to test the model of Blake and Haynes [2]. They computed the relaxation of the contact angle of a spreading drop from its initial value to its final equilibrium value by integrating the Blake–Haynes model and found good agreement with the results obtained directly from their MD simulation.

3 MD simulations

In this study, MD simulations were carried out for drops composed of diatomic molecules. Yang et al. [17] used a similar system in their study of "terraced" drop spreading. An advantage of diatomic molecules is that the vapor pressure is much smaller than that for monatomic molecules [8]; this facilitates the analysis of the drop shape.

Following the procedure described by Yang et al. [17], the MD simulations were performed for diatomic liquid drops in contact with a FCC crystal lattice. All interactions between fluid molecules and other molecules or wall atoms were performed using a "site to site" approach in which the individual atoms interacted with one another. The atoms interacted through modified Lennard–Jones ("LJ") potentials with characteristic energies ε_{ff}, ε_{wf}, and ε_{ww} for the interactions between fluid atoms and other fluid atoms, fluid and wall atoms, and wall atoms and other wall atoms, respectively. The wettability of the surface was varied by choosing different values for ε_{wf}. The LJ potential energy of interaction V between atoms of type i and type j located a distance f apart is given by

$$V(r) = 4\varepsilon_{ij}\left[\left(\frac{\sigma_{ij}}{r}\right)^{12} - \delta_{ij}\left(\frac{\sigma_{ij}}{r}\right)^6\right]. \tag{4}$$

In Equation (4), δ_{ij} was chosen to be unity in the simulations to be reported. The quantities $2^{1/6}\sigma_{ff}$ and $2^{1/6}\sigma_{ww}$ can be interpreted as the atomic diameters for the fluid and wall atoms, respectively. The quantity σ_{ff} will be denoted by σ, ε_{ff} will be denoted by ε, and the mass of a fluid atom will be denoted by m. Dimensionless quantities will be used in the rest of the paper. The reference mass is m, the reference length is σ, and the reference time is $(m\sigma^2/\varepsilon)^{1/2}$. The mass of a wall atom was 5, the values of σ_{ww} and σ_{wf} were $\sqrt{2}/2^{1/6}$ and $(1 + \sigma_{ww})/2$, respectively, and the value of ε_{ww} was 50.

To reduce computation time, periodic boundary conditions were imposed in the direction normal to the wall, which was 5 atomic layers thick [17]. Periodic boundary conditions were also imposed in the transverse directions. The LJ potentials were modified by setting them equal to zero when the distance r was larger than 2.5 for the fluid-fluid interactions, $2.5\sigma_{wf}$ for the fluid-wall interactions, and $1.8\sigma_{ww}$ for the wall-wall interactions.

The equations of motion for the system are a coupled set of nonlinear ODE's for the coordinates and the velocity components of each atom. The "velocity Verlet" algorithm [1] was used to solve the equations. The dimensionless time step in all simulations was 0.005. The Head of Chain-Link List (HOC-LL) method was used to reduce computation time [13]. Similar techniques are described by Allen and Tildesley [1]. The technique exploits the finite range of the atomic interactions to reduce the number of computations per time step to $O(N)$ instead of $O(N^2)$, where N is the number of atoms. The computational domain is divided into a three-dimensional array of identical cells and each atom is assigned to a cell. In searching for interaction partners of a given atom, one considers only atoms in the given atom's cell or in an adjacent cell.

Numerical error and viscous heating can lead to temperature variations. Yang et al. [17] employed a re-scaling of the molecular velocities to solve this problem. In the initial part of a simulation, the velocities of all atoms were rescaled. After this initial part, it was found sufficient to rescale only the velocities of the atoms in the middle layer of the wall. The same approach was used to obtain the results in the present paper.

To perform a simulation, several steps were followed. First, a spherical drop was created. This was done by placing the fluid molecules in a rectangular block that was well separated from the wall, and then performing the computations for 5×10^4 time steps. Then, a small body force was applied to the drop to bring the drop into contact with the wall. Once the drop had begun to wet the wall, the body force was removed. In the simulations of drops on uniform surfaces, the computations were performed for a sufficient number of time steps to allow the drop to reach an equilibrium shape. Typically, this involved more than 10^6 time steps.

In the simulations to be discussed, there were 51,622 diatomic fluid molecules and 50,000 wall atoms. The simulations were performed at a dimensionless temperature equal to 0.7.

Figure 1 shows a side view of an equilibrium drop in contact with a wall for which $\varepsilon_{wf} = 1.3$. The dimensionless periods of the computational domain are 200 in x, 100 in y, and 85 in z, where x and y are measured parallel to the solid surface and z is normal to the solid surface. The equilibrium contact angle was determined to

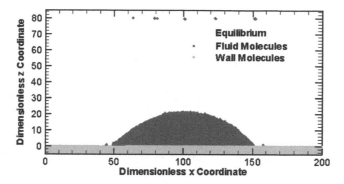

Fig. 1. An equilibrium drop on a uniform surface with $\varepsilon_{wf} = 1.3$.

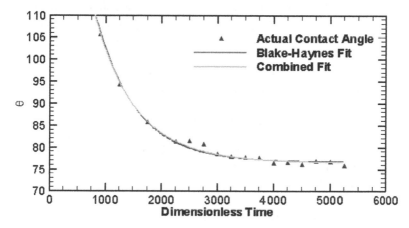

Fig. 2. The dynamic contact angle determined from the MD simulation and the value predicted by the Blake–Haynes theory are plotted as a function of dimensionless time.

be 44° by fitting the drop to a spherical cap shape and excluding the 10 fluid layers closest to the wall as recommended by de Ruijter et al. [6].

4 Determination of friction coefficients

To determine values for the friction coefficient, ς_0, one can use results for the spreading of drops on uniform surfaces. At regular time intervals, the dynamic contact angle was determined by fitting a spherical cap to the instantaneous shape of the drop. The results of this procedure are shown in Figure 2 for a drop spreading on a surface for which $\varepsilon_{wf} = 1.0$. It may be seen that the equilibrium contact angle is approximately 76°. Also shown in Figure 2 are two theoretical curves. To obtain these results, the friction coefficient was varied to obtain the best agreement with the results from the MD simulation. The curve labeled "Blake–Haynes Fit" shows the best fit when

Table 1. Equilibrium contact angles and values of the friction coefficient for different values of the wettability parameter, ε_{wf}.

ε_{wf}	θ_e (degrees)	Advancing ς_0	Receding ς_0
0.4	133	13	31
0.7	104	55	82
1.0	76	83	151
1.3	43	167	NA

viscous dissipation is neglected. The value obtained by this method was $\varsigma_0 = 94$. The "Combined Fit" curve includes both the Blake–Haynes dissipation and viscous dissipation. The values of the viscosity and the interface thickness in Equation (3) were 8.9 and 5.0, respectively. The value obtained by this method was $\varsigma_0 = 83$. In what follows, only the results obtained when viscous dissipation was included will be discussed. Using the same procedure for other wettabilities, it was found that ς_0 varied monotonically from 13 at $\varepsilon_{wf} = 0.4$ to 167 at $\varepsilon_{wf} = 1.3$.

To predict the translation velocity of a drop on a surface with a wettability gradient, one should also determine the friction coefficient for a receding contact line. This was done by starting with an equilibrium drop on a surface and abruptly reducing the value of the wettability (i.e., the value of ε_{wf}). For example, starting with the equilibrium drop in Figure 1, the value of ε_{wf} was changed from 1.3 to 0.7. The value of ς_0 for a receding contact line on a surface with $\varepsilon_{wf} = 0.7$ was thereby found to be 82. For an advancing contact line on the same surface, the friction coefficient is 55. For $\varepsilon_{wf} = 0.4$, the values of ς_0 for advancing and receding contact lines were 13 and 31, respectively. The difference between the values of the friction coefficients for advancing and receding contact lines increases with ε_{wf}.

Table 1 summarizes the values of the friction coefficient for advancing and receding contact lines as well as the equilibrium contact angles for several surfaces. The value of the friction coefficient for a receding contact line on a surface with $\varepsilon_{wf} = 1.3$ is not presently available because of the large amount of computer time needed to compute it. The difficulty is that one must first create an equilibrium drop on a surface with a significantly larger value of ε_{wf} and then abruptly reduce the value of ε_{wf} to 1.3. Unfortunately, the time needed to attain equilibrium increases rapidly with ε_{wf} because of the large resistance to contact line motion.

In all cases, the dynamic angle approached the same steady value for receding contact lines as for advancing contact lines to within the error within which the angle could be determined ($\pm 1°$). This may be because the surface is molecularly smooth.

5 Drop motion on a wettability gradient

When drops are placed on a solid surface with a wettability gradient, they can migrate in the direction of increasing wettability [7, 10]. A wettability gradient was created by setting $\varepsilon_{wf} = 0.4 + 0.006x$. A spherical drop was created and then brought into contact with the solid surface. Figure 3 shows the center of mass x-coordinate of the

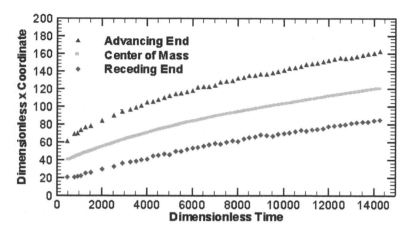

Fig. 3. The x-coordinate of the center of mass and the advancing and receding ends of a drop on a wettability gradient are shown as a function of dimensionless time.

drop as a function of dimensionless time. It may be seen that the velocity gradually decreases with time. Also shown in Figure 3 are the x-coordinates of the leading and trailing ends of the drop as functions of dimensionless time.

The Blake–Haynes theory was used to derive an expression for the migration velocity by assuming that the friction coefficient was relatively uniform around the drop's perimeter. Viscous effects were also included in the calculations by using the drag computed from the Cox wedge solution.

To compute the driving force that causes a drop to migrate, the cosine of the equilibrium contact angle as a function of position is required. Figure 4 shows that the cosine of the equilibrium contact angle is very well approximated by a linear function of the x-coordinate: $\cos(\theta_e(x)) = \cos(\theta_e(x_0)) + \alpha(x - x_0)$, where x_0 is a convenient reference coordinate. Figure 4 was prepared by using the relation between the equilibrium contact angle and ε_{wf} in Table 1; it reflects the fact that the cosine of θ_e is well approximated by a linear function of ε_{wf}. It is interesting to note that Maruyama et al. [14] also found that the cosine of the equilibrium contact angle was well-approximated by a linear function of ε_{wf} in their MD simulations.

Using the Blake–Haynes theory and the viscous drag obtained from the Cox wedge solution, it may be shown that the velocity of the drop is given by the following relationship:

$$v = \frac{\alpha \gamma R}{\left(\varsigma_0 + \frac{\mu \psi \ln(1/\varepsilon)}{\pi}\right)}. \qquad (5)$$

In Equation (5), the quantity ψ is a function of the local equilibrium contact angle that is discussed by Subramanian et al. [16]. At dimensionless time 14250, the friction coefficient at the value of ε_{wf} corresponding to the location of the center of mass was 124. Using this value, the predicted migration velocity is 0.0040. The actual velocity was 0.0031. The difference between the theoretical and observed val-

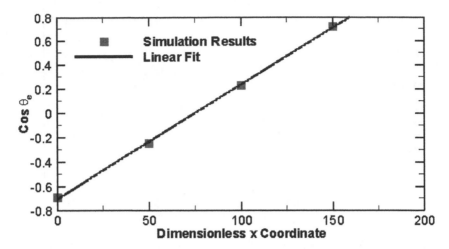

Fig. 4. The cosine of the local equilibrium contact angle is plotted as a function of the dimensionless x-coordinate.

ues is partly due to the fact that the friction coefficient varies significantly over the perimeter of the drop. Part of this variation is due to fact that ε_{wf} depends on x, and part of it is due to the difference between the values of the friction coefficients for advancing and receding contact lines. A crude way of accounting for this difference is to replace the friction coefficient in the expression for the migration velocity by an average of the advancing and receding values evaluated at the center of mass location. With this modification, the predicted migration velocity is approximately 0.0035.

6 Conclusions

Our results provide several insights into the behavior of contact lines. First, in simulations of drops on homogeneous surfaces, we found that a theoretical model that included the Blake–Haynes contact line dissipation and the viscous dissipation computed from the Cox wedge solution described the MD results well. The viscous dissipation was small but not negligible in comparison with the Blake–Haynes dissipation. De Ruijter et al. [6] performed a similar simulation and arrived at the same conclusion. Two differences with the present work are that they used a different approach to modeling the viscous dissipation and their molecules were much larger. We have developed more comprehensive results over a range of partially wetting conditions and correlated the Blake–Haynes friction coefficient with the LJ parameter that controls the wettability. The friction coefficient increases rapidly with increase in the wettability of the surface. We also have presented results for receding contact lines. These results were obtained by starting with equilibrium drops on relatively hydrophilic surfaces and abruptly changing the fluid-wall LJ parameter to make the wall more hydrophobic. We found that the value of the friction coefficient for a receding contact line is significantly larger than the value for an advancing contact line; the ratio of the values for receding and advancing contact lines increases as the surface is made more hydrophilic. At present, one can only speculate about the cause of this difference. When a contact line advances, molecules hop from the liquid surface to adsorption sites on the solid surface. When a contact line recedes, liquid molecules must be pulled from the solid surface. It seems likely that the energies required for the two processes are different.

The results of the above studies were used to model the migration of drops on a wettability gradient. The results suggest that the Blake–Haynes theory can describe the motion of the drops over a broad range of conditions provided that account is taken of the facts that (1) the friction coefficient depends on the local wettability of the surface and (2) the friction coefficient for a receding contact line has a different value from that for an advancing contact line on the same surface.

Acknowledgement

This research was supported by NASA under grant NAG3-2703.

References

1. Allen, M.P. and Tildesley, D.J., 1987, *Computer Simulation of Liquids*, Oxford Science Publications, Oxford.
2. Blake, T.D. and Haynes, J.M., 1969, Kinetics of liquid/liquid displacement, *Journal of Colloid and Interface Science* **30**, 421–423.
3. Brochard, F., 1989, Motions of droplets on solid surfaces induced by chemical or thermal gradients, *Langmuir* **5**, 432–438.
4. Brochard-Wyart, F. and de Gennes, P.G., 1992, Dynamics of partial wetting, *Advances in Colloid and Interface Science* **39**, 1–11.
5. Cox, R.G., 1986, The dynamics of spreading of liquids on a solid surface. Part 1. Viscous flow, *Journal of Fluid Mechanics* **168**, 169–194.
6. de Ruijter, M.J., Blake, T.D. and De Coninck, J., 1999, Dynamic wetting studied by molecular modeling simulations of droplet spreading, *Langmuir* **15**, 7836–7847.
7. Chaudhury, M.K. and Whitesides, G.M., 1992, How to make water run uphill, *Science* **256**, 1539–1541.
8. Freund, J.B., 2003, The atomic detail of a wetting/de-wetting flow, *Physics of Fluids* **15**, L33–L36.
9. Gladstone, S., Laidler, K.J. and Eyring, H.J., 1941, *The Theory of Rate Processes*, McGraw-Hill, New York.
10. Greenspan, H.P., 1978, On the motion of a small viscous droplet that wets a surface, *Journal of Fluid Mechanics* **84**, 125–143.
11. Hocking, L.M., 1981, Sliding and spreading of thin two-dimensional drops, *Quarterly Journal of Mechanics and Applied Mathematics* **34**, 37–55.
12. Hocking, L.M. and Rivers, A.D., 1982, The spreading of a drop by capillary action, *Journal of Fluid Mechanics* **121**, 425–442.

13. Hockney, R.W. and Eastwood, J.W., 1988, *Computer Simulation Using Particles*, Institute of Physics, Bristol.
14. Maruyama, S., Kurashige, T., Matsumoto, S., Yamaguchi, Y. and Kimura, T., 1998, Liquid droplet in contact with a solid surface, *Microscale Thermophysical Engineering* **2**, 49–62.
15. Subramanian, R.S., Moumen, N. and McLaughlin, J.B., 2005, Motion of a drop on a solid surface due to a wettability gradient, *Langmuir* **21**, 11844–11849.
16. Yang, J.-X., Koplik, J. and Banavar, J.R., 1992, Terraced spreading of simple liquids on solid surfaces, *Physical Review A* **46**, 7738–7749.

Fluctuating Immersed Material (FIMAT) Dynamics for the Direct Simulation of the Brownian Motion of Particles

Yong Chen, Nitin Sharma and Neelesh A. Patankar*

Department of Mechanical Engineering, Northwestern University, Evanston, IL 60208, USA

Abstract. In the paper a Direct Numerical Simulation (DNS) scheme, named Fluctuating Immersed MATerial (FIMAT) dynamics, for the Brownian motion of particles is presented. In this approach the thermal fluctuations are included in the fluid equations via random stress terms. Solving the fluctuating hydrodynamic equations coupled with the particle equations of motion results in the Brownian motion of the particles. There is no need to add a random force term in the particle equations. The particles acquire random motion through the hydrodynamic force acting on its surface from the surrounding fluctuating fluid. The random stresses in the fluid equations are easy to calculate unlike the random terms in the conventional Brownian Dynamics (BD) type approaches.

Key words: fluctuating hydrodynamics, mesoscopic scale, Brownian motion, direct numerical simulation (DNS), Distributed Lagrange Multiplier (DLM) method.

1 Introduction

The interaction of sub-micron/nanoscale objects (such as macromolecules or small particles or small devices) with fluids is an important problem in small scale devices. A better understanding of fluid dynamics is critical in e.g. bio-molecular transport, manipulating and controlling chemical and biological processes using small particles. These objects could be moving in an environment with varying temperatures and fluid properties. Thermal fluctuations can influence the motion of such objects.

Direct Numerical Simulations (DNS) of particle motion in fluids is a tool that has been developed over the past twelve years [4, 8, 9, 13, 15, 19, 20]. In this approach the fluid equations are solved coupled with the equations of motion of the particles. DNS allows investigation of a wide variety of problems including particles in Newtonian or viscoelastic fluids with constant or varying properties. DNS can be an excellent tool to investigate the motion of sub-micron particles in varying fluid

---

* Author for correspondence.

S. Balachandar and A. Prosperetti (eds), Proceedings of the IUTAM Symposium on Computational Multiphase Flow, 119–129.

environments. The objective of this work is to find a convenient way to incorporate the effect of thermal fluctuations in the DNS schemes.

A particle suspended in a fluid experiences a hydrodynamic force due to the average motion of the fluid around it. The average motion of the fluid is represented by the continuum equations – the Navier–Stokes equations. In addition to the average force, small particles in fluids experience a random force due to the thermal fluctuations in the fluid. In Brownian dynamic (BD) simulations the principle is to model this thermal force from the fluid in terms of a random force in the particle equation of motion.

The conventional approach to perform Brownian dynamic (BD) simulations is based on the algorithm by Ermak and McCammon [3]. Their numerical method is based on the Langevin equation for particle motion. Properties of the random force in the particle equation of motion depend on the hydrodynamic interactions between the particles. Typically, approximate expressions are used to model the hydrodynamic interactions.

Brady and Bossis [1] presented Stokesian dynamics technique for simulating the Brownian motion of many particles. They also considered the Langevin equations for the motion of the Brownian particles. They computed the hydrodynamic interactions through a grand resistance tensor instead of using approximations as was done by Ermak and McCammon [3]. Using these techniques to objects of irregular shapes and to cases where the fluid exhibits varying properties is not straightforward. This is mainly because the properties of the random force in the particle equations depend on the grand resistance tensor, which in turn depends on the particle positions, shapes and the fluid properties.

In accordance with the BD approach, it is possible to envisage a DNS scheme where the Navier–Stokes equations for the fluid are solved coupled with the Langevin equation (which includes a random force term) for particle motion. Again, as stated above, generation of the random force term is not straightforward because it depends on the particle resistance tensor. A different approach is preferred.

An alternate approach is to model the thermal fluctuations in the fluid (instead of in the particle equations) via random stress terms in its governing equations. A general theory of fluctuating hydrodynamics is given by Landau and Lifshitz [11]. Solving the fluctuating hydrodynamic equations coupled with the particle equations of motion can result in the Brownian motion of the particles. There is no need to add a random force term in the particle equations. The particles acquire random motion through the hydrodynamic force acting on its surface from the surrounding fluctuating fluid. The random stresses in the fluid equations are easy to calculate unlike the random terms in the BD approach. In this paper we present such an approach along with validation.

Ladd [10] presented a Lattice-Boltzmann (LB) method to simulate the Brownian motion of solid particles. They added a fluctuating term in the LB equation for the fluid which was equivalent to the random stress term in the fluctuating hydrodynamic equations of Landau and Lifshitz [11]. Fluctuating LB equations were solved to get results for the decay of an initially imposed translational and rotational velocity of an isolated Brownian sphere in a fluid. The current work in this paper is aimed at adding the random fluctuating terms directly into the Navier–Stokes equations instead of the Lattice-Boltzmann equations. It can therefore be easily incorporated in existing conventional solvers for Navier–Stokes equations to model fluid-particle behavior at small scales.

Zwanzig [22] showed that the motion of an isolated Brownian particle computed using fluctuating hydrodynamic equations is consistent with the traditional Langevin description in the long time (dissipative) limit. Hauge and Martin-Löf [6] showed that the Langevin equation describing the Brownian motion is a contraction from the more fundamental, but still phenomenological, description of an incompressible fluid governed by fluctuating hydrodynamics in which a Brownian particle with no-slip boundary condition is immersed. They showed that the fluctuating hydrodynamics approach captures the algebraic tail ($t^{-3/2}$) in the velocity autocorrelation function consistent with the molecular time correlation functions. The Langevin description gives an exponential tail in the velocity autocorrelation function. Hauge and Martin-Löf [6] also identified conditions under which the classical Langevin description is applicable. These results imply that the simulation of the Brownian motion of particles based on fluctuating hydrodynamic equations is a sound phenomenological approach. In this work, only the long time dissipative limit is considered, which is equivalent to neglecting the inertia terms in the governing equations. As a result, the velocity autocorrelation function will not be considered. However, the method will be tested by comparing the Brownian diffusion (in the long time limit), obtained from the simulations, with known analytic values. The problem involving the solution of the fluctuating hydrodynamic equations including the inertia terms will be considered in future work.

Patankar [14] presented preliminary results for the Brownian motion of a cylinder by solving the fluctuating hydrodynamic equations of the fluid coupled with the particle equation of motion. A 2D problem was considered.

Fluctuating hydrodynamic equations have been solved for a single fluid case by Serrano and Español [16, 17] using a finite volume Lagrangian discretization based on Voronoi tessellation. They obtained a discrete form of the governing equations that satisfied the fluctuation dissipation theorem. They ensured this by casting their discrete equations in the GENERIC (General Equation for Non-Equilibrium Reversible/Irreversible Coupling) structure. The GENERIC structure proposed by Grmela and Öttinger [5, 12] ensures that the equations describing the macroscopic dynamics of a system are thermodynamically consistent and that the fluctuation dissipation theorem is satisfied.

In this paper a DNS technique is used to solve the fluctuating hydrodynamic equations of Landau and Lifshitz [11] coupled with the particle equations of motion. The method used here is based on our earlier work presented in [18]. In this approach the entire fluid-particle domain is considered to be a fluid. It is ensured that the 'fluid' occupying the particle domain moves rigidly by adding a rigidity constraint [4, 15, 18–20]. Solution of this system of equations results in the Brownian motion of the particles. The technique is validated by comparing numerical results with analytic values.

In Section 2 the mathematical formulation of the problem will be discussed. Numerical results are presented in Section 3 and conclusions in Section 4.

2 Mathematical formulation

Let Ω be the computational domain which includes both the fluid and the particle domain. Let P be the particle domain. Assume that the computational domain is periodic in all directions. Consider one particle in the computational domain. The particle can be of any shape. In this paper a sphere and an ellipsoid are considered. The formulation to be presented is not restricted to periodic boundary condition – it can be extended to non-periodic domains. It is assumed that the entire fluid-particle domain is a single fluid governed by [4, 15]

$$\rho \left(\frac{\partial \mathbf{u}}{\partial t} + (\mathbf{u} \cdot \nabla) \mathbf{u} \right) = -\nabla p + \mu \nabla^2 \mathbf{u} + \nabla \cdot \tilde{\mathbf{S}} + \mathbf{f} \quad \text{in } \Omega, \tag{1}$$

$$\nabla \cdot \mathbf{u} = 0 \quad \text{in } \Omega, \tag{2}$$

$$\nabla \cdot (\mathbf{D}[\mathbf{u}]) = \mathbf{0} \quad \text{in } P, \tag{3a}$$

$$\mathbf{D}[\mathbf{u}] \cdot \mathbf{n} = \mathbf{0} \quad \text{on } \partial P, \tag{3b}$$

$$\mathbf{u}|_{t=0} = \mathbf{u}_0(\mathbf{x}) \quad \text{in } \Omega, \tag{4}$$

where ρ is the fluid and particle density (neutrally buoyant particles are considered here; however, the formulation can be easily generalized to heavy or light particles), $\mathbf{u}$ is the fluid velocity, $\mathbf{n}$ is the outward normal on the particle surface, p is the dynamic pressure (i.e. without the hydrostatic component) due to the incompressibility constraint (Equation 2) and μ is the viscosity of the fluid.

Equation (3) represents the rigidity constraint and Equation (2) is the incompressibility constraint. The rigidity constraint, imposed only in the particle domain, ensures that the deformation-rate tensor

$$\mathbf{D}[\mathbf{u}] = \frac{1}{2}(\nabla \mathbf{u} + \nabla \mathbf{u}^T) = \mathbf{0} \quad \text{in } P. \tag{5}$$

Thus the 'fluid' in the particle domain is constrained to move rigidly as required. The viscous stress is zero in the particle domain due to the rigidity constraint [15].

Equation (3) represents three scalar constraint equations at a point in the particle domain. They give rise to a force $\mathbf{f}$ in the particle domain similar to the presence of pressure due to the incompressibility constraint [15]. This is the Distributed Lagrange Multiplier (DLM) approach for particulate flows [4, 15]. $\mathbf{f}$ is zero in the fluid domain.

$\tilde{\mathbf{S}}$ in Equation (1) is the random stress tensor which is computed as proposed by Landau and Lifshitz [11]. $\tilde{\mathbf{S}}$ is included in the Navier–Stokes equations to model the fluid at mesoscopic scales. Hydrodynamics as such is a macroscopic theory. At the macroscopic level the hydrodynamic variables represent an average value over a macroscopic length and time scale. Consequently, information regarding the random fluctuations arising due to the molecular nature of the fluid is lost. $\tilde{S}$ accounts for these fluctuations when modeling flows at mesoscopic scales. By mesoscopic scales we typically imply scales ranging from tens of nanometers to micron, depending on the problem.

$\tilde{S}$ has the following property [11]:

$$\left.\begin{aligned} \langle \tilde{S}_{ij} \rangle &= 0, \\ \langle \tilde{S}_{ik}(x_1, t_1) \tilde{S}_{lm}(x_2, t_2) \rangle &= 2 k_B T \mu (\delta_{il}\delta_{km} + \delta_{im}\delta_{kl}) \delta(x_1 - x_2) \delta(t_1 - t_2). \end{aligned}\right\} \tag{6}$$

where $\langle \rangle$ denotes averaging over an ensemble, k_B is the Boltzmann constant, T is temperature of the fluid and we have used indicial notation. The above equations are in accordance with the fluctuation dissipation theorem for an incompressible fluid [11].

The solution of Equations (1)–(4) and (6) gives the velocity field $\mathbf{u}$ in the entire domain. The particle translational and angular velocities, $\mathbf{U}$ and ω, respectively, can then be computed by

$$M\mathbf{U} = \int_P \rho \mathbf{u} \, dx \quad \text{and} \quad I_P \omega = \int_p \mathbf{r} \times \rho \mathbf{u} \, dx, \tag{7}$$

where $\mathbf{r}$ is the position vector of a point with respect to the centroid of the particle, I_p is the moment of inertia of the particle and M its mass.

In this work the inertia is neglected (which is equivalent to taking the long time limit [18]) and the following Stokes problem, that is driven by the random stresses, is solved

$$-\nabla p + \mu \nabla^2 \mathbf{u} + \nabla \cdot \tilde{S} + \mathbf{f} = 0 \quad \text{in } \Omega, \tag{8}$$

where the time discretized properties of the random stress are given by

$$\left.\begin{aligned} \langle \tilde{S}_{ij} \rangle &= 0, \\ \langle \tilde{S}_{ik}(x_1) \tilde{S}_{lm}(x_2) \rangle &= \frac{2 k_B T \mu}{\Delta t} (\delta_{il}\delta_{km} + \delta_{im}\delta_{kl}) \delta(x_1 - x_2). \end{aligned}\right\}. \tag{9}$$

Solution of the Stokes problem represented by Equations (8), (2), (3) and (9) gives the velocity $\mathbf{u}$ in the entire domain. The velocity field $\mathbf{u}$ in Equation (8) is not the true velocity of the fluid material point at any instant. It is the Brownian diffusion of the fluid material point in time Δt divided by Δt. Hence, the velocity will also be referred to as the apparent velocity in the following discussion. For further discussion see [18] and references therein. The translational and angular velocities of the particle as computed by Equation (7) must be interpreted similarly.

The governing equations to be solved are stochastic. It is known that, for deterministic equations, central differencing ensures second-order accuracy. Discretization of stochastic equations based on central differencing may not be sufficient to obtain thermodynamically consistent discrete equations. A consistent discretization should ensure that the resultant discrete equations satisfy the corresponding fluctuation dissipation theorem (FDT). It must be noted that even if the differential equations satisfy the FDT it does not imply that the corresponding discretized equations based on central differencing will necessarily satisfy the FDT for the discrete equations.

Thermodynamic consistency of the discrete equations can be ensured if we discretize the equations such that they are in the GENERIC form as proposed by Grmela and Öttinger [5, 12]. Serrano and Español [16] and Serrano *et al.* [17] have presented a systematic derivation of two-dimensional discrete equations that obey the GENERIC structure for the case of a fluid (i.e. no particles in the domain). They used a finite volume Lagrangian discretization based on Voronoi tessellation. They showed that simple central differencing does not ensure thermodynamically consistent discretized equations but this can be corrected by adding certain terms to the discrete equations. They also argued that these additional terms may be neglected under certain conditions.

In this paper an Eulerian control (finite) volume discretization based on cubic cells is used. A staggered control volume scheme is used to solve the fluid equations [18, 20]. The momentum equations obtained by simple central difference discretization of Equation (8) did not strictly satisfy the FDT, however, in agreement with the analysis in [17], it was found that the additional term was small. Hence it was neglected. A detailed derivation will be presented elsewhere [2]. It was also found that the solution of the fluid equations without the particles gave the results in close agreement with the FDT [2]. The details of the algorithm to solve the particulate Stokes flow problem are given in [18, 20].

3 Results

The Brownian diffusion of a single sphere, of diameter d, in a fully periodic domain is considered first. A non-dimensionalized problem was solved. The fundamental scales for non-dimensionalization are

$$
\left.
\begin{aligned}
\text{Time} &\rightarrow \sqrt{\frac{\mu \Delta V \Delta t}{k_B T}}, \\
\text{Length} &\rightarrow L, \\
\text{Mass} &\rightarrow L \sqrt{\frac{\mu^3 \Delta V \Delta t}{k_B T}},
\end{aligned}
\right\}
\tag{10}
$$

where $\Delta V = \Delta h^3$ is the volume of the control volume, Δt is the time over which the Brownian diffusion is computed and L is the length of the periodic domain. The scales for velocity, pressure and stress are

$$
\left.
\begin{aligned}
\text{velocity} &\rightarrow L \sqrt{\frac{k_B T}{\mu \Delta V \Delta t}}, \\
\text{stress, pressure} &\rightarrow L \sqrt{\frac{\mu k_B T}{\Delta V \Delta T}}.
\end{aligned}
\right\}
\tag{11}
$$

In terms of the non-dimensionalized variables, the governing equations are

$$
-\nabla p^* + \nabla^2 \mathbf{u}^* + \nabla \cdot \tilde{\mathbf{S}}^* + \mathbf{f}^* = \mathbf{0} \quad \text{in } \Omega,
\tag{12}
$$

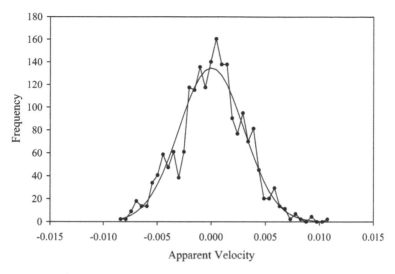

Fig. 1. Comparison of the frequency distribution of the relative apparent velocity of the particle with the analytic Gaussian distribution ($\phi = 0.008$).

$$\nabla \cdot \mathbf{u}^* = 0 \quad \text{in } \Omega, \tag{13}$$

$$\nabla \cdot (\mathbf{D}[\mathbf{u}^*]) = \mathbf{0} \quad \text{in } P \quad \text{and} \quad \mathbf{D}[\mathbf{u}^*] \cdot \mathbf{n} = \mathbf{0} \quad \text{on } \partial P, \tag{14}$$

where the superscript $*$ represents the corresponding non-dimensionalized variable. The above equations were discretized as discussed before. Properties of the random stress $\tilde{\mathbf{S}}^*$ after discretization were obtained by non-dimensionalizing Equation (9) to give

$$\langle \tilde{S}_{ij}^* \rangle = 0 \quad \text{and} \quad \langle \tilde{S}_{ik}^* \tilde{S}_{lm}^* \rangle = 2(\delta_{il}\delta_{km} + \delta_{im}\delta_{kl}). \tag{15}$$

A single sphere was located at the centre of a cubic periodic domain. No body force was applied either in the particle domain or the fluid domain other than the force due to the random stresses. Components of the random stresses at different locations were generated from a Gaussian random number generator with the desired mean and variance. Once the random stresses were generated, the Stokes problem, defined by Equations (12)–(15), was solved [18, 20]. It is ensured that the net momentum in the periodic domain is zero. One simulation was considered as one realization. We solved for several realizations which constituted an ensemble. For each realization a different initial seed was assigned to the Gaussian random number generator for random stresses. This ensured that each realization was different. In a given realization, the apparent velocity $\mathbf{U}^*$ of the sphere was computed according to Equation (7). The variance of the apparent velocity gives the Brownian diffusion D, which in turn is related to the drag coefficient K as follows:

$$D = \frac{\langle |\mathbf{U}|^2 \rangle \Delta t}{6} = \frac{k_B T}{3\pi \mu d K} \Rightarrow K = \frac{6 k_B T}{3\pi \mu d \Delta t \langle |\mathbf{U}|^2 \rangle}. \tag{16}$$

In terms of the non-dimensional variables we get

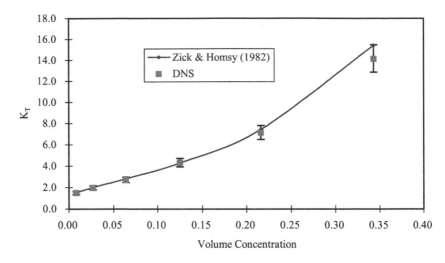

Fig. 2. Plot of the translational drag coefficient as a function of the volume fraction.

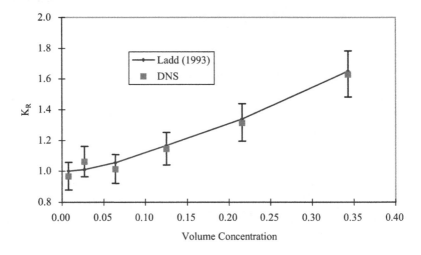

Fig. 3. Plot of the rotational drag coefficient as a function of the volume fraction.

$$K = \frac{6}{3\pi N^3 d^* \langle |\mathbf{U}^*|^2 \rangle}. \qquad (17)$$

Note that $N = L/\Delta h$ gives the information regarding the degree of discretization in the computational domain. Equation (17) was used to compute the drag coefficient from the numerical simulation. This numerical value of the drag coefficient was then compared with the analytic value obtained by Hasimoto [7] and Zick and Homsy [21].

Figure 1 shows a histogram of the apparent velocity based on 900 realizations. It is compared to the analytic Gaussian distribution. The analytic frequency distribution has zero mean and the variance is $6/3\pi d^* N^3 K_A$ (Equation 17), where K_A is the analytic value of the drag coefficient from Hasimoto [7] and Zick and Homsy [21]. Figure 2 shows the translation drag coefficient K_T, calculated according to Equation (17), as a function of the volume fraction. The results are compared with the analytic value of Zick and Homsy [21]. The error bars were drawn based on a chisquare distribution for the variance [18]. Figure 3 shows the results for the rotational drag coefficient K_R, calculated similar to Equation (17) but for the rotational fluctuations. These results are compared to the analytic values by Ladd [10]. In all the cases we see that the agreement is good.

The method was also verified for non-spherical particles. An oblate ellipsoid was considered in a periodic domain at a volume fraction of 0.01676. The long axes were two times the short axis of the ellipsoid. The ellipsoid was placed with the principle axes coincident with the coordinate direction. The translational and rotational drag with respect to the principle axes are different for an ellipsoid. The drag data for ellipsoids are typically represented in terms of an effective radius. For the translational case the effective radius R_{te} is defined by

$$D_t = \frac{\langle u^2 \rangle \Delta t}{2} = \frac{k_B T}{6\pi \mu R_{te}} \Rightarrow R_{te} = \frac{k_B T}{3\pi \mu \langle u^2 \rangle \Delta t}, \qquad (18)$$

where D_t is the translational diffusion. The value of R_{te} is different along different principle directions. E.g. the effective radius with respect to the x direction is obtained if $\langle u^2 \rangle$ is based on the x component of the random velocity of the ellipsoid. Similarly, the rotational drag is represented by the effective radius R_{re} defined by

$$D_r = \frac{\langle \omega^2 \rangle \Delta t}{2} = \frac{k_B T}{8\pi \mu R_{re}^3} \Rightarrow R_{re} = \left(\frac{k_B T}{4\pi \mu \langle \omega^2 \rangle \Delta t} \right)^{1/3}, \qquad (19)$$

where D_r is the rotational diffusion. R_{re} is different for different principle directions. The effective radii were also computed from non-Brownian simulations. In this case a force or torque was applied to the ellipsoid. A Stokes problem was solved to obtain the particle translational and angular velocities [20]. The effective radii are then given by

$$R_{te} = \frac{F}{6\pi \mu U} \quad \text{and} \quad R_{re} = \left(\frac{T}{8\pi \mu \omega} \right)^{1/3}, \qquad (20)$$

where F and T are the applied force and torque with respect to the chosen principle direction. U and ω are the corresponding translational and angular velocities obtained from the Stokes solution of the coupled fluid-particle problem [20]. Table 1 shows the comparison between the effective radii obtained from the Brownian and non-Brownian simulations for the same geometry and volume fraction (0.01676). The agreement is good.

4 Conclusions

A DNS scheme, named FIMAT dynamics, for the Brownian motion of particles is presented. The thermal fluctuations were included in the fluid equations via ran-

Table 1. Effective radii representing the drag on an ellipsoid from non-Brownian and Brownian simulations.

	R_{te}/L Non-Brownian	R_{te}/L Brownian	R_{re}/L Non-Brownian	R_{re}/L Brownian
x-direction	0.2530	0.2682	0.1622	0.1655
y-direction	0.2530	0.2692	0.1622	0.1604
z-direction	0.3289	0.3587	0.1758	0.1801

dom stress terms. Solving the fluctuating hydrodynamic equations coupled with the particle equations of motion resulted in the Brownian motion of the particles. The particles acquired random motion through the hydrodynamic force acting on its surface from the surrounding fluctuating fluid. The random stresses in the fluid equations were easy to calculate unlike the random terms in the conventional Brownian Dynamics (BD) type approaches.

The problem was solved in the long time dissipative limit. Solution of the governing equations gave the Brownian displacements of the particle. The numerical results were used to find the corresponding drag coefficient acting on spheres and ellipsoids. The numerical values of the drag coefficient were compared with analytic values. The agreement was found to be good.

The method can be potentially extended to fluids with varying properties. Application to many particles and the scaling of the computational time with the number of particles is the subject of our future work. Implementation of fast solvers is currently being undertaken after which the computational time for the current method will be compared with the traditional methods for Brownian simulations.

Acknowledgment

This work was supported by the National Science Foundation through the CAREER grant CTS-0134546.

References

1. Brady, J.F. and Bossis, G., 1988, Stokesian dynamics, *Annual Rev. Fluid Mech.* **20**, 111–157.
2. Chen, Y., Sharma, N. and Patankar, N.A., 2005, Fluctuating Immersed Material (FIMAT) dynamics for the direct simulation of the Brownian motion of particles, *J. Fluid Mech.*, submitted.
3. Ermak, D.L. and McCammon, J.A., 1978, Brownian dynamics with hydrodynamic interactions, *J. Chem. Phys.* **69**(4), 1352–1360.
4. Glowinski, R., Pan, T.W., Hesla, T.I. and Joseph, D.D., 1999, A distributed Lagrange multiplier/fictitious domain method for particulate flows, *Int. J. Multiphase Flow* **25**, 755–794.
5. Grmela, M. and Öttinger, H.C., 1997, Dynamics and thermodynamics of complex fluids. I. Development of a general formalism, *Phys. Rev. E* **56**(6), 6620–6632.

6. Hauge, E.H. and Martin-Löf, A., 1973, Fluctuating hydrodynamics and Brownian motion, *J. Stat. Phys.* **7**(3), 259–281.

7. Hasimoto, H., 1959, On the periodic fundamental solution of the Stokes equations and their application to viscous flow past a cubic array of spheres, *J. Fluid Mech.* **5**, 317–328.

8. Hu, H.H., Joseph, D.D. and Crochet, M.J., 1992, Direct numerical simulation of fluid particle motions, *Theoret. Comput. Fluid Dynam.* **3**, 285–306.

9. Hu, H.H., Patankar, N.A. and Zhu, M.Y., 2001, Direct numerical simulations of fluid solid systems using Arbitrary Lagrangian–Eulerian technique, *J. Comput. Phys.* **169**, 427–462.

10. Ladd, A.J.C., 1993, Short time motion of colloidal particles: Numerical simulation via a fluctuating Lattice-Boltzmann equation, *Phys. Rev. Lett.* **70**(9), 1339–1342.

11. Landau, L.D. and Lifshitz, E.M., 1959, *Fluid Mechanics*, Pergamon Press, London.

12. Öttinger, H.C. and Grmela, M., 1997, Dynamics and thermodynamics of complex fluids. II. Development of a general formalism, *Phys. Rev. E* **56**(6), 6633–6655.

13. Patankar, N.A., 2001, A formulation for fast computations of rigid particulate flows, *Center for Turbulence Research, Annual Research Briefs*, 185–196.

14. Patankar, N.A., 2002, Direct Numerical Simulation of moving charged, flexible bodies with thermal fluctuations, in *Technical Proceedings of the 2002 International Conference on Modeling and Simulation of Microsystems*, pp. 32–35.

15. Patankar, N.A., Singh, P., Joseph, D.D., Glowinski, R. and Pan, T.W., 2000, A new formulation of the distributed Lagrange multiplier/fictitious domain method for particulate flows, *Int. J. Multiphase Flow* **26**, 1509–1524.

16. Serrano, M. and Español, P., 2001, Thermodynamically consistent mesoscopic fluid particle model, *Phys. Rev. E* **64**(4), 046115.

17. Serrano, M., Gianni, D.F., Español, P., Flekkøy, E.G. and Coveney, P.V., 2002, Mesoscopic dynamics of Voronoi fluid particles, *J. Phys. A: Math. Gen.* **35**(7), 1605–1625.

18. Sharma, N. and Patankar, N.A., 2004, Direct numerical simulation of the Brownian motion of particles by using fluctuating hydrodynamic equations, *J. Comput. Phys.* **201**, 466–486.

19. Sharma, N. and Patankar, N.A., 2005, A fast computation technique for the Direct Numerical Simulation of rigid particulate flows, *J. Comput. Phys.* **205**, 439–457.

20. Sharma, N., Chen, Y. and Patankar, N.A., 2005, A Distributed Lagrange Multiplier method based computational method for the simulation of particulate Stokes flow, *Comput. Meth. Appl. Mech. Engng.* **194**, 4716–4730.

21. Zick, A.A. and Homsy, G.M., 1982, Stokes flow through periodic arrays of spheres, *J. Fluid Mech.* **115**, 13–26.

22. Zwanzig, R., 1964, Hydrodynamic fluctuations and Stokes' law friction, *J. Res. Natl. Bur. Std. (U.S.)* **68B**, 143–145.

A Novel Definition of the Local and Instantaneous Liquid–Vapor Interface

Gota Kikugawa, Shu Takagi and Yoichiro Matsumoto

Department of Mechanical Engineering, The University of Tokyo, Hongo, Bunkyo-ku, Tokyo 113-8656, Japan; e-mail: gota@fel.t.u-tokyo.ac.jp, takagi@mech.t.u-tokyo.ac.jp, ymats@mech.t.u-tokyo.ac.jp

Abstract. In this paper, we present a new definition of liquid–vapor interface at the molecular level which can capture the local and instantaneous structure of the interface. The new definition is not a thermodynamic definition of the interface, such as the equimolar surface, but is based on the instantaneous particle density distribution of molecules. Applying the new definition of the interface to the MD result of the liquid–vapor interface, we found that our definition of the interface was able to capture the microscopic fluctuation caused by molecular motion. Furthermore, we confirmed that on the longtime average our definition of the interface shows good agreement with the equimolar surface.

Key words: instantaneous interface, molecular dynamics, equimolar surface, level set method.

1 Introduction

The study of liquid–gas or liquid–liquid interface have an important role in the wide range of physical, chemical, and biological processes. Especially the elucidation of the physico-chemical properties of the water–air interface leads to an essential understanding of various interfacial phenomena. Recently, the microscopic phenomena related to the water–air interface have been reported, of which mechanisms have not been understood clearly. For example, nanometer sized bubbles (nanobubbles) stably existing on the hydrophobic surfaces in water, were reported [1–3]. However, it is considered that bubbles of this size in water are inherently unstable due to the strong effect of surface tension. Therefore, the contradiction between the experimental facts and the theoretical predictions about the stability of nanobubbles is recognized as an unresolved problem [4].

So far, a microscopic picture of the liquid–gas or liquid–liquid interface have not fully been understood, although there have been a number of efforts to investigate the interface. For one reason, there is difficulty of the experimental approach to the microscopic features at the liquid interface [5]. On the contrary, our approach to the

S. Balachandar and A. Prosperetti (eds), Proceedings of the IUTAM Symposium on Computational Multiphase Flow, 131–140.

microscopic phenomena is based on the numerical simulation. In this study, we carried out a molecular dynamics (MD) simulation of liquid–vapor coexistence systems of water, and focused on the molecular scale structure of the liquid–vapor interface.

A number of molecular level investigations of the liquid–vapor interface using molecular simulations have been reported (see [6] and references therein). Some of the above studies analyzed the molecular structures of the interface from the viewpoint of both static properties and dynamic properties. Almost all the statistical quantities of molecular properties in these analyses are based on the static interface, that is, physical properties are averaged according to the one dimensional distance from the thermodynamically defined interface.

In contrast, we introduce a new definition of the liquid–vapor interface which can capture the local and instantaneous structure of the interface. We visualized this instantaneous interface and confirmed the interface fluctuations due to molecular motions. To compare with the statistical quantities based on the static interface mentioned above, we obtained new statistical quantities which are calculated according to the fluctuating interface, and found that this definition sheds light on the different physical pictures of the nanometer-scale interface.

2 Molecular dynamics simulations

In order to analyze the liquid–vapor interface, we carried out MD simulations by two computational systems. One is a rectangular cell with a three-dimensional periodic boundary condition (PBC) applied, and water molecules are placed in the middle of the cell. At the equilibrium state, water molecules form the water film which has two flat (in the thermodynamic sense) surfaces. This system is refered to as the planar surface system in this study. The other is a cubic cell also with the three-dimensional PBC, referred to as the bubble system. In this system, we form a nanometer-sized void region by the procedure described in detail later.

We used the water molecules as solvent molecules and 1-heptanol ($C_7H_{15}OH$) as solute molecules to investigate adsorption structure and to verify our newly definition of the interface described in Section 3. The number of each molecule used in the MD simulations is shown in Table 1. In both systems, we performed two calculation sets: pure water and 1-heptanol solution systems.

We used the all-atom model potential which is based on the AMBER force field [7] for the surfactant dynamics, and SPC/E potential [8] for water molecules as a potential function. We adopted a cut-off length of van der Waals interaction; the length is $3.79\sigma_{OO}$ (= 12 Å, σ_{OO} corresponds to σ for oxygen-oxygen pair of water), while the coulomb potential was calculated by the SPME method [9].

The equation of motion was integrated by the r-RESPA method [10] as the Multiple Time Scale (MTS). In this study, the time step for intramolecular motions was 0.2 fs, and that for intermolecular motions was 1.0 fs. The sampling of configuration was taken every 100 fs.

The cell size of the planar surface system is $30 \times 30 \times 120$ Å, while in the bubble system the cell size is varied with each computational set. The initial configurations

Table 1. Simulation conditions about the number of molecules.

Calculation set	# surfactant	# water
Bubble system		
Pure water	0	1024
1-heptanol	15	906
Planar surface system		
Pure water	0	1024
1-heptanol	18	1024

of the bubble systems are achieved by the following procedures. At first, each system is equilibrated at 300 K and 1 atm in the NPT ensemble [11]. Next, the system is expanded isotropically at a certain ratio. In this study, the expansion ratio in volume is 1.20 for the pure water system and 1.11 for the 1-heptanol system. The cell size given by the above procedure was 33.35 Å (side length) for the pure water system, and 33.75 Å for the 1-heptanol system. Snapshots of the planar system and bubble system in the 1-heptanol system are shown in Figure 1, In Figure 1, void regions (white spheres in Figure 1c) are defined by the region where there is no water molecule within a certain radius from the grid point (grid spacing of about 1 Å) in the MD cell [12].

Before the production run, the system temperature is set to 300 K in all systems by velocity scaling [13] for 60 ps, and the system is equilibrated for 240 ps in the NVE ensemble. The production run is implemented in the NVE ensemble for 720 ps. Total MD steps are 720,000 steps for this production run in both the planar surface system and the bubble system, and it typically takes about 1 hour for 10,000 steps (10 ps) using Pentium4 3.2 GHz single processor.

3 Definition of local and instantaneous interface

The Gibbs dividing surface is a mathematical definition of the interface on which interfacial physics has long ago been constructed [14]. Therefore, in the study of interfacial science, investigations are usually based on the Gibbs dividing surface. However, the interface defined in this manner loses the molecular-level fluctuations in the sense of both space and time.

On the contrary in this study, we introduced a new definition of the local and instantaneous interface which is not time- and space-averaged. In this definition, we express the single particle density of solvent molecules, which is described as the summation of the Dirac delta function, as the field quantity. In general, the single particle density in the system is defined as,

$$\rho^{(1)}(\mathbf{r}) = \sum_{i=1}^{N} \delta(\mathbf{r} - \mathbf{r}_i), \tag{1}$$

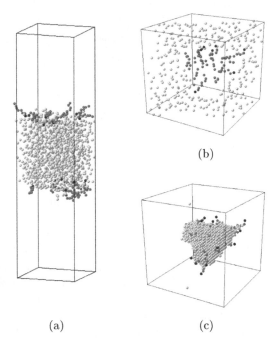

Fig. 1. Snapshots of computational systems of the MD simulation in the 1-heptanol system; (a) planar system, (b) slice view of the bubble system, (c) void view of the bubble system. In (a) and (b), white spheres indicate oxygen atoms of water molecule, and black and dark gray spheres indicate oxygen atoms and carbon atoms of surfactant, respectively. In (c), the water molecules are not shown and the void region is shown as white spheres, and the colors of the surfactant molecules are the same as in (b). The void view shown here is based on the method proposed by Maruyama et al. [12].

where $\mathbf{r}$ is the field position, $\mathbf{r}_i$ is the position of particle i, and N is the number of particles in the system. In order to express the single particle density as the field quantity, we introduce the following smoothed delta function on the center of mass of each solvent molecule instead of the Dirac delta function,

$$
D\left(\mathbf{r} - \mathbf{r}^{(\mathrm{par})}\right) =
\begin{cases}
(2k\Delta x)^{-3} \displaystyle\prod_{i=1}^{3} \left\{1 + \cos\frac{\pi}{k\Delta x}\left(r_i - r_i^{(\mathrm{par})}\right)\right\}, \\
\qquad\qquad \text{if } \left|r_i - r_i^{(\mathrm{par})}\right| < k\Delta x, \\
0, \text{ otherwise},
\end{cases}
\tag{2}
$$

where $\mathbf{r}^{(\mathrm{par})}$ is the position of the particle. The index i of r_i indicates each direction x, y, z of Cartesian coordinates. The Cartesian grids are generated in the MD cell in order to evaluate the density on that grid point, and Δx is a grid spacing. Therefore $k\Delta x$ determines the broadening of the delta function and the arbitrary value k is discussed below. Finally, the smoothed single particle density is described instead of Equation (1) as

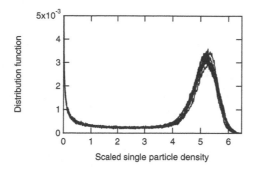

Fig. 2. Time series of the distribution function of scaled single particle density on each grid in the planar surface system of pure water. Lines are drawn every 1.0 ps. The single particle density is scaled by $1/(\Delta x)^3$.

$$\rho_D^{(1)}(\mathbf{r}) = \sum_{i=1}^{N} D(\mathbf{r} - \mathbf{r}_i). \tag{3}$$

An important variable k indicates the broadening of the delta function, and it determines the fluctuation scale of the interface which can be captured. In this study, the value of k was determined so that the broadening of the smoothed delta function approximately coincided with the mean intermolecular distance of water.

In order to decide the position of the interface, we calculated a distribution function of the single particle density on each grid point. The distribution function is defined as the probability which finds the density value ranging from $\rho_D^{(1)}$ to $\rho_D^{(1)} + d\rho_D^{(1)}$ at a grid point. The distribution function is averaged over all the grid points at the indivisual time. As the result, the distribution function typically showed two peaks which correspond to the vapor phase ($\rho_D^{(1)} = 0$) and the liquid phase (Figure 2). Using this information, we determined the density of the interface as the center value between the two density values which gives two peaks of the distribution function. We will discuss whether this definition is proper from the viewpoint of relation with the Gibbs surface later.

After we determine the interface, by applying re-initialization of the level set method [15] we can calculate a distance function from the interface. Thus, once the distance function is defined, we can calculate a normal vector to the interface, mean curvature, and other geometric quantities by using the distance function.

4 Results and discussions

4.1 Result of the new definition of the interface

First, we show the snapshots of the local and instantaneous interface from the MD result (Figure 3). The grid spacing is about 0.6 Å. In both the planar surface system and the bubble system, we can see that our definition of the interface can capture the

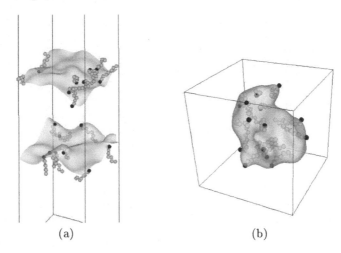

(a)   (b)

Fig. 3. Snapshots of the local and instantaneous interface of each system; (a) planar surface system of the 1-heptanol solution and (b) bubble system of the 1-heptanol solution. In (a) and (b), black and gray spheres indicate oxygen and carbon atoms of surfactant molecules, respectively.

local fluctuation of the surface. Especially, in the bubble system, we found that the shape of the void region did not necessarily hold the spherical shape. Furthermore, we note that in the 1-heptanol system the position of the surface is recognized just at the position of the surfactant head groups (-OH) adsorbed at the surface in accordance with a physical picture of surfactant adsorption. This fact shows the validity of our definition of the interface.

Next, we discuss the relation between our definition of the inteface and the Gibbs dividing surface in the pure water system. At first, we calculated the position of the Gibbs dividing surface, especially the equimolar surface on which the surface excess of number density becomes zero. In the planar surface system, the one-dimensionality can be assumed on long-time average, that is, in the x, y direction (parallel to the surface plane), the system is uniform and physical quantities depend only on the z direction (perpendicular to the surface plane). Therefore, a number density profile of water molecules can be described as [14],

$$\rho\left(z\right) = \frac{\rho_{\text{liq}} + \rho_{\text{vap}}}{2} + \frac{\rho_{\text{liq}} - \rho_{\text{vap}}}{2} \tanh\left(\frac{z - z_{\text{d}}}{2\delta_{\text{d}}}\right), \tag{4}$$

where ρ_{liq} and ρ_{vap} is the number density of bulk liquid and vapor, respectively. z_{d} is the position of the equimolar surface and δ_{d} is the surface thickness. From the MD result, the mean number density is calculated and Equation (4) is fitted to the profile to obtain z_{d} and δ_{d}. On the other hand, in the bubble system one-dimensionality in a radial direction can be assumed, that is, the void region should be a spherical shape thermodynamically. In this study, an instantaneous center of mass of the void region was defined as the center of mass of those grid points with a negative level set

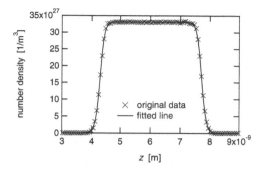

Fig. 4. A number density profile of water in the planar system. Cross marks are the aver-aged raw data, and the solid line is fitted by the tanh function. The center of the cell is $z = 6 \times 10^{-9}$ m.

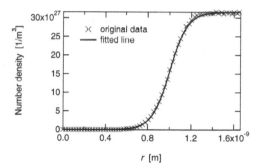

Fig. 5. A number density profile of water in the bubble system. Cross marks are the averaged raw data, and the solid line is fitted by the tanh function.

value (vapor phase), and a number density was averaged according to the radial dir-ection. The number density profile assumed in this system is similar to Equation (4), replacing the coordinate z by the radial coordinate r.

In Figures 4 and 5, the number density profiles in the planar system and the bubble system are drawn with the fitting line mentioned above. In the planar system, the origin of the z coordinate is set at the bottom of the computational domain. From these figures, we found that the density profiles were fitted to Equation (4) well. Then we could obtain the positions of the equimolar surface and the surface thick-ness. These results with the averaged positions of the instantaneous interface in the planar surface system and the bubble system are shown in Tables 2 and 3, respect-ively. From Table 2, we found that in the planar surface system, the positions of the equimolar surface and the averaged positions of the instantaneous surface showed good agreement. In this table, the two positions of the interface correspond to both sides of the surface of the liquid film. Also in the bubble system, from Table 3, we found that the equimolar surface and the averaged position of the instantaneous in-terface showed reasonable agreement.

Table 2. Results of surface positions and surface thickness in the planar surface system

	equimolar surface	instantaneous interface
surface 1	42.92	43.25
surface 2	77.13	77.40

unit: [Å]

Table 3. Results of surface positions and surface thickness in the bubble system

	equimolar surface	instantaneous interface
radius	9.999	9.560

unit: [Å]

4.2 Result of physical quantities based on the instantaneous interface

In this section, we show one example of physical quantities based on the instantaneous interface in the planar surface system, and compare the result with that based on the equimolar surface. Usually, a number density profile is calculated with respect to the z coordinate or the distance function from the equimolar surface as shown in Figure 4. However in this study, we calculated the number density profile with respect to the distance function from the instantaneous interface. This result is shown in Figure 6, and the number density profile with respect to the distance function from the equimolar surface in Figure 7; this figure is basically the same as Figure 4. From these figures, we found that the number density profile based on the equimolar surface changed smoothly from the bulk vapor phase to the bulk liquid phase, while the number density profile based on the instantaneous interface had peaks and an oscillatory profile in the liquid phase. We consider that this oscillation corresponds to the molecular layering at the liquid–vapor interface. This physics has not been so clear from the picture based on the thermodynamic interface as opposed to the case of solid–liquid interface. To our best knowledge, this is the first demonstration of the clear molecular layering at the liquid–vapor interface of water.

5 Conclusions

In this study, we introduced a novel definition of the interface which was able to capture the local and instantaneous structure of the liquid–vapor interface at the molecular level. We consider the density field which is obtained by making the originally discrete single particle density smoothed, and then we can treat the system as a field rather than particles. The interface is defined as the iso-surface which has the same density value. Applying this definition to the MD result of the liquid–vapor equilibrium system, we confirmed that our definition was able to capture the microscopic fluctuations due to the molecular motion.

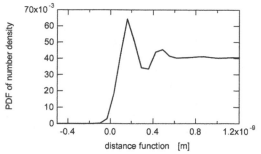

Fig. 6. A probability density of number density according to the distance from the instantaneous interface. A negative value of the distance function corresponds to the vapor phase region, while a positive value of the distance function corresponds to the liquid phase region.

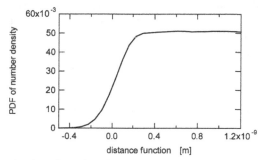

Fig. 7. A probability density of number density according to the distance from the equimolar surface.

In this study, we calculated the physical quantities based on the instantaneous interface proposed here. From the result, we found that the number density across the surface showed the peaks and oscillatory profile because of the molecular layering of water. We can conclude that the statistical average based on the instantaneos interface can extract the characteristic pictures of the interface more fully than that based on the Gibbs dividing surface.

Acknowledgements

This work was supported by Research Fellowships of the Japan Society for the Promotion of Science for Young Scientists (to G.K.) and by NEDO Grant Program under the Project ID 04A20004.

References

1. Ishida, N., Inoue, T., Miyahara, M. and Higashitani, K., 2000, *Langmuir* **16**, 6377–6380.
2. Tyrrell, J.W.G. and Attard, P., 2001, *Phys. Rev. Lett.* **87**, 176104.
3. Yang, J., Duan, J., Fornasiero, D. and Ralston, J., 2003, *J. Phys. Chem. B* **107**, 6139–6147.
4. Attard, P., 2003, *Adv. Colloid Interface Sci.* **104**, 75–91.
5. Richmond, G.L., 2002, *Chem. Rev.* **102**, 2693–2724.
6. Taylor, R.S., Dang, L.X. and Garrett, B.C., 1996, *J. Phys. Chem.* **100**, 11720–11725.
7. Cornell, W.D., Cieplak, P., Bayly, C.I., Gould, I.R., Merz Jr., K.M., Ferguson, D.M., Spellmeyer, D.C., Fox, T., Caldwell, J.W. and Kollman, P.A., 1995, *J. Am. Chem. Soc.* **117**, 5179–5197.
8. Berendsen, H.J.C., Grigera, J.R., Straatsma, T.P., 1987, *J. Phys. Chem.* **91**, 6269–6271.
9. Essmann, U., Perera, L., Berkowitz, M.L., Darden, T., Lee, H. and Pedersen, L.G., 1995, *J. Chem. Phys.* **103**, 8577–8593.
10. Tuckerman, M., Berne, B.J. and Martyna, G.J., 1992, *J. Chem. Phys.* **97**, 1990–2001.
11. Martyna, G.J., Tobias, D.J. and Klein, M.L., 1994, *J. Chem. Phys.* **101**, 4177–4189.
12. Maruyama, S. and Kimura, T., 2000, *Int. J. Heat Technology* **18**, supplement 1, 69–74.
13. Allen, M.P. and Tildesley, D.J., 1987, *Computer Simulation of Liquids*, Clarendon Press, Oxford.
14. Rowlinson, J.S. and Widom, B., 1982, *Molecular Theory of Capillarity*, Clarendon Press, Oxford.
15. Sussman, M., Smereka, P. and Osher, S., 1994, *J. Comput. Phys.* **114**, 146–159.

Part III

Fully-Resolved Multi-Particle Simulations

DNS of Collective Behavior of Solid Particles in a Homogeneous Field

Influence of Reynolds Number, Loading Ratio and Particle Rotation

Takeo Kajishima

Department of Mechanical Engineering, Osaka University, Yamadaoka, Suita, Osaka 565-0871, Japan; e-mail: kajisima@mech.eng.osaka-u.ac.jp

Abstract. To investigate the two-way interaction between solid particles and fluid turbulence, a homogeneous flow field including more than 2000 spherical particles was directly simulated. Since flow around each particle is approximately resolved, no models were used for particle motion or fluid turbulence. A particle settles under gravity with the Reynolds number ranging from 50 to 300, based on diameter and slip velocity. When particle clusters are formed due to the wake attraction, the average settling velocity increases. Thus particular attention was focused on the distribution of particles. The influence of Reynolds number and loading ratio are assessed. It is found that the rotation of particle dominates the cluster dynamics.

1 Introduction

Particle-laden flows are widely observed in nature and industrial applications. Solid particles significantly affect the transfer of momentum, heat and mass in turbulent flows. Extensive research has therefore been conducted for the turbulence modulation by particles.

The interaction takes place through a wide range of scales, such as particle diameter, inter-particle distance, size of particle clusters, wakes from particles as well as clusters. Considering the importance of the energy input at the upstream in the cascade process of turbulence, the interaction at the larger scales could be dominant for turbulence modulation. The largest scale is not close to the particle scale but related to the particle distribution. Particularly the non-uniformity in particle distribution is investigated in this study.

The objective of this study is to clarify factors governing the pattern of particle distribution. Especially, the influence of particle Reynolds number, loading ratio and particle rotation are considered. To this end, direct numerical simulation (DNS) is applied. In DNS, the flow around each particle is calculated to represent the basic physics involved in the turbulence modulation. Therefore, any empirical model such as point-source model for particle motion is not used.

Hereafter particles are assumed to be rigid spheres of uniform diameter. The sphere is the simplest three-dimensional shape, but the flow around it has a wide

S. Balachandar and A. Prosperetti (eds.) Proceedings of the IUTAM Symposium on Computational Multiphase Flow, 143–152.

variation, even for a fixed particle in a uniform stream, as known experimentally [1, 2] and numerically [3, 4]. The Reynolds number ranges attracting researchers' interests have been from 200 to 300 and near 3×10^5, where the Reynolds number is based on the sphere diameter D_p and relative velocity U. The unsteadiness is caused at the Reynolds number higher than the former range. The drag crisis takes place around the latter. This study deals with the Reynolds number range below 400.

The pattern of wake has been observed as follows [1, 4]. A steady and axisymmetric vortex ring attaches at the sphere for Reynolds number less than approximately 210. The vortex ring becomes non-axisymmetric, but steady and plane-symmetric, for Re between 210 and 270. For Reynolds numbers greater than approximately 270, unsteady vortex shedding takes place. But there remains some diversity in critical Reynolds number. The orientation and period of vortex shedding seems nearly constant for $Re \simeq 300$ and then they become more random for higher Reynolds number. These patterns have also been successfully reproduced by our numerical method [5, 6].

This paper describes DNS result of homogeneous flow laden by particles falling gravitationally. First, the numerical method for full-scale computation from particle scale to particle-induced turbulence is outlined. Then DNS results are discussed from the viewpoint of particle distribution. Particular attention is focused on the influence of particle Reynolds number, loading ratio and particle rotation.

2 DNS method

2.1 Numerical scheme

A DNS method [5, 6] has been developed for the full-scale simulation of flow including more than 1000 particles moving with vortex shedding, up to Reynolds numbers of the order of $\mathcal{O}[10^2]$. In this method, flow around each particle is resolved and the force on particle is evaluated based on the surface integral of fluid stress. For such a sense, we believe it may be used for complete DNS of particle-laden flows, rather than the previous ones using point-source model.

The Cartesian coordinate system is selected for the DNS of particle-laden turbulence due to the following reasons. The most accurate and efficient method for computation of flow around a sphere may be to use a spherical coordinate system attached to the body. But the cost increases significantly for many spheres in relative motion. In addition, considering the objective of the present simulation, uniform and isotropic resolution of the computational grid is desired for turbulent flows.

The computational mesh, cubic in this study, does not fit the surface of spherical particles. The volume fraction of the particle in the cell, including solid-fluid interface, is taken into account [6].

The volume-weighted average velocity

$$\boldsymbol{u} = \alpha \boldsymbol{u}_p + (1 - \alpha)\boldsymbol{u}_f \tag{1}$$

is introduced for two-way coupling between solid particles and fluid turbulence. In Equation (1), α represents the volumetric fraction of the solid in the computational cell, u_f the fluid velocity, $u_p (= v_p + \omega_p \times r)$ the velocity in the solid body moving with velocity v_p and angular velocity ω_p. Based on the Navier–Stokes equation and continuity equation for u_f

$$\frac{\partial u_f}{\partial t} = -\nabla \frac{p}{\rho_f} - u_f \cdot \nabla u_f + \nu_f \nabla^2 u_f, \quad \nabla \cdot u_f = 0, \tag{2}$$

a governing equation for u is given as

$$\frac{\partial u}{\partial t} = -\nabla \frac{p}{\rho_f} - u \cdot \nabla u + \nu_f \nabla^2 u + f_p, \quad \nabla \cdot u = 0, \tag{3}$$

where ρ_f is the fluid density and ν_f the kinematic viscosity. The additional term

$$f_p = \alpha(u_p - \hat{u}_f)/\Delta t \tag{4}$$

is given at the cell that includes the solid-fluid boundary. The meaning of f_p can be explained through the time-marching procedure as follows. First, the unsteady equations for fluid flow (2) is calculated, as if the field was occupied by fluid. This result is once expressed by $\hat{u}_f$. Next, f_p modifies $\hat{u}_f$ to u using $u = \hat{u}_f + \Delta t f_p$. Considering the difference between Equation (2) and Equation (3), f_p is interpreted as the momentum exchange between the phases [6], which is meaningful in a cell of $\alpha > 0$.

The surface integral of the fluid stress in the equations for the particle can therefore be replaced by the volumetric integral of f_p as

$$\frac{d(m_p v_p)}{dt} = \int_{V_p} f_p dV + g_p, \quad \frac{d(I_p \cdot \omega_p)}{dt} = \int_{V_p} r \times f_p dV + h_p, \tag{5}$$

where m_p denotes the mass of the particle, I_p the inertia tensor, and r the relative position from the center of rotation. The last two terms, g_p and h_p, are external force and moment: respectively $g_p = -[(\rho_p - \rho_f)/\rho_p]m_p g e_z$ and $h_p = 0$ in this computation. The domain V_p is slightly larger than the particle, including all its interfacial cells. Since the grid for the fluid-flow simulation is used for the volumetric integral in Equation (3), there is no residual in the momentum exchange between the two phases.

We apply our DNS method to homogeneous turbulence including spherical particles. Grid points for the fluid turbulence simulation are distributed uniformly in a periodic computational domain. The spatial derivative is approximated by a central finite-difference method of fourth-order accuracy. Second-order schemes are applied for time marching, namely the Adams–Bashforth method for the equations of motion for fluid and solid particle and the Crank–Nicholson method for the displacement of particle position. The SMAC method is used for velocity-pressure coupling in Equation (3).

The above-mentioned method is one of the simplest immersed boundary techniques, in which the fortified Navier–Stokes approach [7] is extended for the realization of solid boundaries.

2.2 Computational setup

Periodic boundary conditions are applied in all directions assuming homogeneity of the flow field. Each computational cell is cubic. The numbers of them are $N_x = N_y = 512$ in the horizontal directions and $N_z = 1024$ in the vertical direction. The ratio of particle diameter to grid spacing is $D_p/\Delta = 10$, which allowed sufficient accuracy for vortex shedding at the particle Reynolds number range of interest [5, 6]. Four steps for the number of solid particles are calculated: $N_p = 256, 512, 1024$ and 2048.

The number of particles is limited so that the volumetric fraction Φ is 0.4%. In such a relatively dilute loading, inter-particle collisions could occur, but are unlikely to dominate the particle distribution and the flow field. Thus, elastic collisions for simplicity are assumed in this study.

Initially, both the uniformly distributed particles and the fluid are at rest. The density ratio between solid and fluid is $\rho_p/\rho_f = 8.8$. Hence, particles settle due to gravity. To keep the mass flow rate of the mixture to zero, we adjusted the vertical gradient of pressure in the equation of fluid motion. The Reynolds number is adjusted as $Re_{ps} = 50, 100, 200$ and 300 by changing the fluid viscosity so that the gravity and drag are in balance. This is based on the particle diameter and the terminal velocity when a particle falls in a stationary and infinite domain. The drag is estimated by the standard Re_{ps}-C_D curve for a fixed sphere in a uniform flow [8]. On the other hand, shown later is based on the average slip velocity between falling particles and fluid in a periodic domain. This is different from Re_{ps}. The setups correspond to copper particles having diameters in the range from 0.34 mm ($Re_{ps} = 50$) to 0.85 mm ($Re_{ps} = 300$) in water.

One can manipulate a particular parameter in the numerical experiment. In this study, the value of ω_p is forced to be 0 as a virtual situation, for the purpose of investigating the effect of particle rotation. Namely, the angular momentum, governed by the second Equation (5), is ignored in such a situation. Hereafter, such a hypothetical particle is expressed by 'irrotational', while 'rotational' means a realistic particle.

3 Results and discussion

3.1 Effects of Reynolds number and particle rotation

First, the Reynolds number dependence together with the influence of particle rotation is considered at constant loading-ratio ($\Phi = 0.1\%$, $N_p = 512$). Figure 1 shows the time evolution of the averaged Reynolds number Re_p and rotation intensity, $\omega_{ph} = (\omega_{px}^2 + \omega_{py}^2)^{1/2}$ of particles. Figure 2 compares the influence of rotation on the inter-particle (center-to-center) distances.

As for the lower Reynolds number particles $Re_{ps} = 50$ and 100, they fall with almost the same velocity as a single particle ($Re_p \approx Re_{ps}$) as shown in Figure 1(a). The influence of particle rotation is insignificant. The particles moving with low Reynolds number have almost axisymmetric vortex rings in their wake. Since the

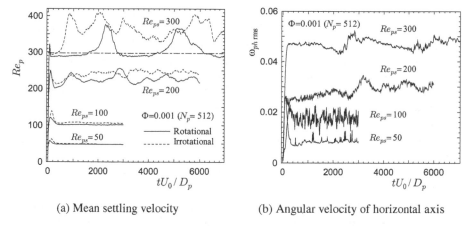

(a) Mean settling velocity   (b) Angular velocity of horizontal axis

Fig. 1. Time evolution of settling velocity and intensity of particle rotation.

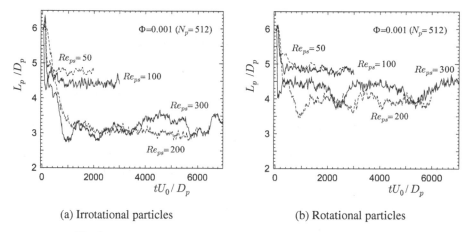

(a) Irrotational particles   (b) Rotational particles

Fig. 2. Time evolution of average distance to the nearest particles.

interaction between them through their wakes is not evident, they tend to move individually. The inter-particle distances are therefore almost constant in these cases as shown in Figure 2. Particle rotation is weak as shown in Figure 1(b). The time evolution of rotation intensity takes spiky profile when a few particles move due to rotation in a disturbed flow field.

For the intermediate Reynolds number $Re_{ps} = 200$, the particles exhibit collective behavior as shown in Figure 2. Accordingly, the settling velocity becomes larger than for a single particle of the same property, $Re_p > Re_{ps}$, as shown in Figure 1(a). The experimental evidence is that flow around a particle fixed in a uniform stream at this Reynolds number is axisymmetric. But this axisymmetric state is unstable unlike that at the lower Reynolds number. Thus particles of $Re_{ps} = 200$ moving in a disturbed field are likely to have continuous rotation.

For particles with higher Reynolds number, $Re_{ps} = 300$, vortex shedding takes place from each particle. In this case, the above-mentioned tendency becomes more evident. Moreover, the influence of rotation causes qualitative difference in collective motion of particles. Clearly from Figures 1(a) and 2, fluctuation of Re_p and L_p shows negative correlation. It suggests that the falling velocity increases when particles form clusters. In this paper, the high-density region of particles is denoted by 'cluster'. The drag on a particle trapped in the wake of another becomes smaller and it reaches the other. Such a 'wake-attraction' is the mechanism of clustering in our case. Since many particles have less drag in the cluster, the average Reynolds number increases.

The correlation between particle distribution and fluid turbulence is as follows [9, 10]. Velocity fluctuation in the vertical direction increases with slight delay of increase in particle Reynolds number due to clustering. Then the horizontal component follows with additional delay. As a consequence, clusters break up mainly due to horizontal fluctuations. This is the life cycle of a cluster. Through some cycles, the particle-laden flow field develops to statistically steady state. At this stage, there seems to be quite an important difference between actual particles and non-rotating particles. Irrotational particles keep cluster structure with weak fluctuations. On the other hand, particles with rotation cause formation/break-up of clusters with long period.

As shown in Figure 1(b), for $Re_{ps} = 300$, the rotation increases gradually through the non-dimensional time between 2000 and 3000 and then between 5000 and 6000. They are the period of clustering as suggested in Figures 1(a) and 2(b). Particles acquire angular momentum within the high shear region around clusters. Thus continual lift force due to rotation acts on the particles and they tend to slide horizontally. The direction of lift by rotation is outward from clusters. As a result, clusters are reproduced periodically but become weaker due to the inertia of rotation.

3.2 Effects of loading ratio and particle rotation

The collective motion of particles in the self-induced turbulence is strongly affected by the particle distribution as discussed in the former section. Thus it may be also influenced by the mean loading ratio of solid particles. Here, loading-ratio dependence in couple with the influences of particle rotation are considered at the highest Reynolds number ($Re_{ps} = 300$).

Figure 3 compares the time evolution of averaged Reynolds number Re_p. Figure 4 shows time evolutions of mean distance to the nearest particles L_p normalized by $L_m = (V_c/N_p)^{1/3}$, where V_c is the volume of computational domain. Clearly from the comparison of Figure 4 with Figure 3, variations of Re_{ps} and L_p show a negative correlation. It means the falling velocity becomes larger when particles form clusters. Settling velocity of irrotational particles is larger than for a single by approximately 20%, regardless of the loading ratio. Clusters of irrotational particles seem to be maintained continuously.

The behavior of actual particles, on the other hand, is affected by the mean loading ratio. For 0.05% volume loading, the average fall velocity is reduced. In this case,

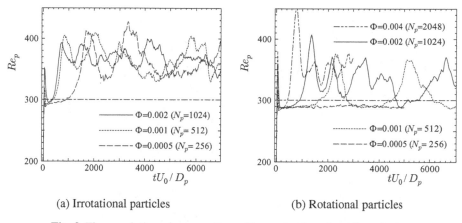

(a) Irrotational particles   (b) Rotational particles

Fig. 3. Time evolution of average Reynolds number based on slip velocity.

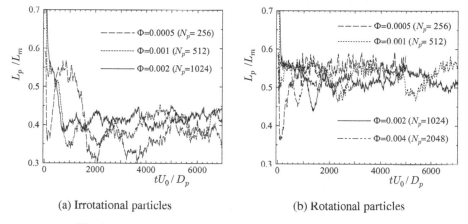

(a) Irrotational particles   (b) Rotational particles

Fig. 4. Time evolution of average distance to the nearest particle.

particles do not form clusters. Thus the drag increases because they move in the self-induced turbulence. As the loading ratio increases from 0.1% to 0.2%, clusters are formed intermittently. Furthermore, for 0.4% loading, the period becomes shorter and more irregular. The mean drag coefficient decreases as a consequence of the development of clusters.

Figures 3 and 4 show a clear difference between actual particles and rotation-ignored particles. Irrotational particles keep cluster structure. On the other hand, rotational particles cause quasi-periodic formation and break-up of clusters. Such a collective behavior is affected by the loading ratio.

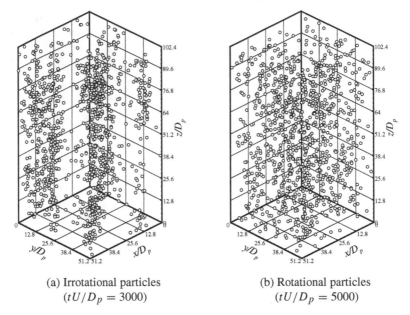

(a) Irrotational particles
$(tU/D_p = 3000)$

(b) Rotational particles
$(tU/D_p = 5000)$

Fig. 5. Typical example of instantaneous distribution of particles ($N_p = 1024$, $Re_{ps} = 300$, $\rho_p/\rho_f = 8.8$.)

3.3 Instantaneous flow fields

Figure 5 shows examples of the instantaneous flow field including 1024 particles ($\Phi = 0.2\%$) for the case of $Re_{ps} = 300$ [10]. Each particle sheds hairpin-type vortices similar to the observation in experiments [1, 2].

Irrotational (hypothetical) particles form vertically elongated clusters. From movies one can observe that irrotational particles, even if they once drop out of a cluster, they return to it or go into another cluster. Clusters behave somewhat dynamically but the structure is maintained, as observed in Figures 3(a) and 4(a). Particles in the high concentration region fall faster than average. The flow field including rotational (realistic) particles, on the other hand, fluctuates intensely as observed in Figure 3(b).

When the particle rotation is accounted for, firstly developed clusters have similar size and strength with those of irrotational ones shown in Figure 5(a). But they are broken completely. At the next step, regenerated clusters become larger but the particle density is lower. During the iteration of this process, the particle rotation grows as shown in Figure 1(a). Consequently, the high concentration clusters such as initial ones are never reproduced as shown in Figure 5(b). From movies one can confirm the sliding motion of particle in the horizontal plane due to the Magnus lift force.

The difference between rotational and irrotational particles is due to the difference in the direction of lift. Clusters cause faster downward current of fluid and make high-shear layer around them. In this region, particles without rotation receive the lift force to the direction of cluster center, resulting in the absorption. This corresponds to an experimental finding [11], in which they reported the lift is in the direction to less relative velocity side for Reynolds number greater than 60. When particles receive angular momentum in the shear region, the Magnus lift force is in the direction outward from cluster center. In addition, the rotation tends to sustain the sliding motion because the orientation of unsteady vortex shedding tends to be fixed. Then, rotational particles have continuous horizontal motion, which is originally in the direction dropping out from clusters. As a consequence, they travel randomly and unlikely form high concentration clusters until their rotation is damped.

4 Conclusion

The interaction in larger scale is more important for turbulence modulation. The largest scale is not close to the particle size but it is related to the particle distribution. So the non-uniformity in particle distribution was particularly discussed in this study, in the multiple-scale interactions between particles and turbulence. Thus, the dominant factors for the collective behaviors of particles, especially, the influence of Reynolds number, loading ratio and rotation of particles were considered by means of DNS.

Particles in clusters fell faster than average because of the smaller fluid drag in the wake of other particles. It is due to the high density in fluid-particle mixture. Then clusters made high-shear region around them resulting in turbulence production. The effect of particle rotation manifested itself after such a process. Irrotational (hypothetical) particles maintained cluster structure while rotational (realistic) ones moved randomly in the horizontal direction. Moreover, the collective behavior of rotational particles was affected by the loading ratio. The major reason for the difference was the direction of lateral component of the fluid force, that is the lift force. A dominant factor of rotational particles was the Magnus lift force, which was given by the fluid shear and preserved due to the inertia of rotation.

Recently, Nishino and Matsushita found a vertically elongated structure in particle distribution by experiment and named it 'columnar particle accumulation' (CPA) [12]. Our DNS method was applied to a condition corresponding to the experiment: $Re_{ps} = 200$ and $\rho_p/\rho_f = 2.5$. The simulated and measured flow field is in qualitative agreement [13].

Acknowledgements

The author tenders his acknowledgment to Dr. Satoshi Takiguchi (Mitsubishi Heavy Industries, Ltd.) who made important contribution to developing the numerical scheme. This work was partially supported by a Grant-in-Aid Scientific Research on Priority Areas (B) No. 12125202 from the Ministry of Education, Culture, Sports, Science and Technology of Japan.

References

1. Achenbach, E., 1974, *J. Fluid Mech.* **62**, 209–221.
2. Sakamoto, H. and Haniu, H., 1990, *J. Fluids Eng.* **112**, 386–392.
3. Shirayama, S., 1992, *AIAA J.* **30**, 349–358.
4. Johnson, T.A. and Patel, V.C., 1999, *J. Fluid Mech.* **378**, 19–70.
5. Takiguchi, S., Kajishima, T. and Miyake, Y., 1999, *JSME Int. J. Ser. B* **42**, 411–418.
6. Kajishima, T., Takiguchi, S., Hamasaki, H. and Miyake, Y., 2001, *JSME Int. J. Ser. B* **44**, 526–535.
7. Fujii, J., 1995, *J. Comput. Phys.* **118**, 92–108.
8. Clift, R., Grace, J.R. and Weber, M.E., 1978, *Bubbles, Drops and Particles*, Academic Press, New York.
9. Kajishima, T. and Takiguchi, S., 2002, *Int. J. Heat and Fluid Flow* **23**, 639–646.
10. Kajishima, T., 2004, *Int. J. Heat and Fluid Flow* **25**, 721–728.
11. Kurose, R. and Komori, S., 1999, *J. Fluid Mech.* **384**, 183–206.
12. Nishino, K. and Matsushita, H., 2004, in *Proc. 5th Int. Conf. Multiphase Flow*, Yokohama, No. 248 (CD-ROM).
13. Kajishima, T., 2004, in *Proc. 5th Int. Conf. Multiphase Flow*, Yokohama, No. K03 (CD-ROM).

Proteus – A New Computational Scheme for Deformable Particles and Particle Interaction Problems

Zhi-Gang Feng and Efstathios E. Michaelides

Mechanical Engineering and Southcentral Regional Center of the National Institute for Global Environmental Change, Tulane University, New Orleans, LA 70118, USA; e-mail: zfeng@tulane.edu, emichael@tulane.edu

Abstract. *Proteus*[1] is a new code that utilizes elements of the Immersed Boundary (IB) and Lattice Boltzmann Method (LBM) as well as a Direct Forcing (DF) scheme. As a computational method, it is very flexible and it appears to be ideal in solving fluid-particle interaction problems including problems with deformable boundaries. Proteus uses a regular Eulerian grid for the flow domain and a regular Lagrangian grid to follow particles that are contained in the flow field. The rigid body conditions for the fluid and the particles are enforced by applying the external force acting on the boundary of particles. A penalty method is used, which assumes that the particle boundary is deformable with a high stiffness constant. The velocity fields for the fluid and particles are solved by incorporating a force density term into the lattice Boltzmann equation. This force term is determined by using a technique that is based on the direct forcing scheme. *Proteus* preserves all the advantages of LBM in tracking a group of particles and, at the same time, provides an alternative and better approach to treating the solid-fluid boundary conditions. Because of this it provides for a smooth boundary interface, with only a few nodes assigned for the size of particles. This new method also solves the problems of fluctuation of the forces and velocities on the particles when the "bounceback" boundary conditions are applied. The method has the capability to simulate deformable particles and fluid-structure deformation. The results of the Proteus code have been validated by comparison with results from other computational methods as well as experimental data. Some of the validation results will be given in the presentation of this paper.

1 Introduction

Ladd [8, 9] successfully applied the Lattice Boltzmann Method (LBM) to particle-fluid suspensions. The LBM overcame the limitations of the conventional Finite Volume and Finite Element Methods (FEM) by using a fixed, non-adaptive (Eulerian) grid system to represent the flow field. Since then, the LBM has proven to

---

[1] In the Greek mythology, *Proteus* is one of the many mythical heroes. He was the son of God Poseidon and was granted by Zeus the abilities to change shapes, to take different forms at will and to correctly predict the future. One cannot ask for better attributes from a computer code.

S. Balachandar and A. Prosperetti (eds), Proceedings of the IUTAM Symposium on Computational Multiphase Flow, 153–161.
© 2006 *Springer. Printed in the Netherlands.*

be a robust and efficient method to accurately simulate particulate flows with a large number of particles. When the LBM is used to simulate particle-fluid interaction problems, the no-slip condition on the particle-fluid interface is treated by the so-called "bounce-back" rule [9], and the particle surface is represented by the so-called "boundary nodes", which are essentially a set of mid-points of the links between two fixed grids, in which one of the grid points is within the fluid domain and the other is within the solid domain. This arrangement causes the computational boundary of a particle to be defined by step-wise scheme. In order to represent a smooth boundary and accurately represent the shape of any particle, it is necessary to use a large number of lattice points. In addition, when a particle moves, its computational boundary varies in each time step and this introduces fluctuations in the forces that act on the particle. This limits the ability of LBM to solve particle-fluid interaction problems at very high Reynolds numbers. Peskin [11] developed the immersed boundary method (IBM) in order to model the flow of blood in the heart. This method uses a fixed Cartesian mesh for the fluid, which is composed of Eulerian nodes. However, for the solid boundaries, which are immersed in the fluid, the IBM uses a set of Lagrangian boundary points, which are advected by the fluid-solid interactions. This method is especially suitable for the simulation of the effect of deformed immersed boundaries and has been widely used in biological fluid dynamics. Höfler and Schwarzer [7] presented a finite-difference method for particle-laden flows by adding a constraint force into the Navier-Stokes equations to enforce particle rigid motions, with the constraint force being determined by a penalty method. Goldstein et al. [6] used a so-called adaptive or feedback forcing scheme to model the no-slip conditions on a stationary boundary. This technique necessitates the use of two free parameters that must be chosen, based on the flow conditions. In the recent years, the concept of IBM has been employed into the FEM. Glowinski et al. [4, 5] developed the Ficti- tious Domain Method (FDM) by using Lagrange multipliers to enforce the no-slip boundary conditions between the particle surfaces and the fluid. They were able to apply this method in order to simulate a flow system with 1024 spherical particles [5]. Ten Cate et al. [13] used an adaptive-forcing scheme with the LBM to simu- late the sedimentation of a single sphere in an enclosure. Feng and Michaelides [2] combined the IBM and the LBM by computing the force density through a penalty method in the simulations of particulate flows. This method has the disadvantage that it requires a priori selection of the stiffness parameter based on the specific problem to be solved.

The key point of the success of both LBM and IBM is that instead of re-meshing the fluid domain, they both use a fixed mesh to represent the fluid field. In the LBM, the moving boundaries are approximated by the fixed points on the grid. These are essentially the midpoints of the boundary links if the bounce-back rule is used to implement the no-slip boundary condition. Hence, the moving boundaries are de- scribed by Eulerian points. In the IBM, the moving boundaries are represented by a set of Lagrangian boundary points, which are advected by the fluid. In this paper, we develop a new computational method called *Proteus*, which combines the direct forcing and the lattice Boltzmann methods. *Proteus* makes use of Eulerian lattice nodes for the fluid flow field and Lagrangian boundary points to represent particles or moving-boundary surfaces. Unlike the penalty method we have employed in our previous study [2], this method applies the direct forcing scheme, which was originally proposed by Mohd-Yusof [10] for fixed complex boundaries. This eliminates the need for the determination of the free parameter for the stiffness coefficient and makes the method much more straightforward and efficient. In addition, *Proteus* allows us to implement the rigid-body conditions inside a particle in a more convenient manner.

2 *Proteus* – Description of the numerical method

Proteus resolves the no-slip boundary conditions by adding a force density term in the Navier–Stokes equations. It computes the force density via a direct forcing scheme, and then solves the flow field by using the LBM. The basic idea of the LBM is to decompose the flow domain into a regular lattice grid and model the fluid as a group of fluid particles that are only allowed to move between lattice nodes or stay at rest. To apply the conventional LBM to particulate flow, the boundary of solid particles is enforced using the bounce-back rule, according to which, the fluid particles will bounce back when they run into a solid boundary. However, the difficulty with this approach is that it uses boundary nodes, which are the midpoints of boundary links. We call these nodes Eulerian boundary nodes to differentiate them from the Lagrangian boundary points/nodes that represent the surface of the particles as the particles move inside the fluid. This representation of a surface causes significant fluctuations of the computational boundaries, especially when a relatively small number of lattice nodes are used to represent the surface of the particles. The determination of the Eulerian boundary nodes is a non-trivial task especially when the particles do not have simple shapes. The worst disadvantage of the "bounce-back" rule is that it either fails to achieve accurate results or it could not yield converge results at high Reynolds number flow, which requires finer updating scheme for particle velocity and position during one lattice time step. To resolve this problem one may use the IB-LBM [2] and represent the particulate surfaces by using a set of independent Lagrangian boundary points that are attached to the boundary. The main advantages of doing this is that the computational boundary of the particles will be smooth and that the exact locations of the Lagrangian boundary points may be easily determined if we keep a track of the transformation matrix.

Let us consider a particle with a boundary surface, Γ, immersed in a three-dimensional incompressible viscous fluid with a domain, Ω. The particle boundary surface, Γ, is represented by the Lagrangian parametric coordinates, $\mathbf{s}$, and the flow domain, Ω, is represented by the Eulerian coordinates $\mathbf{x}$. Hence, any position on the particle surface may be written as $\mathbf{x} = \mathbf{X}(\mathbf{s}, t)$. Let $\mathbf{F}(\mathbf{s}, t)$ and $\mathbf{f}(\mathbf{x}, t)$ represent the particle surface density and the fluid body force density. The no-slip boundary condition is satisfied by enforcing the velocity at all boundaries to be equal to the velocity of the fluid, $\mathbf{u}$, at the same location:

$$\frac{\partial \mathbf{X}(\mathbf{s}, t)}{\partial t} = \mathbf{u}(\mathbf{X}(\mathbf{s}, t), t). \tag{1}$$

The governing equations for the fluid-particle composite as follows:

$$\rho \left(\frac{\partial \mathbf{u}}{\partial t} + \mathbf{u} \cdot \nabla \mathbf{u} \right) = \mu \nabla^2 \mathbf{u} - \nabla p + \mathbf{f}, \tag{2}$$

$$\nabla \cdot \mathbf{u} = 0, \tag{3}$$

$$\mathbf{f}(\mathbf{x}, t) = \int_{\Gamma} \mathbf{F}(\mathbf{s}, t) \delta(\mathbf{x} - \mathbf{X}(\mathbf{s}, t)) \, d\mathbf{s} \tag{4}$$

and

$$\frac{\partial \mathbf{X}}{\partial t} = \int_{\Omega} \mathbf{u}(\mathbf{x}, t) \delta(\mathbf{x} - \mathbf{X}(\mathbf{s}, t)) \, d\mathbf{x}, \tag{5}$$

where $p(\mathbf{x}, t)$ is the fluid pressure, ρ is the fluid density and μ the fluid viscosity. Equations (2) and (3) are the Navier–Stokes equations of a viscous incompressible flow. Equation (4) shows how the force density of the fluid, $\mathbf{f}(\mathbf{x}, t)$, may be obtained from the immersed boundary force density, $\mathbf{F}(\mathbf{s}, t)$ through the integration over the immersed boundary. Equation (5) is essentially the no-slip condition at the interface, since the particle moves at the same velocity as the neighboring fluid. In the numerical implementation of the IBM the whole fluid domain, including the parts that are occupied by immersed bodies, is divided into a set of fixed regular nodes. Since these fluid nodes are not moving with the flow, we will call them Eulerian nodes. The immersed boundary is discretized into a group of boundary points that move under the action of the moving fluid. We will call these boundary nodes Lagrangian nodes. It must be pointed out that in the IBM, the Lagrangian nodes do not necessarily coincide with the Eulerian nodes.

To solve the fluid field with a body force density, $\mathbf{f}(\mathbf{x}, t)$, the LBM equation is modified by adding a term to the collision function. The details of the implementation may be found in [1].

In the *Proteus* method we use a set of Lagrangian boundary points to describe the particle boundary. Equation (2) is also valid at these Lagrangian boundary points. Assuming that the velocity and pressure fields at the time step $t = t_n$ are known, we have an explicit scheme to determine the force term at these Lagrangian boundary points at time $t = t_{n+1}$:

$$f_i^{(n+1)} = \rho \left(\frac{u_i^{(n+1)} - u_i^{(n)}}{\Delta t} + u_j^{(n)} u_{j,i}^{(n)} \right) - \mu u_{i,jj}^{(n)} + p_{,i}^{(n)}. \tag{6}$$

The Einstein notation for subscripts and derivatives is used in the last equation. In order to impose the boundary condition that at $t = t_{n+1}$, the velocity on the immersed Lagrangian boundary points is equal to the velocity of the particle at the same point, $U_i^{P(n+1)}$, the density force at these points should be given by the following expression:

$$f_i^{(n+1)} = \rho \left(\frac{U_i^{P(n+1)} - u_i^{(n)}}{\Delta t} + u_j^{(n)} u_{j,i}^{(n)} \right) - \mu u_{i,jj}^{(n)} + p_{,i}^{(n)}. \tag{7}$$

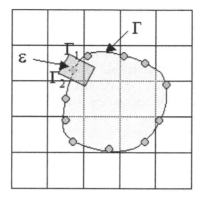

Fig. 1. A set of Lagrangian boundary points for a two-dimensional particle.

The above equation is called direct forcing, since it may be used to evaluate the force density at the Lagrangian boundary points without introducing any pre-defined parameters.

Figure 1 shows the boundary points for a two-dimensional particle boundary. The velocities at these Lagrangian points at the current time step $t = t_n$ may be computed by a bilinear interpolation using the velocity values of four neighboring grid points for two-dimensional flows or eight neighboring grid points for three-dimensional flows. However, the calculated force density using Equation (7) is at a Lagrangian boundary point, and we have to spread it into the neighboring Eurlerian nodes using a spreading function. In the adaptive forcing scheme, the fluid density force is also computed at boundary points. In this work, we use a delta function to spread the force density to the nearby Eulerian nodes.

The use of the force spreading technique is consistent with the theory of the immersed boundary method. The procedure employed may be explained as follows: For simplicity, we consider a two-dimensional problem with a particle boundary Γ. We also consider a small area ε around a Lagrangian boundary point s_i, as shown in Figure 1 and by integrating the force density within the small area, we obtain:

$$\int_{\varepsilon} \mathbf{f}(\mathbf{X})\mathbf{x}, t)) \, dA = \int_{\Gamma} \mathbf{F}(\mathbf{s}, t) \left(\int_{\varepsilon} \delta(\mathbf{x} - \mathbf{X}(\mathbf{s}, t)) \, dA \right) ds = \int_{\Gamma_{\varepsilon}} \mathbf{F}(\mathbf{s}, t) \, ds. \quad (8)$$

The relation between the flow force density, $\mathbf{f}(\mathbf{X}(\mathbf{s}, t))$, and the surface force density, $\mathbf{F}(\mathbf{s}, t)$, of Equation (4) is used. Equation (8) implies that the flow force density integral for a small area ε is equal to the boundary force density integral over the boundary element Γ_{ε}, which is the intersection of this small area and the particle boundary Γ (from Γ_1 to Γ_2 in Figure 1). For a uniform grid with spacing $\delta x = \delta y = \delta$, it is reasonable to assume that a Lagrangian boundary node s_i carries an area of $\delta s \delta$ (where δs is the length of the small boundary element Γ_{ε}). The force acting on this area is $\mathbf{f}(\mathbf{X}(\mathbf{s}_i, t))\delta s \delta$. Hence, the integral of the surface force density over the small area can be approximated as:

$$\int_{\Gamma_\varepsilon} \mathbf{F}(\mathbf{s}, t) \, ds \approx \mathbf{f}(\mathbf{X}(\mathbf{s}_i, t)) \delta_s \delta f. \tag{9}$$

For the three-dimensional case, one needs to consider a small volume around a Lagrangian point. Hence, the line integral should be replaced by a surface integral over a small boundary surface area intersected by a small volume. Also, δs should be replaced by δA, the area of this boundary surface element. By using Equation (9) as an approximation for the surface force density integral, the flow force density at each Eulerian node may be determined from Equation (4). This is the same procedure as the force spreading if a delta function is used for the spreading of the force. Equation (9) also provides an approach for the computation of the force acting on the particles, since the total force acting on a particle is equal to the sum of the forces acting on each surface element. The direct-forcing method combined with LBM is to be implemented in three-dimensional particulate flows. Details of the method are given by Feng and Michaelides [3].

3 Modeling the inter-particle collisions

In any type of particulate flows collisions between particles are unavoidable, especially when the flow is dense and the particles move at high Reynolds numbers. The correct handling of these collisions in any direct numerical simulation (DNS) is very important for the study of all particulate processes. Generally, the grid used in a DNS study is not fine enough to handle the lubrication force that develops between the particles or between particles and a solid boundary. Therefore, an artificial mechanism is necessary to be introduced in the numerical scheme in order to account for the repulsive force during collision processes. Without such a mechanism, it is likely that the particles will penetrate significantly into each other's computational boundary, thus, rendering the results meaningless.

We introduce a repulsive force when the gap between two particles is lower than the "safe zone". This artificial short-range repulsive force is added as an external force, with the functional form that was developed by Glowinski et al. [5]. This collision technique allows particles to overlap when the stiffness parameter is very large, that is, when the particles undergo "soft" collisions. For "soft" collisions, the partial overlapping of particles may be significant when a large number of particles undergo a packing process. The particles at the bottom, which have to bear the load of the particles above, will exhibit the maximum overlapping. To counteract this overlapping, we choose the repulsive force by considering the following situations: before the two particles contact, a repulsive force developed by Glowinski et al. [5] is used; when the two particles start to overlap, a higher spring force is applied. This force is proportional to the overlapping distance of two particles, and typically is much larger than the repulsive force with no overlapping. The advantage of the present collision scheme is that it enables us to use a small repulsive force for particles sedimentation before packing and the larger spring force for particles in the packing process. When particles start the sedimentation and before packing, the small repulsive force will be enough to repel the two particles and this will reduce unwanted side effects by large repulsive force used otherwise; when the particles start packing, the spring force will take effect to prevent penetration.

The *Proteus* numerical scheme has been validated by comparison with several sets of experimental and computational data. Details of the validation as well as more details of the method are given by Feng and Michaelides (2005).

4 Results – Sedimentation of 1232 spherical particles in a shallow box

We simulated the sedimentation process of 1232 spherical particles in a narrow enclosure. The group of particles is initially packed in a closed three-dimensional box, 3.125 cm long, 3.125 cm high and 0.09375 cm wide. The diameter of the particles is $d = 0.0625$ cm and, hence, the width of the box is $1.5d$. As in the previous case, the fluid density is $\rho_f = 1000$ kg/m^3, and the particle/fluid density ratio is 1.01. The dynamic viscosity of the fluid is 0.001 kg/ms. The "safe zone" between particles is equal to $d/8$. The stiffness parameters for the collisions are $\varepsilon_P = 0.25$, $E_P = 0.02$ and $\varepsilon_W = 0.5\varepsilon_P$. From the results presented in the last section, it is reasonable to conclude that *Proteus* will yield accurate results with the parameters chosen for this simulation. Initially, both the fluid and particles are stationary in an arrangement similar to closely-packed spheres, with the heavier particles on top of the fluid.

The numerical simulation box is $400 \times 12 \times 400$ in lattice units, and the diameter of each particle is equal to 8 lattice units. As in the last simulation, two boundary conditions will be examined in the width direction: (a) solid walls with no-slip boundary condition and (b) periodic boundary conditions, which imply an infinite array of identical particles. The boundaries in all the other directions are solid boundaries with no-slip velocity conditions. The relaxation time for the first case is $\tau = 0.9915$ and each lattice time step corresponds to a physical time of 0.001 s; the relaxation time for the second case $\tau = 0.74576$ and each lattice time step corresponds to a physical time of 0.0005 s. All the simulations were conducted on a SGI Onyx 3500 machine. In the case of sedimentation with 1232 particles, the time to complete a single iteration is about 15.8 s. This results in approximately 4.3 hours computational time to simulate 1 s of physical time for the first case, or about 11 days to complete a simulation of 60 s of physical time without any paralleling. More details of the *Proteus* code and more results are given by Feng and Michaelides [3].

5 Conclusions

An efficient three-dimensional computational method, *Proteus*, based on the LBM and the direct forcing scheme, has been developed for use with large groups of particles. This method computes the force term directly and does not require the use of any other coefficients as additional parameters. Compared with the conventional LBM, this method provides a smooth computational boundary. While it has

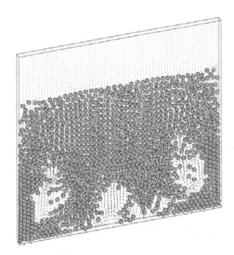

Fig. 2. Particle positions at $t = 25$ s.

the same order of accuracy as the LBM, *Proteus* is capable of achieving results at higher Reynolds numbers and, if needed, to easily enforce the rigid body motion in the interior of the particles. *Proteus* is also easier and more efficient to be used when the particles do not have a simple shape. The method has been validated by comparison of results from the simulations of the motion of single spheres settling in an enclosure with analytical and experimental results that were derived in the past and has been successfully applied to the sedimentation problem of an arrangement of spheres.

References

1. Feng, Z.-G. and Michaelides, E.E., 2002, Inter-particle forces and lift on a particle attached to a solid boundary in suspension flow, *Phys. Fluids* **14**, 49–60.
2. Feng, Z.-G. and Michaelides, E.E., 2004, An immersed boundary method combined with lattice Boltzmann method for solving fluid and particles interaction problems, *J. Comput. Phys.* **195**, 602–628.
3. Feng, Z.G. and Michaelides, E.E., 2005, *Proteus* – A direct forcing method in the simulation of particulate flows, *J. Comput. Phys.* **202**, 20–51.
4. Glowinski, R., Pan, T.-W., Hesla, T.I. and Joseph, D.D., 1999, A distributed Lagrange multiplier/fictitious domain method for particulate flows, *Int. J. Multiphase Flow* **25**, 755–794.
5. Glowinski, R., Pan, T.-W., Hesla, T.I., Joseph, D.D. and Periaux, J., 2001, A fictitious domain approach to the direct numerical simulation of incompressible viscous flow past moving rigid bodies: Application to particulate flow, *J. Comput. Phys.* **169**, 363–426.
6. Goldstein, D., Handler, R. and Sirovich, L., 1993, Modeling a no-slip flow boundary with an external force field, *J. Comput. Phys.* **105**, 354–366.

7. Höfler, K. and Schwarzer, S., 2000, Navier–Stokes simulation with constraint forces: Finite-difference method for particle-laden flows and complex geometries, *Phys. Rev. E* **61**, 7146–7160.

8. Ladd, A.J.C., 1994, Numerical simulations of particulate suspensions via a discretized Boltzmann equation. Part I. Theoretical foundation, *J. Fluid Mech.* **271**, 285–310.

9. Ladd, A.J.C., 1994, Numerical simulations of particulate suspensions via a discretized Boltzmann equation. Part II. Numerical results, *J. Fluid Mech.* **271**, 311–339.

10. Mohd-Yusof, J., 1997, Combined immersed boundaries/B-splines methods for simulations of flows in complex geometries, *Annual Research Briefs, Center for Turbulence Research*, Stanford University.

11. Peskin, C.S., 1977, Numerical analysis of blood flow in the heart, *J. Comput. Phys.* **25**, 220–252.

12. Peskin, C.S., 2002, The immersed boundary method, *Acta Numerica* **11**, 479–517.

13. Ten Cate, A., Nieuwstad, C.H., Derksen J.J. and Van den Akker, H.E.A., 2002, Particle imaging velocimetry experiments and lattice-Boltzmann simulations on a single sphere settling under gravity, *Phys. Fluids* **14**, 4012–4025.

An Explicit Finite-Difference Scheme for Simulation of Moving Particles

A. Perrin and H.H. Hu

Department of Mechanical Engineering and Applied Mechanics, University of Pennyslvania, 29 Towne Building, 220S. 3d Street, Philadelphia, PA 199USA

Abstract. We present an explicit finite-difference scheme for direct simulation of the motion of solid particles in a fluid. The method is based on a second-order MacCormack finite-difference solver for the flow, and Newton's equations for the particles. The fluid is modeled with fully compressible mass and momentum balances; the technique is intended to be used at moderate particle Reynolds number. Several examples are shown, including a single stationary circular particle in a uniform flow between two moving walls, a particle dropped in a stationary fluid at particle Reynolds number of 20, the drafting, kissing, and tumbling of two particles, and 100 particles falling in a closed box.

1 Introduction

This paper develops an explicit finite difference scheme for direct numerical solution of particles in a nearly incompressible Newtonian fluid. It is hoped that the present scheme will find success in regimes where conventional methods work awkwardly. Finite element methods, such as the Particle-Mover arbitrary Lagrangian–Eulerian (ALE) method of Hu [2–4] or Johnson and Tezduyar's stabilized space-time method [5], are efficacious for small numbers of particles at moderate Re but have prohibitive computational requirements when particles fill the domain. When these methods arc used to simulate closely-space particles, the mesh may need to be refined excessively. This gives rise to memory and/or processing issues due to the remeshing and projection procedures. A fixed, uniform grid becomes an attractive choice. Additionally, the time step must be smaller than the intrinsic time scale of the particle motion in order to properly capture the physics. For dense particle configurations this timescale may be comparable to the largest stable time step of an explicit scheme, which would make the explicit scheme competitive.

In solving the incompressible Navier–Stokes equations using the primitive variables (velocity and pressure), one numerical difficulty lies in the continuity equation. The continuity equation can be regarded either as a constraint on the flow field to determine the pressure or the pressure plays the role of the Lagrange multiplier to satisfy the continuity equation. In a flow field, the information (or disturbance)

S. Balachandar and A. Prosperetti (eds) Proceedings of the IUTAM Symposium on Computational Multiphase Flow, 163–172.
© 2006 *Springer. Printed in the Netherlands.*

travels with both the flow and the speed of sound in the fluid. Since the speed of sound is infinite in an incompressible fluid, pressure disturbances are propagated instantaneously throughout the domain. In many numerical schemes for solving the incompressible Navier-Stokes equations, the pressure is obtained by solving a Poisson equation. The Poisson equation may occur in either continuous form or discrete form. Solving the pressure Poisson equation is often the most costly step in these schemes.

One technique to surmount the difficulty of the incompressible limit is to introduce an artificial compressibility (AC) as Chorin did [1]. This formulation is normally used for steady problems with a pseudo-transient formulation. In the formulation, the continuity equation is replaced by

$$\frac{\partial p}{\partial t} + c^2 \nabla \cdot \mathbf{u} = 0, \tag{1}$$

where c is an arbitrary constant and could be the artificial speed of sound in a corresponding compressible fluid with the equation of state $p = c^2 \rho$. The formulation is called pseudo-transient because (1) does not have any physical meaning before the steady state is reached. However, when c is large, (1) can be considered as an approximation to the unsteady solution of the incompressible Navier–Stokes problem [1]. Nourgaliev et al. [8] have pointed out that the AC method is both easily parallelized and economically coded.

The present work is closest in spirit to Norgaliev's Numerical Acoustic Relaxation (NAR) method [8]. Our scheme is fully explicit and second order in both time and space. Rather than use Equation (1), we instead use the fully compressible continuity and momentum equations, but with Chorin's artificial equation of state. For the particle interface, we redraw the particle at each time-step and handle change of phase in a manner similar to Udaykumar et al. [10]. The force and torque are calculated by integration over the interface, and the particles are then explicitly moved according to Newton's third law.

2 Explicit MacCormack scheme

Instead of using the artificial continuity equation of (1), one may start with the exact compressible Navier–Stokes equations with the artificial equation of state. In Cartesian coordinates, the component form of the continuity equation and compressible Navier–Stokes equation in two dimensions can be written as

$$\frac{\partial \rho}{\partial t} + \frac{\partial (\rho u)}{\partial x} + \frac{\partial (\rho v)}{\partial y} = 0, \tag{2}$$

$$\frac{\partial}{\partial t}(\rho u) + \frac{\partial}{\partial x}(\rho u^2) + \frac{\partial}{\partial y}(\rho v u) = \rho g_x - \frac{\partial p}{\partial x} + \mu \nabla^2 u + \frac{\mu}{3}\frac{\partial}{\partial x}\left(\frac{\partial u}{\partial x} + \frac{\partial v}{\partial y}\right), \tag{3}$$

$$\frac{\partial}{\partial t}(\rho v) + \frac{\partial}{\partial x}(\rho u v) + \frac{\partial}{\partial y}(\rho v^2) = \rho g_y - \frac{\partial p}{\partial y} + \mu \nabla^2 v + \frac{\mu}{3}\frac{\partial}{\partial y}\left(\frac{\partial u}{\partial x} + \frac{\partial v}{\partial y}\right), \tag{4}$$

with the equation of state $p = c^2\rho$, where c is the speed of sound in the medium. As long as the flows are limited to low Mach numbers and the conditions are almost isothermal, the solution to this set of equations should approximate the incompressible limit (see [6]). The physical continuity equation is preferred because the range valid Mach number of (1) is unknown, and there is no computational reason to make an approximation.

The explicit MacCormack scheme, after [7], is essentially a predictor-corrector scheme, similar to a second-order Runge–Kutta method commonly used to solve ordinary differential equations. For a system of equations of the form

$$\frac{\partial \mathbf{U}}{\partial t} + \frac{\partial \mathbf{E}\,(\mathbf{U})}{\partial x} + \frac{\partial \mathbf{F}\,(\mathbf{U})}{\partial y} = 0, \tag{5}$$

the explicit MacCormack scheme consists of two steps,

$$\mathbf{U}^*_{i,j} = \mathbf{U}^n_{i,j} - \frac{\Delta t}{\Delta x}(\mathbf{E}^n_{i+1,j} - \mathbf{F}^n_{i,j}) - \frac{\Delta t}{\wedge y}(\mathbf{F}^n_{i,j+1} - \mathbf{F}^n_{i,j}), \quad \text{(Predictor)}$$

$$\mathbf{U}^{n+1}_{i,j} = \frac{1}{2}\left[\mathbf{U}^n_{i,j} + \mathbf{U}^*_{i,j} - \frac{\Delta t}{\Delta x}(\mathbf{E}^*_{i,j} - \mathbf{E}^*_{i-1,j}) - \frac{\Delta t}{\Delta y}(\mathbf{F}^*_{i,j} - \mathbf{F}^*_{i,j-1})\right]. \quad \text{(Corrector)}$$

The vector $\mathbf{U} = (\rho, \rho\mathbf{u}, \rho\mathbf{v})$ contains the update variables. The vectors $\mathbf{E}$, and $\mathbf{F}$ are functions of the update variables and some of their spacial derivatives. Notice that the spatial derivatives in (5) are discretized with opposite one-sided finite differences in the predictor and corrector stages. The star variables are supposed to be evaluated at time level t_{n+1}. This scheme is second order accurate in both time and space.

Applying the MacCormack scheme to the compressible Navier–Stokes equations (2)–(4) and replacing the pressure with $p = c^2\rho$, we have the predictor step

$$\rho^*_{i,j} = \rho^n_{i,j} - c_1\left[(\rho u)^n_{i+1,j} - (\rho u)^n_{i,j}\right] - c_2\left[(\rho v)^n_{i,j+1} - (\rho v)^n_{i,j}\right], \tag{6}$$

$$(\rho u)^*_{i,j} = (\rho u)^n_{i,j} - c_1\left[(\rho u^2 + c^2\rho)^n_{i+1,j} - (\rho u^2 + c^2\rho)^n_{i,j}\right] \tag{7}$$

$$- c_2\left[(\rho u v)^n_{i,j+1} - (\rho u v)^n_{i,j}\right] + \frac{4}{3}c_3(u^n_{i+1,j} - 2u^n_{i,j} + u^n_{i-1,j})$$

$$+ c_4(u^n_{i,j+1} - 2u^n_{i,j} + u^n_{i,j-1})$$

$$+ c_5(v^n_{i+1,j+1} + v^n_{i-1,j-1} - v^n_{i+1,j-1} - v^n_{i-1,j+1}),$$

$$(\rho v)^*_{i,j} = (\rho v)^n_{i,j} - c_1\left[(\rho u v)^n_{i+1,j} - (\rho u v)^n_{i,j}\right] \tag{8}$$

$$- c_2\left[(\rho v^2 + c^2\rho)^n_{i,j+1} - (\rho v^2 + c^2\rho)^n_{i,j}\right] + c_3(v^n_{i+1,j} - 2v^n_{i,j} + v^n_{i-1,j})$$

$$+ \frac{4}{3}c_4(v^n_{i,j+1} - 2v^n_{i,j} + v^n_{i,j-1})$$

$$+ c_5(u^n_{i+1,j+1} + u^n_{i-1,j-1} - u^n_{i+1,j-1} - u^n_{i-1,j+1}).$$

and likewise for the corrector. The coefficients are defined as

$$c_1 = \frac{\Delta t}{\Delta x}, \quad c_2 = \frac{\Delta t}{\Delta y}, \quad c_3 = \frac{\mu \Delta t}{(\Delta x)^2}, \quad c_4 = \frac{\mu \Delta t}{(\Delta y)^2} \text{ and } c_5 = \frac{\mu \Delta t}{12 \Delta x \Delta y}. \quad (9)$$

In both the predictor and corrector steps the viscous terms (the second-order derivative terms) are discretized with centered-differences to maintain second-order accuracy. For brevity, body force terms in the momentum equations are neglected here.

Tannehill et al. [9] give the following semi-empirical stability criterion for the explicit MacCormack scheme:

$$\Delta t \leq \frac{\sigma}{(1 + 2/\text{Re}_\Delta)} \left[\frac{|u|}{\Delta x} + \frac{|v|}{\Delta y} + c \sqrt{\frac{1}{\Delta x^2} + \frac{1}{\Delta y^2}} \right]^{-1}, \quad (10)$$

where σ is a safety factor (≈ 0.9), $\text{Re}_\Delta = \min(\rho|u|\Delta x/\mu, \rho|v|\Delta y/\mu)$ is the minimum mesh Reynolds number. This condition is quite conservative for flows with small mesh Reynolds numbers. We find that at moderate flow Re (Re = 10 to 500) the CFL condition gives results that more closely resemble the actual stability limits of our scheme,

$$\Delta t \leq \min \left(\frac{0.5 \Delta x}{c}, \frac{0.5 \Delta y}{c} \right). \quad (11)$$

To resolve the motion of the particles in particulate flows, time step is usually limited by the distance the particle is allowed to move during each step. If this distance is one grid spacing in the simulation, this restriction represents a similar condition for the time step based on the particle velocity. The condition in (11) is not very restrictive for such flows, since the ratio of the time step in (11) to the time step for capturing the particle motion is just the Mach number. For simulating incompressible behavior of the particulate flows, we limit the Mach numbers to be small, say around 0.1, but not too small, to avoid excessively tiny time steps.

3 Flow over a circular cylinder between sliding walls

To test the scheme for the case of flow over an immersed body, we simulated an infinite (i.e. two dimensional) circular cylinder in a channel with two moving side walls and a uniform inlet velocity profile (Figure 1). This problem is mathematically equivalent to the problem of a cylinder moving in a fluid at constant speed down a channel with stationary walls, although the numerical treatment of these two cases differ.

To apply the no-slip condition exactly at the surface of the particle, one can Taylor expand from the boundary gridpoint to the true surface. It is natural to enforce no-slip on the point on the cylinder surface nearest to each boundary point, because it minimizes the truncation error in the Taylor series.

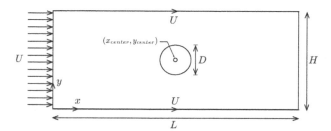

Fig. 1. Flow over a circular cylinder between two sliding walls. Inlet and wall velocities are both U, channel length is L, channel width is H, cylinder diameter is D, and the center of the cylinder is at (x_{center}, y_{center}).

The continuity equation provides one possible boundary condition for density (or pressure, since they are proportional here). When this condition is used on the cylinder, it results in a jagged pressure distribution on the surface. The reason for this is not yet understood. A boundary condition based on the momentum equation gave better results. This condition can be derived by taking the component of the pressure (density) normal to the cylinder surface, which is $\hat{n} \cdot \nabla p$, and combining with Equations (3) and (4). The result is

$$\frac{\partial \rho}{\partial n} = \frac{\mu}{c^2} \left(\frac{\hat{n}_x}{3} \left(4\frac{\partial^2 u}{\partial x^2} + 3\frac{\partial^2 u}{\partial y^2} + \frac{\partial^2 v}{\partial x \partial y} \right) + \frac{\hat{n}_y}{3} \left(3\frac{\partial^2 v}{\partial x^2} + 4\frac{\partial^2 v}{\partial y^2} + \frac{\partial^2 u}{\partial y \partial x} \right) \right)$$
$$- \frac{\rho}{c^2} \left(\hat{n}_x \left(\frac{\partial u}{\partial t} + u\frac{\partial u}{\partial x} + v\frac{\partial u}{\partial y} \right) + \hat{n}_y \left(\frac{\partial v}{\partial t} + u\frac{\partial v}{\partial x} + v\frac{\partial v}{\partial y} \right) \right), \qquad (12)$$

where the equation of state $p = c^2 \rho$ has been used to eliminate pressure and $\hat{n}_x$ and $\hat{n}_y$ are the x- and y-components of the surface normal. Each term can be evaluated using one-sided second-order differences for the derivatives.

In this section, the MacCormack scheme for flow over a cylinder is validated. In the tests that follow, $L = 35D$, $H = 4D$, and $\Delta x = \Delta y$ in all cases. The cylinder center is 15.5 diameters from the inlet. All lengths have been non-dimensionalized using D, and the pressure and shear stress with ρU^2. We examine lift and drag coefficients (C_L and C_D) as functions of time, the convergence of C_D at two Reynolds and Mach numbers, and the pressure and shear distributions at three Re and M = 0.05. When calculating the pressure and shear stress on the particle surface, the density and velocity gradient terms were Taylor expanded from the boundary point to the surface in the same manner described above for the velocity. Lift and Drag coefficients were determined by numerical integration of the pressure and shear on the particle surface using the trapezoidal rule.

The lift and drag coefficients as a function of time are shown for Re = 100 (M = 0.05) in Figure 2b. The oscillation of the lift coefficient, $C_L = 2F_y/(\rho U^2 D)$, is induced by vortex shedding.

The pressure distribution on a fine mesh (40 grid spacings across the cylinder diameter) is shown in Figure 2a. It is worth noting that the drag coefficient at Re = 100

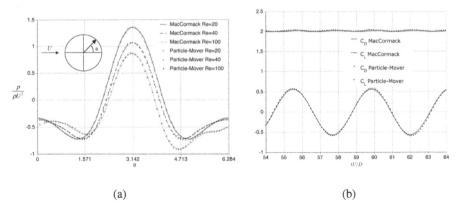

(a)   (b)

Fig. 2. (a) Non-dimensional pressure distribution. (b) C_D and C_L vs. time at Re = 100. There are 40 grid spacings across the cylinder diameter, and the Mach number is 0.05. The Strohl number for the MacCormack scheme is 0.233, and for Particle-Mover it is 0.227. Note that the walls are only $1.5D$ away from the cylinder.

changed by just 2% when the mesh was set to 20 grid spacings per diameter, so one need not use such a fine mesh in most instances. The non-dimensional pressure distribution plot also includes the results of a Particle-Mover finite element calculation with 20 elements across the cylinder diameter. At Re = 100, the flow is unsteady, so the pressure distribution is a snapshot at non-dimensional time $t = 77.36$. Time was non-dimensionalized by L/U.

4 Circular cylinder translating at constant speed

Next we consider the case of a cylinder dragged at constant speed through a stationary fluid in a channel. This introduces some new implementation issues. The boundary gridpoints must be picked anew each time the cylinder moves, resulting in points that were formerly inside the cylinder leaking from the rear. In the previous case of the stationary cylinder, these interior points did not need values assigned to them – they were invisible to the fluid. Since this is no longer the case, the velocity and density must now be chosen for these points.

Our method of dealing with the "leaking" of interior points into the bulk fluid is to assign them reasonable values before the cylinder is moved. Only the first layer of interior points needs to be considered, since the particle always moves less than one grid spacing per time step due to the CFL condition. Expanding in a Taylor series from the boundary point to each associated interior point, density and the velocity components are assigned to the interior points. (The derivatives of the velocity and density at the boundary point are already known from the boundary point update.)

The drag and lift coefficients for the translating cylinder, shown in Figures 3a and 3b, demonstrate that the results obtained for a moving cylinder in a stationary

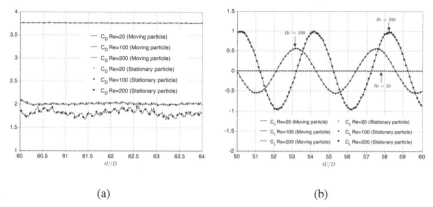

(a)   (b)

Fig. 3. Lift and drag coefficients vs. non-dimensional time. The thin lines represents the moving cylinder, and the dots represents the stationary cylinder. (a) The drag coefficient C_D. (b) The lift coefficient C_L.

fluid agree with those for a stationary cylinder in a moving fluid. This is expected, but given that the boundary points representing the cylinder are changing with each iteration, direct confirmation is important. In both cases, the Mach number was 0.05. For the stationary case, the cylinder center was 15.5 diameters from the inlet, and the channel length was 35 diameters. The channel length for the moving cylinder was 70 diameters. In both cases, the cylinder was centered horizontally in the channel, which was 4 diameters wide. For the case of the stationary cylinder, the outflow boundary condition was used. The channel for the moving cylinder was closed (velocity set to zero) on all boundaries. In both cases there were 40 grid spacings across the cylinder diameter, and the time step was chosen according to Equation (11) with a safety factor of 0.5.

5 Freely falling cylinders

The translating cylinder simulation may now be modified to deal with the case of freely falling cylinders by adding a collision scheme. A collision scheme is necessary because under most circumstances the lubrication forces will only become large enough to prevent collisions in a time that is much smaller than Δt. It will also be necessary to integrate the equations of motion for the cylinders, but this is done as a part of the collision scheme. For details on the collision scheme, see [4]. The equations of motion, which are solved as part of the collision scheme, are the result of a force and moment balance on each particle.

For a Reynolds number of 20 based on the terminal velocity and diameter, the particle remained in the center of the channel. (There was some off-center wandering due asymmetries in the MacCormack discretization, but in all cases, the wandering was by less than a single grid spacing.) Results for the velocity versus time are shown

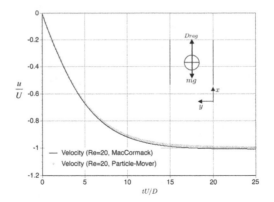

Fig. 4. Velocity (non-dimensionalized by U) versus non-dimensional time tU/D for a single falling particle. U is the terminal velocity.

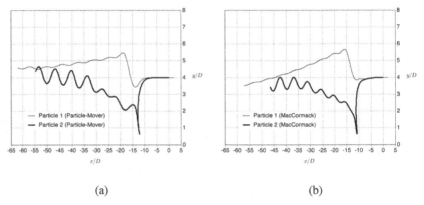

(a)   (b)

Fig. 5. The paths of the center of each particle in the drafting-kissing-tumbling sequence. For the MacCormack result, $\delta/D = 7.5\%$, and for Particle-Mover it was $\delta/D = 3\%$. (a) Particle-Mover finite element result. (b) MacCormack result. In both cases, the thick line corresponds to the particle initially on the bottom.

in Figure 4. The simulation parameters are the same as in the previous translating cylinder tests.

If two particles in a fluid are initially placed one above the other and released, drafting, kissing, and tumbling occurs. This behavior was observed in the MacCormack simulation when the two particles were initially separated by two diameters (measured center to center). The locus of particle positions in the laboratory reference frame is shown in Figure 5 alongside the results from Particle-Mover. The density ratio of the solid phase to the fluid was 1.04. There were 30 grid spacings across the diameter of each particle. The channel was 8 diameters wide, and there were 40 diameters between the top and bottom (the channel length is infinite). The minimum distance between two particles, δ, was set to three grid spacings, or

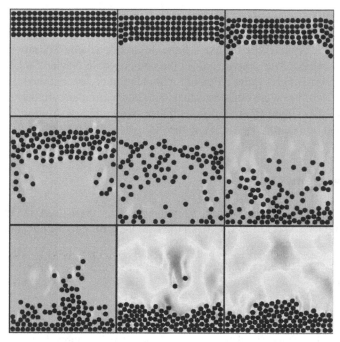

Fig. 6. One hundred particles sedimenting in a closed box. The time increases from left to right and top to bottom. $\rho_s/\rho_f = 2$. The box side length is 24.2 particle diameters, and there are 20 grid spacings across each particle. The safety zone thickness $\delta = 3\Delta x$.

a tenth of the particle diameter. The maximum Reynolds number was around 80. The results of the finite element solution and the MacCormack agree only qualitatively because the details of the collision scheme differ. The safety zone in the case of the MacCormack results was 7.5% of the particle diameter, compared to 3% for Particle-Mover. A thick safety zone was required here by the use of one-sided derivatives in the boundary conditions, however, improvement can be explored in future work. After the tumble, wiggles in the particle path appear due to vortex shedding.

For a given resolution, the computational effort required to compute any number of particles hardly changes, provided the collision scheme is efficient. One hundred particles are shown sedimenting in a closed box in Figure 6. The sides of the box are 24.2 particle diameters long (a non-integer number of particle diameters is required because the particles must be separated by at least a safety zone thickness δ). The solid phase is twice as dense as the fluid. There are 20 grid spacings across each particle.

6 Concluding remarks

A second-order explicit MacCormack finite difference scheme for moving particles has been described. The scheme solves the compressible Navier–Stokes equations on a uniform grid, and is quite efficient for simulation of concentrated suspensions.

Several examples were demonstrated, including a circular cylinder translating at constant speed, a single freely falling cylinder, two cylinders drafting, kissing, and tumbling, and one hundred particles sedimenting in a closed box.

Acknowledgement

A. Perrin acknowledges the support from the Graduate Assistance in Areas of National Need (GAANN) fellowship from the U.S. Department of Education.

References

1. A.J. Chorin, 1967, A numerical method for solving incompressible viscous flow problems, *Journal of Computational Physics* **2**, 12–26.
2. H.H. Hu, 1996, Direct simulation of flows of solid-liquid mixtures, *International Journal of Multiphase Flow* **22**(2), 335–352.
3. H.H. Hu and D.D. Joseph, 1992, Direct simulation of fluid particle motions, *Theoretical and Computational Fluid Mechanics*, **3**, 285–306.
4. H.H. Hu, N.A. Patankar and M.Y. Zhu, 2001, Direct numerical simulations of fluid-solid systems using the arbitrary Lagrangian-Eulerian technique, *Journal of Computational Physics* **169**, 427–462.
5. A.A. Johnson and T.E. Tezduyar, 1996, Simulation of multiple spheres falling in a liquid-filled tube, *Comput. Methods Appl. Mech. Engrg.* **134**, 351–373.
6. P.K. Kundu and I.M. Cohen, 2004, *Fluid Mechanics*, Elsevier Academic Press, 3rd edition, pp. 686–687.
7. R.W. MacCormack, 1969, The effect of viscosity in hypervelocity impact cratering, Technical Report 69-354, AIAA.
8. R.R. Nourgaliev, T.N. Dinh and T.G. Theofanous, 2004, A pseudocompressibility method for the numerical simulation of incompressible multifluid flows, *International Journal of Multiphase Flow* **30**, 901–937.
9. J.C. Tannehill, D.A. Anderson and R.H. Pletcher, 1997. *Computational Fluid Mechanics and Heat Transfer*. Taylor & Francis, Washington, DC, 2nd edition.
10. H.S. Udaykumar, R. Mittal and W. Shyy, 1999, Computation of solid-liquid phase fronts in the sharp interface limit on fixed grids, *Journal of Computational Physics* **153**, 535–574.

Use of Variable-Density Flow Solvers for Fictitious-Domain Computations of Dispersed Solid Particles in Liquid Flow

John C. Wells[1], Hung V. Truong[1] and Gretar Tryggvason[2]

[1]*Department of Civil & Environmental Engineering, Ritsumeikan University, Noji Higashi 1-1-1, Kusatsu 525-8577 Japan; e-mail: jwells@se.ritsumei.ac.jp*
[2]*Department of Mechanical Engineering, Worcester Polytechnic Institute, USA*

1 Introduction

Our group aims to simulate "bedload transport" of sediment by a turbulent flow over a sediment bed, and thence to investigate how bed shear drives sediment flux, how sand becomes suspended, how bed particles are sorted by size or density, how bed ripples form, and so on. Passing up through the bedload layer, the mechanism resisting flow changes rapidly from interparticle contacts and collisions to turbulent momentum transport. This juxtaposition necessitates both (1) tracking $O(1000)$ 3D particles in the manner of Particle Dynamics Simulations, with appropriate modeling of contact forces, and (2) resolving most of the scales of the turbulence for long enough to measure third moments of fluctuating quantities. We are interested in the scientifically important transition from hydraulically smooth to rough beds, which corresponds to particle Reynolds numbers, based on relative velocity, from 10 to a few hundred; in the upper half of this range, the above considerations set a lower limit on resolution of $O(10)$ grid points per diameter. Current processing power sets an *upper* limit that is not much greater, while imposing "fictitious-domain" techniques in which particles, moving through a fixed Cartesian grid, are "rigidified" by artificial forcing. The simple scheme due to Kajishima et al. [1] remains the only such scheme to have calculated the motion of 1000 or more solid 3D particles in a turbulent flow, with particle Reynolds number in the range 10~300, and for this reason our efforts focussed from an early stage on their method. Following a standard "fluid" step, the motion inside particles is forced to a "target velocity" which is predicted from the *prior motion* of the particle (see [1, 2, 3] for details). As an "explicitly-coupled" method, the particle acceleration is not treated as an unknown while calculating the fluid-particle interaction. Also, the method is built on a constant-density N-S solver, so the fluid acceleration calculated within the particle during the "fluid" step is that based on the fluid's density, not the solid's. Thus there is no guarantee that the added-mass reaction is predicted correctly by the fluid solver. Similar comments apply to

S. Balachandar and A. Prosperetti (eds), Proceedings of the IUTAM Symposium on Computational Multiphase Flow, 173–183.

other explicitly-coupled methods (see, for example, Uhlmann and Pinelli [4] in the present volume for a review of such methods).

For the particular case of neutral solid density, Patankar [5] improved the method of Kajishima et al. [1] so as to achieve implicit interphase coupling; the target (angular) velocity to which particles are forced at a new time step is treated, analogously to new fluid velocities, as an *unknown*, and is accordingly calculated by averaging the (angular) momentum, following a "fluid step", over the grid cells occupied by the particle. The net forcing applied to recover rigid motion thus sums to zero within each particle. By building on a *variable-density* incompressible flow solver, the present paper has adapted Patankar's [5] idea to cover non-neutrally buoyant cases.[1] Because the density of the virtual fluid equals that of the particle, the particle acceleration, and thus the added-mass force, is calculated correctly. By design, this method does not require any separate equations of motion for the particles, and as such is an "implicitly-coupled" method. This property will be extremely important in future calculations of bedload sediment transport, when the particles can collide anywhere and at any instant.

While the computational cost of previously reported implicitly-coupled methods, e.g.[6], [7] and references therein, has so far prevented their application to 3D problems with thousands of particles, the marginal computational load associated with the present method of rigidifying particles is, like in [8], only a few percent. Exploiting efficient multigrid codes for the pressure in a variable-density flow solver, processing power should not be a major obstacle to simulating 3D problems of the scale reported in [1]. As a first step in that direction, the present work has tested the method in 2D in the context of a single disk: (1) dropped from rest due to gravity at terminal Reynolds numbers of 13 and 310, and (2) in a Couette channel flow. Particle trajectories compare favorably with reliable benchmark data [6] and with the exact solution, respectively.

2 Numerical algorithms

2.1 Variable density-based implicit volumetric forcing method "VIV"

The basic finite-difference Navier–Stokes solver, for incompressible but variable-density flow, is a very standard one based on a staggered grid, forward Euler stepping in time, central differences in space, and a SMAC treatment of pressure. Steps therein are denoted below by **F#**). **P#**) denotes steps treating the particle phase, notably "rigidification" where the velocity field within particles is forced to rigid motion and details of which are given in the following subsection.

---

[1] During the late stages of the present work, Sharma and Patankar [8] reported an equivalent extension of Patankar [5] to non-neutral cases based on a variable-density Navier–Stokes solver. Flowing from the common seminal ideas [1, 5], the core of the method reported here and in our preliminary report [9] (Equations (8) and (9) herein) corresponds to Equations (32) through (35) in their work.

F1) With the Navier–Stokes equations for momentum conservation in an incompressible velocity field u in fluid of variable density ρ written as

$$\frac{D(\rho u)}{Dt} = -\nabla p + \Pi + \rho g \equiv F, \tag{1}$$

calculate the intensive fluid force F^n, i.e. F at the n–th timestep t^n, in each momentum cell. Π is the viscous term and the "gravity" g can be any conservative body force.

P1) Update particle positions to t^{n+1}, then calculate the new density field ρ^{n+1}.

F2) Subtracting the momentum advection term A, obtain the "fractional-step" momentum density $\rho^{n+1}\tilde{u}$:

$$\rho^{n+1}\tilde{u} = \rho^n u^n + \Delta t \left(F^n - A^n \right) \tag{2}$$

and thence the fractional step velocity $\tilde{u}$.

F3) Project $\tilde{u}$ to obtain a solenoidal velocity field

$$\tilde{\tilde{u}} = \tilde{u} - \frac{\Delta t}{\rho^{n+1}} \nabla_h \phi, \tag{3}$$

where ϕ satisfies the following elliptic equation (cf. [10]):

$$\nabla_h \cdot \left(\frac{1}{\rho^{n+1}} \nabla_h \phi \right) = \frac{\nabla_h \cdot \tilde{u}}{\Delta t}. \tag{4}$$

The pressure field is then updated by $p^{n+1} = p^n + \phi$.

P2) Within each particle, redistribute momentum to yield a rigidified momentum density $\rho^{n+1}u^{n+1}$, and reset the particle's (angular) velocity accordingly.

    The pressure gradient is a conservative body force, applied to satisfy incompressibility. Analogously, our term "rigidification" means constraining the "flow" within particles to rigid-body motion, by applying *non-conservative* body forces [5].

2.2 Details of steps treating the particle phase

P1) We use forward Euler steps to update the position each particle's center:

$$x_m^{n+1} = x_m^n + v_m^n \Delta t, \tag{5}$$

where m indexes the particles, taken herein to be spherical. The new solid volume fractions corresponding to each particle are calculated by the following smoothed distribution:

$$\alpha_m^i = 1 \qquad\qquad\qquad\qquad r_m^i \le (a - \delta): \qquad \text{core region}$$

$$= \frac{1}{2}\left(1 + \cos\left(\pi\frac{r_m^i - (a - \delta)}{2\delta}\right)\right) \quad a - \delta < r_m^i < a + \delta: \text{"surface cells"} \quad (6)$$

$$= 0 \quad \text{if} \quad r_m^i \ge a + \delta,$$

where r_m^i is the distance from the center of cell i to the center of particle m, a is particle radius, and δ is the half-width of the smoothing fringe. The new density and viscosity fields, ρ^{n+1} and μ^{n+1}, are then calculated in proportion to the new solid volume fractions. We typically take δ to be one-half to two grid spacings, so that particles can interpenetrate to some degree. These solid volume fractions are assigned to pressure nodes, and the corresponding density ρ_i in that node is taken to be

$$\rho_i = \sum_{m=0}^{np} \alpha_m^i \rho_m, \tag{7}$$

where ρ_m is the density of particle m, with $m = 0$ referring to the fluid, and np is the number of particles contributing mass to the cell. When solid volume fraction is required in momentum cells, it is simply taken to be the average of the two pressure nodes straddled by the momentum node.

To avoid constraining the time step any more than necessary, the viscosity employed for the solid phase is usually set proportional to density, so that the kinematic viscosity is constant; results of the test cases described below were insensitive to this choice.

P2) The average velocity is calculated by

$$v_m^{n+1} = \sum_i \rho_m \alpha_m^i \tilde{u}_i \Big/ \sum_i \rho_m \alpha_m^i. \tag{8}$$

This average is then assigned as the particle's tentative velocity, in addition to determining the rigid-motion velocity field in (12) below. The tentative angular velocity of the particle is calculated by

$$\Omega_m = \sum_i \rho_m \alpha_m^i r_i \times u_i^n \Big/ \left(\sum_i \rho_m \alpha_m^i (x_i - x_c)^2 + \sum_i \rho_m \alpha_m^i (y_i - y_c)^2\right). \tag{9}$$

The momentum density field is rigidified according to

$$\rho^{n+1} u_i^{n+1} = \sum_{m=0}^{np} \alpha_m^i \rho_m u_m^i, \tag{10}$$

where the forcing velocity of the fluid and solid phases are

$$u_m^i = \tilde{u}_i \qquad\qquad\qquad \text{if } m = 0 \quad \text{(fluid; unforced)}$$
$$= v_m^{n+1} + \Omega_m^{n+1} \times r_m^i \quad \text{if } m \ge 1 \quad \text{(solid).} \tag{11}$$

This step introduces a modification to the velocity field *after* the pressure correction, generally introducing some divergence into the velocity field. This could be remedied by iterating over steps F3) and P2). Our tests with a single such iteration in the case of a freely falling particle indicated negligible effect on the particle trajectories.

2.3 A constant density-based explicit volumetric forcing method "CEV"

For comparison, we have also implemented a method named Constant density-based Explicit Volumetric Forcing (CEV). This represents our best effort to implement Kajishima et al.'s [2] volumetric forcing idea to a constant density solver in a way that is stable and conserves momentum. In this implementation, **P1)** of Section 2.1 is followed by **F1)**, **F2)** and **F3)**, with however ρ_f replaced everywhere by ρ. This is followed by

P2/CEV) Set "target velocity" u_p^{n+1} within each particle based on most recent particle (angular) velocity

$$u_p^{n+1} = v_p^n + \Omega_p^n \times r_i^n, \tag{12}$$

evaluate the body force f_p^{n+1} required to impose target velocity in proportion to volume fraction α

$$f_p^{n+1} = \alpha^{n+1} \rho_f \frac{u_p^n + 1 - \tilde{\tilde{u}}}{\Delta t}, \tag{13}$$

and apply with particles

$$u^{n+1} = \tilde{\tilde{u}} + \frac{\Delta t f_p^{n+1}}{\rho_f} = (1 - \alpha)\tilde{\tilde{u}} + \alpha u_p^{n+1}, \tag{14}$$

update particle velocity by

$$M_p U_p^{n+1} = M_p v_p^n + \Delta t \left(\Delta M_p g - h^2 \sum_i f_p^{n+1} \right) \tag{15}$$

and analogously for angular velocity.

It would appear desirable at this point to update the velocity field u^{n+1} within the particle to match the rigid velocity field $u_p^{n+1} = v_p^{n+1} + \Omega_p^{n+1} \times r_i^{n+1}$, and indeed Kajishima et al. [2] did precisely that "at every grid point inside the particle ($\alpha = 1$)". However we have not found a simple way to perform such an update and still conserve momentum, even approximately, so no such adjustment is done here.

As a result, the updated particle velocity u_p does not affect the velocity field u until Equation (14) in **P2/CEV**, nearly a full step later. Consider a gravitational force that is applied, starting in step 1, to a particle initially at rest in still fluid. This will affect the velocity field inside the particle only in substep **P2/CEV** of step 2, and the velocity field outside the particle remains unaffected until step 3. This very loose

Table 1. Numerical and computational parameters in the simulation. Boundary conditions are 4 walls for Case D# and periodic in x for Case C. The last column reports the terminal Reynolds number observed in [6].

Case	Domain	x_c	y_c	d	ν	ρ_p/ρ_f	$\lvert g \rvert$	Re_s	Re_t
D13	$[0, 2] \times [0, 4]$	1.0	3.5	0.25	0.1	1.25	981	-	13.83
D310	$[0, 2] \times [0, 6]$	"	5.5	"	0.01	1.5	"	–	310.75
C	$[0, 8] \times [0, 4]$	4.0	2.0	1.0	0.01	1.0	0.0	40.0	–

nature of the fluid-particle coupling in our CEV scheme will strongly influence the particle trajectories, especially those presented in Section 3.1.[2]

3 Results for 2D test cases

We have implemented the above implicit and explicit schemes for cylindrical "particles" in two 2D test problems: a disk dropped in a quiescent fluid (Cases D#) and a single freely translating,though non-rotating, disk in Couette channel flow (Case C); conditions are summarized in Table 1. Further details of the test conditions are provided in the following subsections.

3.1 Circular disk dropped from rest in a quiescent fluid

In the first test problem we consider the motion of a rigid disk dropped from rest in an incompressible viscous Newtonian fluid. Parameters in our simulation, which match those reported in [6], are given in Table 1 (Cases D13 and D310). Terminal Reynolds number is defined by $Re_t = dV_t/\nu$, where d is circular disk diameter and V_t is terminal velocity. All Reynolds numbers given in Table 1 correspond to reference data.

At the initial state, a disk of diameter d is submerged in the fluid domain $[0, \text{Width}] \times [0, \text{Height}]$ at the location $[x_c, y_c]$ and everything is at rest. When dropped, the disk accelerates until its immersed weight balances the fluid drag force. Figure 1 shows histories of velocity, for a nominal half-width of the density fringe of $2h$ and grid spacings of $1/32$, $1/64$ and $1/128$ ($d/8$, $d/16$, $d/32$) calculated by the CEV and VIV methods together with the reference data [6] for Cases D13 and D310. The calculated terminal velocities are given in Table 2.

In Figure 1 the curves calculated by the VIV method are closer to the reference curve than those calculated by the CEV method during the initial acceleration phase. We believe that the way of calculating the particle target velocity in the CEV method is the main source of the discrepancy. Because the particle's target velocity used to force the velocity field has a one-time step lag, its acceleration is under-predicted.

---

[2] As explained in [3], it is conceptually sounder to replace M_p with the excess mass ΔM_p in the non-gravity terms of Equation (15), but the resulting scheme is only stable for density ratios greater than about 1.9 [11], which is greater than the test cases considered herein.

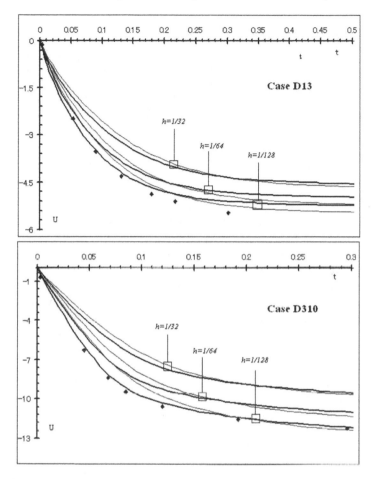

Fig. 1. Calculated velocity histories. Symbols are reference data in [6], thick curves are VIV calculation and thin curves are CEV calculation. Numbers indicate grid spacing.

3.2 A freely translating disk in Couette channel flow

In this test problem we simulate a single neutrally buoyant circular disk placed initially in the centerline of a Couette channel flow that is taken periodic in x. The channel dimensions, material properties and calculation parameters, which match those in [12], are summarized in Case C, Table 1. Re_s is shear Reynolds number defined by $Re_s = LV_w/\nu$ in where L is the shear gap between walls, and V_w is the wall velocity. The disk's initial velocity is set to the average of the two walls. The initial velocity field, including that within the circular disk, is set to the undisturbed velocity that would occur with no particle; the solver corrects this violently to satisfy the rigid motion inside the particle. Galilean invariance then dictates that the disk should continue thereafter at the initial velocity, and such a behavior was indeed observed when the walls moved in opposite directions at equal speeds; the disk remained in

Table 2. Comparison between our calculated terminal velocities with reference data [6] for Cases D13 and D310 at varying grid resolutions h.

Case	h	Terminal velocity V_t			Relative error (%)	
		Ref. [6]	CEV	VIV	CEV	VIV
D13	1/32		4.67	4.57	15.55	17.36
	1/64	5.53	5.19	4.98	6.15	9.95
	1/128		5.45	5.23	1.45	5.42
D310	1/32		10.28	9.96	17.29	19.87
	1/64	12.43	11.86	11.39	4.58	8.36
	1/128		12.71	12.34	-2.25	0.72

place. However, with the upper wall fixed and the lower one moving at a speed of 0.1 to the right, the disk's velocity was calculated to decrease and the disk migrated toward the fixed wall to an incorrect equilibrium position, as shown in Figure 2a.

Figure 2b shows the "lead velocity" of the particle, i.e. the difference between the particle's velocity and the undisturbed fluid velocity at the height of the particle center, as a function of x. For all cases, the particle initially leads the fluid, but soon starts to lag more and more until reaching a steady value. The early overshoot presumably results from the sudden adjustment to the initially uniform shear prescribed inside the particle. Normally, one would like to initialize simulations less violently, but the present method serves nicely to contrast the behavior of the two methods, as follows.

The peak lead velocities observed for VIV are less than 1/5 that for CEV at the same resolution, and the duration of the leading period, which is roughly independent of resolution, is more than twice as long for CEV as for VIV. It would appear that the response time of the former method is about twice that of the latter. This is not surprising; recall that in CEV there is a one step lag between update of particle velocity and the consequent forcing of the velocity field inside the particle domain, and an additional step before any influence is felt by the fluid velocity. More surprising than the relative durations are the absolute durations, which seem inexplicably long given that the particle is neutrally buoyant.[3]

As shown in Table 3, when increasing the grid resolution, the VIV method converges to the exact solution faster than the CEV method. At the finest resolution, the error of the CEV method is still quite large compared with that of the VIV method.

4 Conclusion

In the present paper, we have presented a Variable density Implicit Volumetric forcing (VIV) method, and compared with a Constant density Explicit Volumetric forcing (CEV) method [3], modified from [2], for a single disk: (1) dropped from rest

---

[3] The plots of lag velocity suffer from oscillations in x, of period 8 and perfectly aligned from one curve to another, which implicates our handling of the periodic boundary condition. Despite strenuous efforts, we have not been able to identify the cause of these oscillations.

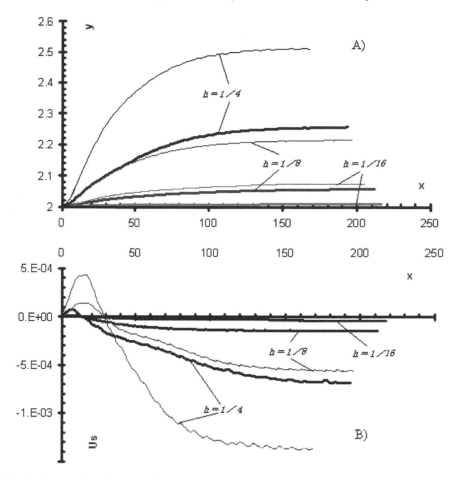

Fig. 2. Trajectories of a neutrally buoyant non-rotating disk in Couette flow (Case C) at different grid resolutions. Thin curves: CEV; thick curves: VIV. Numbers indicate grid spacing. Near-equilibrium part of trajectory for CEV at $h = 1/16$, not shown, is barely distinguishable from that for VIV at $h = 1/8$.

due to gravity at terminal Reynolds numbers of 13 and 310, and (2) in a Couette channel flow. The results for the dropped disks show that the VIV method is better than the CEV method in treating the particle's instantaneous acceleration. The inferior performance of the CEV method may result from the fact that the momentum of the coupled solid-fluid system is not conserved within one time step. The disk's target velocity is calculated by separate equations, so generally the net forcing added to the domain is not zero. This non-zero forcing is recorded and re-injected to the velocity field through Equation (15) at the *next* time step. The error due to this delayed re-injection is relatively small during one time step but its accumulated error is a potential source of non-physical behavior.

Table 3. Comparison between our calculated equilibrium positions of the moving disk for Case C at different grid resolutions h with the exact solution.

Case	h	Equi. position			Relative error		
		Exact	CEV	VIV	CEV(%)	VIV(%)	VIV/CEV
C	1/4	2.0	2.51	2.26	25.5	13.0	0.51
	1/8		2.22	2.06	11.0	3.0	0.27
	1/16		2.07	2.01	3.5	0.50	0.14
	1/32		2.01	2.00	0.50	0.00	0.00

Trajectories in the Couette flow revealed a lateral motion of the disk that was much greater for the CEV method. Such non-physical behavior would be especially unacceptable in the calculations of the highly sheared region near a particle bed that we are targeting.

Acknowledgments

We are grateful for discussions with Dr. T. Kajishima and Dr. M. Uhlmann. We thank Dr. N. Patankar for pointing out the prior use of momentum-neutral "rigidification" in his work [5]. This work was supported by a grant from the MEXT Ministry of Japan, and by the Enviromental Flows Research Group at Ritsumeikan University.

References

1. T. Kajishima, S. Takiguchi and Y. Miyake, 1999, Modulation and subgrid scale modeling of gas-particle turbulent flow, in *Recent Advances in DNS and LES*, D. Knight and L. Sakell (eds), Kluwer Academic Publishers, pp. 235–244.
2. T. Kajishima, S. Takiguchi, H. Hamasaki and Y. Miyake, 2001, Turbulence structure of particle-laden flow in a vertical plane channel due to vortex shedding, *JSME Int. J. Ser. B* **44**(4), 526–530.
3. H.V. Truong, J.C. Wells and G. Tryggvason, 2005, Explicit vs. implicit particle-liquid coupling in fixed-grid computations at moderate particle Reynolds number, FEDSM2005-77206, ASME FED Summer Meeting, June 19–23, 2005, Houston, TX
4. M. Uhlmann and A. Pinelli, 2006, Performance of various fluid-solid coupling methods for DNS of particulate flow, in *Proceedings of the IUTAM Symposium on Computational Multiphase Flow*, S. Balachandar and A. Prosperetti (eds), Springer, Dordrecht, pp. 215–223.
5. N.A. Patankar, 2001, A formulation for fast computations of rigid particulate flows, *Ann. Res. Briefs, Center for Turbulence Research*, pp. 185–196.
6. R. Glowinski, T.W. Pan, T.I. Hesla, D.D. Joseph and J. Periaux, 2001, A fictitious domain approach to the direct numerical simulation of incompressible viscous flow past moving rigid bodies: Application to particulate flow, *J. Comput. Phys.* **169**, 363–426.
7. A. Vikhansky, 2003, A new modification of the immersed boundaries method for fluid-solid flows: Moderate Reynolds numbers, *J. Comput. Phys.* **191**, 328–339.

8. N. Sharma and N.A. Patankar, 2005, A fast computation technique for the direct numerical simulation of rigid particulate flows, *J. Comput. Phys.* **205**, 439–457.

9. J.C. Wells, G. Tryggvason and H.V. Truong, 2004, A new algorithm for simulating dispersed solid bodies in turbulent flow, in *Memoirs of the Institute of Science and Engineering*, Ritsumeikan University, No. 62, Kusatsu, Japan, pp. 33–41.

10. G. Tryggvason, B. Brunner, A. Esmaeeli, D. Juric, N. Alrawashi, W. Tauber, J. Han, S. Nas and Y.J. Jan, 2001, A front tracking method for the computations of multiphase flow, *J. Comput. Phys.* **169**, 708–759.

11. M. Uhlmann, First experiments with the simulation of particulate flows, Technical Report No. 1020, CIEMAT, ISSN 1135-9420.

12. J. Feng, H.H. Hu and D.D. Joseph, 1994, Direct simulation of initial value problems for the motion of solid bodies in a Newtonian fluid. Part 2. Couette and Poiseuille flows, *J. Fluid Mech.* **277**, 271–301.

3D Unsteady Simulation of Particle Sedimentation towards High Regimes

G. Pianet, E. Arquis and S. Vincent

*Laboratoire TrEFlE, UMR 8508, site ENSCPB, 16 Av. Pey-Berland 33607, Pessac, France;
e-mail: gregoire.pianet@free.fr*

1 Introduction

This paper reports on an original computational method for simulating dynamics of liquid-solid mixtures. Our purpose is to validate fully-resolved unsteady simulations with a view to built confidence about the implicit tracking of fluid-solid interfaces, via a viscosity-based penalty method called the 1-Fluid method (1F). This method is generally used for simulating bubbles, drops and free-surface flows. For examples we refer the reader to [2, 5, 10, 11]). The 1 fluid method has been recently adapted by Caltagirone [3] who was inspired by volume penalty methods in fictitious domains and interface reconstruction. The term "1-Fluid" means that fluid and solid phases are considered as a single equivalent fluid in the sense of the Navier–Stokes equations. The equivalent fluid requires equivalent densities and viscosities depending on a phase function. The latter indicates whether the considered grid node belongs to fluid or solid phase, and drives the coupling of the Navier–Stokes equations with the interface transport equation. The following features make the 1F method very attractive for simulating particles in fluid: (i) using phase function allows to use computationally practical fixed Cartesian grids, (ii) the same set of equations is solved in both dispersed and continuous phases and the interface between the phases is only tracked implicitly; as a result the computational effort does not scale with the number of particles, (iii) the tensorial penalty method [15] used to ensure both incompressibility constraint in the fluid phase and undeformability constraint in the solid is just based upon two parameters defined from the characteristic flow scales, and (iv) for solving the velocity-pressure coupling, the augmented Lagrangian algorithm [6] is a stable and efficient technique.

The 1F method was tested successfully in academic configurations. The goal of this work is to study and extend the scope of our approach to flows dominated by inertia.

The global convergence and validation of the method have been carried out by comparison with PIV's measurements from Ten Cate et al. [4] concerning a sphere settling under gravity with Reynolds numbers based on particle diameter (Re_p) ran-

S. Balachandar and A. Prosperetti (eds), Proceedings of the IUTAM Symposium on Computational Multiphase Flow, 185–196.

ging from 1.5 to 31.9. Early results are presented in this paper for a sphere settling in a rectangular tank at $Re_p = 280$.

2 Numerical method

2.1 Principles of the 1-Fluid model

The global technique consists of adding extra-terms to Navier–Stokes equations so that local modifications of equations are induced through local viscosities and densities $\mu(x, y, z, t)$ and $\rho(x, y, z, t)$. In that sense, the 1F method has similarities with the fictitious domain approach of Glowinski et al. [7] and with the Immersed Boundary Method of Peskin [12].

Great advantages are found by using a single Cartesian grid and by introducing the phase function $C(x, y, z, t)$ that matches the multiphase topology of the flow. So the dispersed phase is characterized by $C = 1$ and the continuous one by $C = 0$. It is consequently possible to express the global properties of the mixture as function of C, it is said basically : $\mu = \mu_f(1-C) + \mu_p C$ and $\rho = \rho_f(1-C) + \rho_p C$, where μ is the dynamic viscosity, ρ the density. The subscripts f and p pointing out respectively the fluid and solid phases. The latter one being considered like a fluid, it becomes necessary to set its viscosity as $\mu_p \rightarrow \infty$ so as to ensure a solid behavior. The final equations set compiles the Navier–Stokes equations (1), an advection equation on the phase function (2), and the incompressibility constraint (3):

$$\rho\left(\frac{\partial \vec{u}}{\partial t} + (\vec{u}.\nabla)\vec{u}\right) = \rho\vec{g} - \nabla p + \nabla.(\mu(\nabla\vec{u} + \nabla^t\vec{u})), \qquad (1)$$

$$\frac{\partial C}{\partial t} + \vec{u}.\nabla C = 0, \qquad (2)$$

$$\nabla.\vec{u} = 0. \qquad (3)$$

2.2 Computational methodology

Solving Equation (2) requires transport schemes to handle correctly the high gradients located on liquid-solid interfaces. Consequently Volume of Fluid schemes (see the VOF-PLIC method of Youngs et al. [17]) were chosen and validated.

Concerning the resolution of the Navier–Stokes equations, and in particular the velocity-pressure coupling, the method of the Augmented Lagrangian [6] based on an Uzawa optimization method has been adapted.

In order to satisfy the fluid incompressibility this technique have been designed for multiphase flows by Vincent et al. [16]. It has been generalized to ensure the solid phase undeformability. The augmented Lagrangian-like methods are based on the principle of adding specific implicit terms to the conservation equations. By this way some specific constraints such like incompressibility and solid behavior could

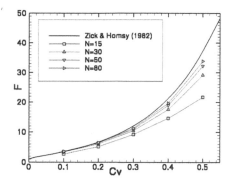

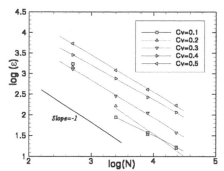

Fig. 1. Non-dimensional drag force F Vs. solid volume fraction C_v at increasing spatial resolution N.

Fig. 2. Convergence rate of the 1F method as a function of N.

be imposed locally. For a complete discussion about the formulation, the implementation and the validation of this method, we refer the reader to [15, 16]. Equations (1) to (3) are approximated by means of a finite volume method on a staggered mesh. The momentum equation is linearized with respect to inertia. The associated time discretization is fully implicit including inertial, augmented Lagrangian and viscous terms. All of the terms in Equation (1) are spatially discretized via a second order accurate centered scheme. We use a second order accurate Gear scheme for discretization in time. The algebraic system, resulting from the discretization of the various equations, is inverted by the iterative method BiCG-STAB.

2.3 Validations and numerical settings

Several academic configurations were used to set the numerical parameters and check the numerical consistency of the 1F method.

Convergence in space

Simulations of Stokes flow through simple cubic arrays of fixed spheres are compared to the results from Zick and Homsy (see [13]). The non-dimensional drag force F of the array is modulated through changes in the solid volume fraction C_v.

The simulation domain is a periodic box and the sphere stands on the box center. With N defined as the number of points along the cube side, Figures 1 and 2 demonstrate the spatial convergence of the 1F method. The latter is shown to be first order accurate in space, but the $N = 15$ case show that a good physical consistency is still observed for very low resolutions.

Numerical settings for liquid-solid coupling

Concerning moving particles (see [15]), the deformation rate must tend to zero as a rigid motion is constrained in the inclusion. The main difficulty lies in setting a con-

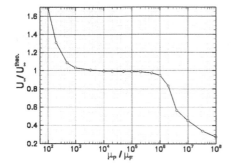 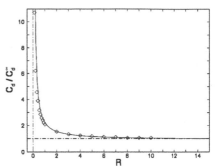

Fig. 3. Impact of viscosity ratio k on the non-dimensional settling velocity U_∞/U_∞^{th} of a sphere.

Fig. 4. Dimensionless drag coefficient Vs. particle/wall gap, 1F method(–), Theory [1] ($\Diamond$).

venient viscosity μ_p in the dispersed phase which does not alter the flow solution. This paragraph details the impact of the solid to liquid viscosity ratio $k = \mu_p/\mu_f$ on the coupling between phases. Theoretical corrections of terminal settling velocity for a sphere sedimenting in infinite cylindrical tanks are used (Haberman exact theory [8]). Three-dimensional simulations of this case are performed and the results are depicted in Figure 3 where different values of U_∞/U_∞^{th} are obtained through changes in the ratio k. It is made clear that there is a range $k \in [10^3, 10^6]$ where the liquid-solid coupling is efficient. For $k \in [1, 10^3]$, the particle is subjected to strong deformations. For viscosity ratios higher than 10^6 (with a fixed number of iteration steps for both the linear system inversion and the augmented Lagrangian), increasing discrepancies are found. This is a numerical artifact due to the deterioration of the linear system conditioning subjected to the viscosity ratio. For $k > 10^6$, accurate results require nonlinearly-increasing costs that make such simulations impossible to lead in practice.

Particle-wall interactions

In this section, our goal is to check the lubrication effects involved in the motion of a sphere perpendicular to a plane wall. We use well-known theoretical results[1] for which the non-dimensional drag coefficient C_d/C_d^∞ should scale as $1/R = 1/(y_c/a - 1)$, where a is the sphere radius and y_c its vertical coordinate. For reducing computational costs, the flow was simulated in an axisymmetric domain. The sphere is initially placed 10 diameters away from the top and bottom walls and 20 diameters from the side wall. Concerning spatial resolution, approximatively 18 grid points per diameter are set here. A physical analysis is done to obtain $Re_p = 0.01$ in the steady state for an unbounded fluid. In the present simulation, the error in the terminal settling velocity with respect to Stokes' velocity does not

---

[1] Limited to creeping regimes and a single unlimited wall, see Brenner [1].

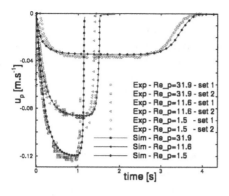
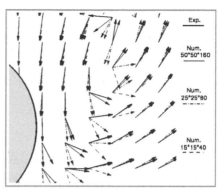

Fig. 5. Particle time series $U_p(t)$, 1F simulations, experiments from Ten Cate et al. [4]. $Re_p = 1.5, 11.6, 31.9$.

Fig. 6. Spatial convergence of the velocity direction fields. 1F method ($d_p = 15, 7.5, 4.5$ grid points), PIV fields [4].

exceed 1%. As the dimensionless gap R tends to zero, Figure 4 shows that the normalized drag is very well reproduced. Then some significant over-prediction appear as $R < 0.2$. This means that while the gap R exceeds approximatively two grid points, no sub-grid lubrication model is required which is quite a good result for a DNS method.

3 Simulations

In Section 3.1, our method is compared to experimental data from Ten Cate et al. [4] related to the case of a single sphere settling under gravity at three different particle Reynolds numbers, $Re_p = 1.5, 11.6, 31.9$. Section 3.2 deals with the simulation of a sphere settling at $Re_p = 280$, with a view to reproduce the experimental data from Mordant et al. [14].

3.1 Simulations at $Re_p \in [1.5, 31.9]$

In this section we use a full set of PIV experiments reported by courtesy of Ten Cate and Derksen [4]. Designed for the validation of Lattice-Boltzmann direct numerical simulations, a wall-bounded geometry is set, and the particle containment (0.15) is relatively important. The container sizes $0.1 * 0.1 * 0.16$ [m^3], the latter dimension being parallel to the gravity direction. A sphere of radius $a = 0.0075$ [m] and density $\rho_p = 1120$ kg/m^3 is initially placed at mid-distance from the spanwise walls. The gap separating sphere's bottom from the container's is initially set to 8 diameters. PIV measurements are performed in a slice placed at mid-distance from the two opposite side-walls. Data are available for three flow regimes representing $Re_p = 1.5, 11.6, 31.9$, and corresponding respectively to dynamic viscosities $\mu_f = 0.373, 0.113, 0.058$ [Pa.s] and fluid densities $\rho_f = 970, 962, 960$ kg/m^3.

Concerning the simulations, a three-dimensional domain is defined. A way to reduce computational costs lies in using the two perpendicular symmetry planes expected in the flow, according to the moderate particle Reynolds numbers. The other boundaries are modelled by no-slip conditions. The particle time series plotted in Figure 5 denote that simulations match the experiments as well for the terminal settling velocity as for transient velocity.

Numerical consistency

Convergence in time and space have been checked for $Re_p = 31.9$. The convergence in time was found to be first-order accurate. We use second-order time schemes but the impact of the augmented Lagrangian method causes the order to decrease. For particle relaxation time $T_a = 2(\rho_p + .5\rho_f)a^2/9\mu_f$ and time step Δt, it has been found that the ratio $T_a/\Delta t$ must be $\mathcal{O}(10^3)$ for well-resolved transient behavior. Figure 6 has been obtained by plotting on the same mesh the experimental velocity field then the interpolation of three simulated fields with spatial resolutions respectively set at 160.50.50, 80.25.25 and 48.15.15 grid points. It is shown that in the low-sheared zones the results are physically consistent even with extremely weak resolutions in the particle (i.e. $d_p = 4.5$ grid points). In high-sheared zones it is shown that the simulation accuracy improves monotonically toward the experimental field. Following simulations have been performed with $d_p = 15$ grid points.

Direction and magnitude of the velocity field

Figure 7 depicts slices of the flow field at particle Reynolds numbers 1.5 and 31.9 respectively (on each of the six pictures, the left part is numerical, the right one is experimental). By representing the flow field direction (vectors) and the normalized velocity magnitude (contour levels from 0 to 100% of the maximal settling velocity in an unbounded medium), one must notice that the fluid and particle behaviors are correctly predicted.

Far from the sphere, the PIV fields are dominated by experimental noise which explains discrepancies in the comparison of velocity directions. Time series in fluid are also checked on a monitor point situated one diameter away from the bottom wall and one diameter away from the tank vertical axis. It is roughly the point where the recirculation next to the sphere is passing through during the deceleration of the sphere. Figure 8 shows that this results in strong velocity gradients. Here is shown that these gradients are accurately predicted by the 1F method.

3.2 Simulations at $Re_p = 280$

In this section we are interested to reproduce the motion of a sphere settling at a particle Reynolds number $Re_p = 280$ and a Stokes number $St = Re_p\rho_p/9\rho_f = 240$. The reference we use lies in one of the experimental results[2] from Mordant et

---

[2] Experimental data: tank of dimensions $H * D * W = 0.75 * 1.1 * 0.65$ [m] filled with water; particle: steel bead $\rho_p = 7710$ [kg/m³], $u_\infty = 0.316$ [m/s].

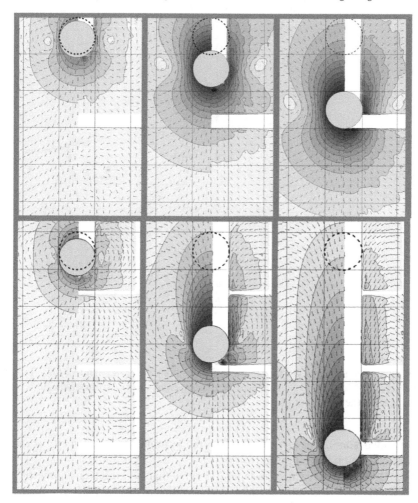

Fig. 7. Comparison of 1F simulation and PIV slices at specific times. Re_p numbers are respectively 1.5 and 31.9 from top to bottom. Times are respectively $t \simeq 0.1$, $t \simeq 0.5s$ and $t \simeq 1s$ from left to right. On each of the 6 pictures, the left part is numerical, the right one is experimental. Contour levels of normalized velocity magnitude $|u|/u_\infty$, direction of velocity field.

al. [14]. The effects of the containment on particle's motion must be negligible, so the computational domain is taken as a box of dimensions $H * D * W = 0.064 * 0.004*0.004$ [m]. Simulations are performed here using the full sphere. The temporal resolution is set as $T_a/\Delta t = 5000$. Three simulations have been performed with increasing resolutions in space successively defined as $d_p = 7$, 8 and 10 grid points. There is no significant changes between $d_p = 8$ and 10. At increasing resolutions, the convergence of our method to experimental data is made clear in Figure 9 (see

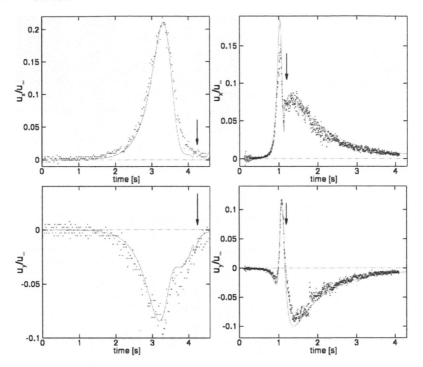

Fig. 8. Fluid time series for normalized velocity components u_x/u_∞ and u_y/u_∞. From left to right: $Re_p = 1.5$ and 31.9. 1F (lines) & experimental (dots). Arrows indicate when the particle reaches the bottom of the tank.

the magnified zone). The error realized on the terminal settling velocity is roughly 2% with our method (see Figure 10).

The flow fields represented in Figure 11 show that the initial axisymmetric topology is not broken. Regarding the high density ratio ($\rho_p/\rho_f = 7.71$), our results can be evaluated by assuming the analogy with fixed-sphere data from literature. The flow characteristics such as the recirculation size or the maximum velocity in the fluid are found consistent with the literature. But at $Re_p = 280$, we should obtain a transition from steady planar-symmetric to unsteady planar-symmetric topologies. So we try here to detect such a transition at a lower level, so our investigation is focused on the out-of-plane velocity field $V_z(x, y, 0)/u_\infty$ depicted in Figure 12 with a zero-centered logarithmic scale. Different zones appear successively as coherent 'V-shape' structures.

This oscillating behavior makes it possible to evaluate the Strouhal number which has been found to be of the same order as those reported in the literature. It is made clear that much more simulations are required for being confident about the results, specially concerning the particle containment, suspected here to damp the growing oscillations.

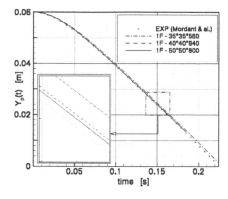

Fig. 9. Particle trajectory, $Re_p = 280$, $d_p = 7$ (dash-dotted line), $d_p = 8$ (dashed line), $d_p = 10$ (plain line), Experiment [14] (symbols).

Fig. 10. Particle velocity, $Re_p = 280$, 1F method $d_p = 10$ (plain line), Experiment [14] (line with symbols).

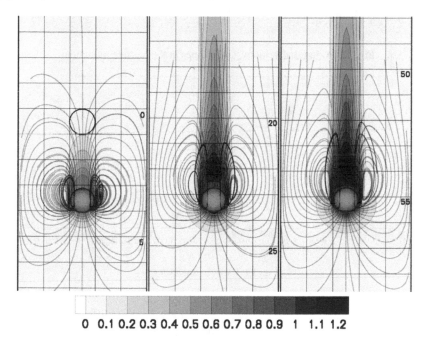

0 0.1 0.2 0.3 0.4 0.5 0.6 0.7 0.8 0.9 1 1.1 1.2

Fig. 11. Flow field slices at times $t = 0.028, 0.091, 0.175$ [s] from left to right. Streamlines and contour levels for $|u|/u_\infty$. The indexes refer to the dimensionless vertical coordinate $(y - y_p(0))/d_p$ and the black circle to the release position.

Recent parallelization efforts has opened new perspectives for investigating this problem. The particle containment is now reduced from 0.2 to 0.125, and the mesh size is $1000 \times 80 \times 80$ with $d_p = 10$ points across the particle diameter. Lateral

-5.0E-04 -5.0E-05 -5.0E-06 -5.0E-07 -5.0E-08 -5.0E-09 .0E+00 1.0E-08 5.0E-07 5.0E-06 5.0E-05 5.0E-04

Fig. 12. Out-of-plane velocity $V_z(x, y, 0)/u_\infty$.

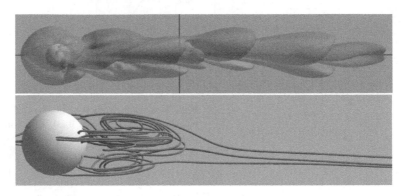

Fig. 13. Azimuthal velocity, streamlines in the particle referential.

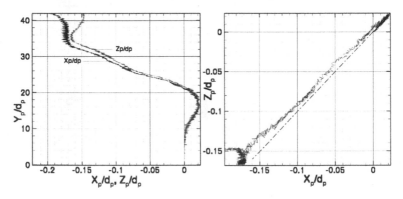

Fig. 14. Particle trajectory.

walls are replaced by periodic boundary conditions. Results differ sensitively from preceding simulations as the sidewall effects are reduced. Very similar 'V-shaped' structures are plotted in 3D in Figure 13 but the wake action on the particle behavior is now distinct from numerical noise. It can be seen in Figure 14 that a very weak lateral deviation occurs. As the regime increases, the particle trajectory decomposes into three successive stages : vertical, oblique, oblique and oscillating. In the range $Re_p \in [270, 300]$, very similar behavior was shown in a recent paper from Jenny et al. [9]. During the oblique stage, a plane distinct from principal directions is selected.

Related streamlines are plotted in the referential moving with the particle. At long times the vertical velocity slowly decays with no visible oscillations and shows discrepancies with the experimental terminal velocity. Clearly much more work needs to be done at this level. Further investigations will require more points for defining the particle and its boundary layer. Nevertheless, numerical experiments leaded with the 1F method are qualitatively consistent with the actual knowledge about the onset of instabilities.

4 Conclusions

We have described the 1F method as an efficient way to simulate liquid-solid mixtures. Designed for being totally implicit, the method results in excellent numerical stability properties. Several benchmarks and fully-resolved simulations demonstrated a good physical consistency. We have shown that the accuracy level remains unexpectedly high for very low spatial resolutions. In spite of important requirements in terms of computing time and memory inherent to the present DNS methods, its particular advantage is that it uses fixed grids and that in the case of several inclusions of equivalent densities, the computing time and the amount of particles are independent.

References

1. Brenner H., 1961, The slow motion of a sphere through a viscous fluid towards a plane surface, *Chem. Eng. Sci.* **16**, 242–251.
2. Bunner, B. and Tryggvason, G., 2003, Effect of bubble deformation on the properties of bubbly flows, *J. Fluid Mech.* **495**, 77–118.
3. Caltagirone, J.P. and Vincent S., 2001, Sur une méthode de pénalisation tensorielle pour la résolution des équations de Navier–Stokes, *C.R. Acad. Sci. Paris* **329**(2b), 607–613.
4. Ten Cate, A., Nieuwstad, C.H., Derksen, J.J. and Van den Akker, H.E.A., 2002, Particle imaging velocimetry experiments and lattice-Boltzmann simulations on a single sphere settling under gravity, *Phys. Fluids* **14**(11), 4012–4025.
5. Enright, D., Fedkiw, R., Ferziger, J. and Mitchell, I., 2002, A hybrid particle level set method for improved interface capturing, *J. Comput. Phys.* **183**, 83–116.
6. Fortin, M. and Glowinski, R., 1982, *Méthode de lagrangien augmenté. Application à la résolution numérique des problèmes aux limites*, Collection méthodes mathématiques de l'informatique, Dunod, Paris.
7. Glowinski, R., Pan, T.W., Hesla, T.I., Joseph, D.D. and Periaux, J., 2001, A fictitious domain approach to the direct numerical simulation of incompressible viscous flow past moving rigid bodies: Application to particulate flow, *J. Comput. Phys.* **169**(1), 363–427.
8. Happel, J. and Brenner, H., 1983, *Low Reynolds Number Hydrodynamics*, Kluwer Academic Publishers, Dordrecht.
9. Jenny, M., Dusek, J. and Bouchet, G., 2004, Instabilities and transition of a sphere falling or ascending freely in a Newtonian fluid, *J. Fluid Mech.* **508**(1), 201–239.
10. Josserand, C. and Zaleski, S., 2003, Droplet splashing on a thin liquid film, *Phys. Fluids* **15**, 1650–1657.

11. Liovic, P., Lakehal, D. and Liow, J.G., 2004, LES of turbulent bubble formation and breakup by use of interface tracking, in *Direct and Large-Eddy Simulation V*, Proceedings of DLES-5, R. Friedrich, B.J. Geurts and O. Métais (eds), ERCOFTAC Series, Vol. 9, Kluwer Academic Publishers, Dordrecht.
12. Peskin, C.S., 1977, Numerical analysis of blood flow in the heart, *J. Comput. Phys.* **25**, 220–252.
13. Hill, R.J., Koch, D.L. and Ladd, A.J.C., 2001, The first effects of fluid inertia on flows in ordered and random arrays of spheres, *J. Fluid Mech.* **448**, 213–241.
14. Mordant, N. and Pinton, J.F., 2000, Velocity measurement of a settling sphere, *Eur. Phys. J., Ser. B* **18**, 343–352.
15. Randrianarivelo, N., Pianet, G., Vincent, S. and Caltagirone, J.P., 2005, Numerical modelling of the solid particles motion using a new penalty method, *Int. J. Numer. Meth. Fluids* **47**, 1245–1251.
16. Vincent, S., Caltagirone, J.P., Lubin, P. and Randrianarivelo, N., 2004, An adaptive augmented Lagrangian method for three-dimensional multi-material flows, *Comput. & Fluids* **33**, 1273–1289.
17. Youngs, D.L., Morton, K.W. and Baines, M.J., 1982, Time-dependent multimaterial flow with large fluid distortion, in *Numerical Methods for Fluid Dynamics*, Academic Press, New York.

Microstructural Effects in a Fully-Resolved Simulation of 1,024 Sedimenting Spheres

Z.Z. Zhang[1], L. Botto[1] and A. Prosperetti[1,2]

[1] Department of Mechanical Engineering, Johns Hopkins University,
Baltimore, MD 21218, USA; e-mail: zhang@jhu.edu, lorenz o@jhu.edu, pros peretti@jhu.edu
[2] Faculty of Applied Science and Burgerscentrum, University of Twente, 7500 AE Enschede,
The Netherlands

Abstract. The results of a fully-resolved simulation of 1,024 particles settling under gravity in a periodic domain are described and analyzed. The particle volume fraction is about 13% and the single-particle terminal Reynolds number is about 10. Collisions are modelled as completely elastic. The results show that the formation of nearly-horizontal particle pairs is an important phenomenon which affects the mean settling velocity as well as the velocity fluctuations.

1 Introduction

Most of the substantial literature on fluid flow with suspended particles is based on an Eulerian–Lagrangian treatment of point particles. While much has been learned from these studies, there are many situations which cannot be modelled in this way, such dense systems, particles larger than the smallest flow scales, and liquid-solid systems. For all these situations it is necessary to account for the finite size of the particles. Furthermore, when the ratio of the fluid to the particle density is not very small, considerable uncertainties exist as to the parameterization of the fluid forces, which may introduce errors of uncontrollable and unknown magnitude in the results.

For these reasons, a considerable effort has been devoted to developing methods capable of solving the coupled fluid-particle equations without approximations other than those inherent in the discretization of the continuum equations. In addition to the papers included in this volume, recent representative work is that of Patankar et al. [1], Singh et al. [2] and Dong et al. [3], who used a fixed finite-element mesh, and that of Johnson and Tezduyar [4], who used a moving finite-element mesh. Several finite-difference methods have also been developed. Interesting examples are given in [5–12]; one may also cite the Cartesian grid method of Udaykumar et al. [13]. A great impulse to this type of studies has been given by the advent of the lattice-Boltzmann method [14–16], which has also been combined with the immersed boundary approach in [17, 18].

Each one of these methods has strengths and weaknesses the analysis of which is the object of current research. In this study we briefly describe another method,

S. Balachandar and A. Prosperetti (eds), Proceedings of the IUTAM Symposium on Computational Multiphase Flow, 197–206.

PHYSALIS, and illustrate its application to the sedimentation of 1,024 spherical particles in a viscous fluid. A significant advantage of this method, which is more fully described in [19–21], is the spectral decrease of the error as the number of degrees of freedom assigned to each particle increases. As the method is based on a fixed grid, the computational effort is determined mostly by the extent of the computational domain rather than by the number of particles. Furthermore, the force and torque on each particle are found directly as a by-product of the calculation, thus avoiding the difficulty encountered with some other methods which require extrapolation to obtain the fluid stress distribution on the particle. These advantages come at the price of the restriction of the method – at least as currently formulated – to spherical particles.

2 Mathematical-numerical method

The analytical basis and numerical implementation of the method used in this work are described in several other publications [19–21] and a brief description will be sufficient here.

Due to the no-slip condition, at the surface of each particle the fluid velocity field has a rigid-body character. By continuity, the deviation of the fluid velocity from this rigid-body velocity field in the immediate neighborhood of the particle will be small and its square can therefore be neglected. On the basis of this remark, after some transformation, it is possible to introduce an auxiliary divergenceless velocity field which to an excellent approximation satisfies the Stokes equations very near the particle surface. It should be stressed that this procedure is not a linearization of the Navier–Stokes equation about 0, as the original Stokes equation, but about the rigid-body motion of each particle. Furthermore, the approximation is used only up to a distance of the order of one mesh size from the particle surface and, therefore, the magnitude of the error can be controlled by controlling the discretization.

The general solution of the Stokes equations near a spherical boundary was given by Lamb [22, 23] in terms of three scalar potentials P, Φ, X. Each potential is a harmonic function and can therefore be expressed as the superposition of solid harmonics. For example, $P = \sum_1^\infty (p_n + p_{-n-1})$ consists of an infinite series of harmonics both regular, p_n, and singular, p_{-n-1}, at the particle center. The generic regular harmonic is written as

$$p_n = \left(\frac{r}{a}\right)^n \sum_{m=0}^n \left[P_{nm} \cos m\varphi + \tilde{P}_{nm} \sin m\varphi \right] P_n^m (\cos\theta), \tag{1}$$

where P_{nm} and $\tilde{P}_{nm}$ are dimensionless coefficients, P_n^m is an associated Legendre function, and r, θ, and φ are spherical coordinates centered at the particle center. The other harmonics ϕ_n, χ_n entering the definition of Φ and X are written in a similar way with other dimensionless coefficients. The regular harmonics represent the "incident" flow and the singular ones the disturbance induced by the particle. The coefficients of the singular harmonics can be readily related to those of the regular harmonics by the boundary conditions at the particle surface. The pressure and vorticity fields can also be represented in a similar fashion in terms of the same set of coefficients.

At the same time as we consider these local spectral representations of the solution valid near each particle, we construct by finite-differences a solution of the full Navier–Stokes equations in the interstitial fluid. The advantage of this two-fold representation of the fields is that, rather than imposing the boundary conditions on the finite-difference solution at the particle surfaces, we demand that the finite-difference solution match the local solution at the points of a suitable *cage* of nodes surrounding each particle. The nodes are part of the finite-difference grid and, therefore, the geometric complexity that arises from the mismatch between the regular grid and the particle boundaries is avoided.

The finite-difference grid is constructed to cover the entire domain, irrespective of the presence of the bodies. A standard staggered grid arrangement is used with pressure at cell centers and velocities at the midpoint of cell sides. Each particle is surrounded by a cage of cells with the respective grid nodes for velocity, pressure, and vorticity.

For the solution of the Navier–Stokes equations we use methods of first- or second-order accuracy in time. The former one is described in [19], while the latter one, based on the work of Brown et al. [24], is described in [21]. Spatial discretization is second-order accurate. The Poisson equation(s) are solved by a multigrid method.

After truncating the infinite summations in Lamb's potentials to a finite number of terms N_c, we calculate the flow by an iterative procedure. By matching the pressure and vorticity of a provisional estimate of the fields at the cage nodes, we obtain a first estimate of the coefficients P_{nm} etc. for each particle. These coefficients are then used to calculate the velocity at the cage nodes and the flow fields are updated by solving the discretized Navier–Stokes equations using these velocities as boundary conditions. The process is then repeated until convergence.

3 Sedimentation velocity

We conducted the present simulation in a domain of size $(32a)^3$, where a is the particle radius, discretized with 128^3 cells. With 1,024 spheres, this translates into a volume fraction β of $\pi/24 \simeq 13.1\%$. The spheres were 50% denser than the fluid and the Reynolds number $Re = 2aw_t/\nu$ based on the terminal velocity w_t of a single particle in an unbounded domain, was approximately 10.1. Particle collisions were treated as fully elastic. The infinite summations of the Lamb potentials were truncated to $N_c = 1$, which corresponds to retaining 10 coefficients per particle. First-order-accurate time stepping was used.

The initial configuration of the particles was generated starting from a regular arrangement and subjecting each sphere to a large number of random displacements as described, e.g., in [25, 26]. The spheres were released from rest.

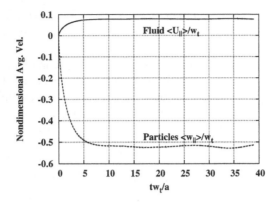

Fig. 1. Normalized average vertical velocity of the fluid (upper line) and particles in a frame of reference in which the mean vertical volumetric flux vanishes.

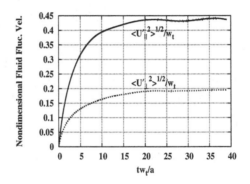

Fig. 2. Evolution of the fluid velocity fluctuations vs. nondimensional time.

The time development of the mean velocities of fluid and particles in a frame of reference in which the total volumetric flux vanishes is shown in Figure 1. Here and in the following the subscript $\parallel$ refers to velocity components in the positive z-direction, which is parallel and opposite to the direction of gravity, and the angle brackets denote the volume average. The velocities are normalized by the single-particle sedimentation velocity w_t.

At steady state, the Richardson–Zaki correlation [27] predicts that the relative particle-fluid velocity should be reduced by a factor $(1-\beta)^N$, with $N = 4.45\,Re^{-0.1}$, with respect to the single-particle value. With $Re = 10.1$ we have $N = 3.531$ and $(1-\beta)^N \simeq 0.61$. According to more recent analyses (e.g. [28]), this value should be reduced by a factor k estimated between 0.8 and 0.9. Upon taking $k = 0.85$, we find $\langle w_\parallel \rangle = 0.52$ in very good agreement with the computed value. The corresponding single-particle Reynolds number is approximately 5.2.

The time evolution of the normalized velocity fluctuations is shown in Figures 2 and 3 respectively. For the fluid, the vertical fluctuations are defined as

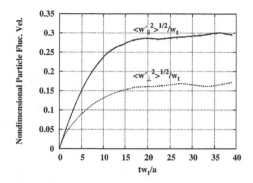

Fig. 3. Evolution of the particle velocity fluctuations vs. nondimensional time.

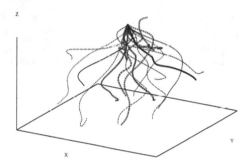

Fig. 4. Sample particle trajectories; the paths are shown as departing from a common origin.

$$\langle U'_{\parallel} \rangle = \langle (U_{\parallel} - \langle U_{\parallel} \rangle)^2 \rangle^{1/2} \tag{2}$$

while the horizontal fluctuations are defined as

$$\langle U'_{\perp} \rangle = \left[\frac{1}{2} \langle (U_x - \langle U_x \rangle)^2 + (U_y - \langle U_y \rangle)^2) \rangle \right]^{1/2}. \tag{3}$$

The volume average is calculated by summing over the fluid nodes and dividing by their number. The definitions of the particle velocity fluctuations are analogous. These results are comparable to those reported in [29] for the case where the distance between the walls included in that simulation is large. These velocity fluctuations exhibit a marked anisotropy, with fluctuations in the vertical direction (upper lines) about twice as large as those in the horizontal direction. A similar result was reported in [30] for a comparable single-particle Reynolds number, but at smaller concentrations.

A small number of randomly chosen sample particle trajectories is shown in Figure 4. The trajectories are shown departing from a common origin to illustrate the diffusive nature of the particle motion. The complexity of these paths is a clear illustration of the strong interactions among the particles. It is interesting to notice that one particle, evidently trapped in a relatively fast upward-moving fluid mass, moves upwards against the direction of gravity.

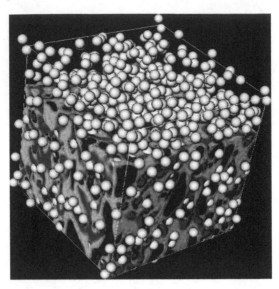

Fig. 5. Snapshot of the particle arrangement and fluid kinetic energy at nondimensional time $tw_t/a = 16$.

4 Microstructural effects

The results shown in Figures 1 to 3 exhibit some quite noticeable fluctuations which can be connected with the evolving microstructure of the system.

Figure 5 shows a snapshot of the settling particles taken at the nondimensional time $tw_t/a = 16$. Fluid from the top 1/4 of the computational domain has been removed to show the particle arrangement. A striking feature of this image is the presence of several clusters of two or more particles which may be observed everywhere in the domain.

A first quantification of this feature may be found by calculating the distance of the center of each particle to that of its nearest neighbor and averaging over all the particles. This quantity, normalized by the particle radius, is plotted in Figure 6 as a function of the nondimensional time. For the initial condition used, the particles are already close at the beginning of the simulation, but their distance decreases further with time stabilizing around $r/a = 2.3$, suggesting a tendency to cluster. This effect may be attributed to the energy loss during collision. Even though collisions are modelled elastically, there is an energy loss due to viscosity as the particles approach and separate before and after contact. A similar clustering was observed in the Stokes flow simulation of Wylie and Koch [31] without a mean relative flow at a particle volume fraction of 20%.

While these considerations explain the large number of close pairs observed in Figure 5, they do not account for the fluctuations of the mean values. For this issue we turn to the study of the orientation factor which we define as (see e.g. [32])

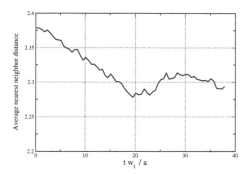

Fig. 6. The average distance to the nearest neighbor vs. nondimensional time.

$$A(t; r_{\min}, r_{\max}) = \frac{1}{N_p(N_p - 1)} \int_{r_{\min}}^{r_{\max}} \int d\Omega \sum_{\alpha=1}^{N_p} \sum_{\beta=1, \beta \neq \alpha}^{N_p} \delta(\mathbf{r} - \mathbf{r}_{\alpha\beta}) P_2(\cos \theta).$$

(4)

Here $N_p = N_p(r_{\min}, r_{\max})$ is the number of particles in the shell $r_{\min} \leq r \leq r_{\max}$, Ω the solid angle, $\mathbf{r}_{\alpha\beta}$ the separation vector between particles α and β, and θ the angle to the vertical. The radial integration limits $r_{\min}$ and $r_{\max}$ will be chosen so as to explore different ranges. Due to the form of the second-order Legendre polynomial $P_2 = \frac{1}{2}(3 \cos^2 \theta - 1)$, A will be positive if there is a prevalence of particle pairs with $\cos \theta \geq 1/\sqrt{3}$ (corresponding to an angle smaller than approximately 54.7°) and negative in the opposite case; A would vanish for a completely isotropic system.

The time evolution of A is plotted in Figure 7 for different values of $r_{\min}$ and $r_{\max}$. To investigate the orientation of closely spaced particles we take $r_{\min} = 2$ and $r_{\max} = 2.5$ (solid line). For particles with an intermediate spacing we take $r_{\min} = 2.5$ and $r_{\max} = 4$ (dashed line) while, for widely separated particles, $r_{\min} = 4$ and $r_{\max} = 6$ (dotted line). All three lines start close to 0, which indicates that no significant asymmetry is present in the initial conditions. The line for widely separated particles (dotted) remains close to zero for all times, indicating that such particle pairs do not exhibit any special orientation. In marked contrast, the line for close particles (solid) quickly becomes negative and, while fluctuating, remains negative for the entire simulation. Particle pairs at an intermediate distance (dashed) also exhibit a weak horizontal preferential orientation.

The curve for close particle pairs ($2 \leq r/a \leq 2.5$) presents two peaks at nondimensional times around 17 and 33. At these instants, then, the probability of vertical arrangements increases, even though horizontal pairs are still more frequent. From Figure 1 it is seen that, at these same times, the mean particle velocity exhibits minima (i.e. maxima in absolute value). To establish a correlation between these two observations, it is instructive to consider the mean vertical velocity of "horizontal" and "vertical" particle pairs.

Figure 8 shows the mean center of mass of velocity of such pairs. For the purpose of this figure, only particle separations in the range $2 \leq r/a \leq 2.5$ were considered. Pairs with a separation vector inclined between 0 and $\pi/6$ or between $5\pi/6$ and

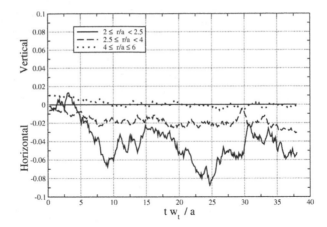

Fig. 7. Orientation factor defined in (4) vs. nondimensional time.

π were defined as "vertical", while pairs with a separation vector inclined between 0.45π and 0.55π were defined as "horizontal". These choices define two regions with the same volume in the three-dimensional configuration space. It is seen that, on average, vertical doublets fall appreciably faster than horizontal ones, with a velocity difference of the order of 20%. While this phenomenon is well known in the case of two particles (see e.g. [33]), our results prove that it also occurs in a concentrated dispersion. From these results it is evident that a temporary increase in the number of vertically-oriented particle pairs will lead to an increase of the mean particle settling velocity, as observed in Figure 1.

The minimum value of A for close pairs is approximatively -0.09, with oscillation around -0.05. Perfect horizontal orientation would correspond to $A = -0.5$. This suggests that while preferential orientation is a weak phenomenon in the present sedimentation process, it is large enough to explain the observed mean-velocity fluctuations.

5 Summary and conclusions

We have described some aspects of a fully resolved simulation of 1,024 particles settling under gravity in a Newtonian fluid and presented some results on the spatial structure of the system and its effects on the mean phase velocities.

In agreement with experiment (e.g., [30]), we have found a marked anisotropy in the velocity fluctuations of both phases. An analysis of the average distance between neighboring particles reveals that the particles tend to cluster due to hydrodynamic interactions. We have shown that the weak maxima in the particle mean settling velocity correlate with an increase in the number of vertically-oriented particle pairs, which fall faster than horizontal ones. The analysis of the orientation of particle

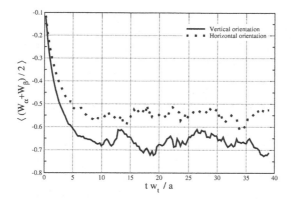

Fig. 8. Center-of-mass vertical velocity vs. nondimensional time for vertical and horizontal particle pairs.

pairs reveals a preference for a horizontal arrangement when the distance between the surfaces of the two particles is less than a particle radius.

These results have been obtained with a newly developed computational method, PHYSALIS, which has lived up to the expectations of robustness and efficiency suggested by earlier calculations [21]. While several aspects of the method can be improved (e.g., a more faithful representation of the flow near colliding particles, acceleration of the iterative procedure, more flexible parallelization), the code can already generate a wealth of interesting results.

Acknowledgments

Support from NSF under grant CTS-0210044 and DOE under grant DE-FG02-99ER14966 is gratefully acknowledged.

References

1. Patankar, N., Singh, P., Joseph, D., Glowinski, R. and Pan, T.W., 2000, *Int. J. Multiphase Flow* **26**, 1509–1524.
2. Singh, P., Hesla, T. and Joseph, D., 2001, *Int. J. Multiphase Flow* **29**, 495–509.
3. Dong, S., Liu, D., Maxey, M.R. and Karniadakis, G.E., 2004, *J. Comput. Phys.* **195**, 695–717.
4. Johnson, A., Tezduyar, T., 1997, *Comp. Meth. Appl. Mech. Engng.* **145**, 301–321.
5. Takiguchi, S., Kajishima, T. and Miyake, Y., 1999, *JSME Int. J.* **B42**, 411–418
6. Fadlun, E.A., Verzicco, R., Orlandi, P. and Yusof, J.M., 2000, *J. Comput. Phys.* **161**, 35–60.
7. Li, Z. and Lai, M.C., 2001, *J. Comput. Phys.* **171**, 822–842.
8. Kim, J., Kim, D. and Choi, H., 2001, *J. Comput. Phys.* **171**, 132–150.
9. Gilmanov, A., Sotiropoulos, F. and Balaras, E., 2003, *J. Comput. Phys.* **191**, 660–669.

10. Tseng, Y.H. and Ferziger, J., 2003, *J. Comput. Phys.* **192**, 593–623.
11. Balaras, E., 2004, *Comput. &luids* **33**, 375–404
12. Gilmanov, A. and Sotiropoulos, F., 2005, *J. Comput. Phys.* **207**, 457–492.
13. Udaykumar, H.S., Mittal, R., Rampunggoon, P. and Khanna, A., 2001, *J. Comput. Phys.* **174**, 345–380.
14. Chen, S. and Doolen, G., 1998, *Annu. Rev. Fluid Mech.* **30**, 329–364.
15. Aidun, C.K., Lu, Y. and Ding, E.J., 1998, *J. Fluid Mech.* **37i3**, 287–311.
16. Hill, R.J. and Koch, D.L., 2002, *J. Fluid Mech.* **465**, 59–97.
17. Feng, Z.G. and Michaelides, E.E., 2004, *J. Comput. Phys.* **195**, 602–628.
18. Feng, Z.G. and Michaelides, E.E., 2005, *J. Comput. Phys.* **202**, 20–51.
19. Takagi, S., Oğuz, H., Zhang, Z. and Prosperetti, A., 2003, *J. Comput. Phys.* **187**, 371–390.
20. Zhang, Z. and Prosperetti, A., 2003, *J. Appl. Mech.* **70**, 64–74.
21. Zhang, Z. and Prosperetti, A., 2005, *J. Comput. Phys.* **210**, 292–324.
22. Lamb, H., 1932, *Hlrodynamics* , 6th edition, Cambridge University Press, Cambridge.
23. Kim, S. and Karrila, S., 1991, *Microhydrodynamics*, Butterworth-Heinemann, Boston.
24. Brown, D., Cortez, R. and Minion, M., 2001, *J. Comput. Phys.* **168**, 464–499.
25. Marchioro, M., Tanksley, M. and Prosperetti, A., 2000, *Int. J. Multiphase Flow* **26**, 783–831.
26. Ichiki, K. and Prosperetti, A, 2004, *Phys. Fluids* **16**, 2483–2496.
27. Richardson, J. and Zaki, W., 1954, *Trans. Inst. Chem. Engrs.* **32**, 35–53.
28. Di Felice, R. and Kehlenbeck, R., 2000, *Chem. Eng. Technol.* **23**, 1123–1126.
29. Kuusela, E., Lahtinen, J.M. and Ala-Nissila, T., 2004, *Phys. Rev. E* **69**, 066310.
30. Parthasarathy, R. and Faeth, G., 1990, *J. Fluid Mech.* **220**, 515–537.
31. Wylie, J.J. and Koch, D.L., 2000, *Phys. Fluids* **12**, 964–970.
32. Ladd, A., 1997, *Phys. Fluids* **9**, 491–499.
33. Wu, J. and Manasseh, R., 1998, *Int. J. Multiphase Flow* **22**, 285–306.

A DNS Approach Dedicated to the Analysis of Fluidized Beds

Tseheno N. Randrianarivelo[1], Stéphane Vincent[1], Olivier Simonin[2] and Jean-Paul Caltagirone[1]

[1]Laboratory Transfert, Ecoulements Fluide Energétique (TREFLE), UMR 8508, ENSAM/ENSCPB/Bordeaux 1, Pessac, France; e-mail: randrian@enscpb.fr, vincent@enscpb.fr, calta@enscpb.fr
[2]Institut de Mécanique des Fluides de Toulouse (IMFT), UMR 5502, CNRS/INPT/UPS, Toulouse, France; e-mail: olivier.simonin@imft.fr

1 Motivation

The numerical modelling of fluidized bed is widely encountered in fundamental research and industrial applications in the field of chemistry, energy or material processes. The problem is complicated by the presence of different length and time scales describing the particles and the reactors. Several modellings exist for simulating particulate flows. In averaged Eulerian models where the size of the particles is small compared to the continuous medium characteristic scale, fluid-particle interactions are taken into account using constitutive laws. In the present study, a numerical model is proposed to describe the fluid-particle interactions at the particle scale, based on fixed Eulerian structured grids, penalty methods and Volume Of Fluid (VOF) techniques. Our interest is to build a general model and approximation methods efficient enough for leading Direct Numerical Simulations of particulate flows without any requirements of physical parametrization such as particle-particle interactions. In addition, our motivation is to analyze the results of the numerical experiments in terms of equivalent macroscopic quantities such as velocity fluctuations, granular pressure or solid fraction.

Fluidized beds are characterized by thousands of particles interacting in a moving fluid. Solving such fluid/particle motion at the particle scale can be achieved thanks to two main approaches. First, the conservation equations can only be solved in the fluid phase and the grid is adapted in a Lagrangian manner at each calculation step to account for the particle shape. In three dimensional unsteady problems, this method is too expansive in calculation time and difficult to implement. The second approach consists in solving the fluid-solid two-phase flow on a fixed Eulerian grid. This method is easy to develop for complex and evolving topologies of the two phases. However, the conservation equations must take into account the discontinuous and unsteady behavior of the flow characteristics.

S. Balachandar and A. Prosperetti (eds), Proceedings of the IUTAM Symposium on Computational Multiphase Flow, 207–214.

A brief presentation of the Direct numerical Simulation (DNS) formulation, in terms of generalized Navier–Stokes equations for multi-material flows, is reported. Physical validations are proposed involving particle group effects and particle interactions, meaningfull for fluidized bed applications. A 2D fluidized bed involving 2860 particles in a liquid is also considered. Time and space statistics on macroscopic and phase average variables are analyzed. Perspectives for 3D numerical experiments and improvements of existing modelling and empirical laws are finally drawn.

2 Numerical model

An Eulerian modelling is chosen. The solid and fluid zones Ω_s and Ω_f are distinguished by using a phase function C defined as $C = 1$ in Ω_s and $C = 0$ in Ω_f. After convolution of the Navier–Stokes equations by C and integration over all the phases, the equations of motion are written as:

$$\nabla \cdot \mathbf{u} = 0,$$
$$\rho \left(\frac{\partial \mathbf{u}}{\partial t} + \mathbf{u} \cdot \nabla \mathbf{u} \right) = -\nabla p + \rho \mathbf{g} + \nabla \cdot \mu [\nabla \mathbf{u} + \nabla^T \mathbf{u}], \tag{1}$$
$$\frac{\partial C}{\partial t} + \mathbf{u} \cdot \nabla C = 0,$$

where $\mathbf{u}$ is the velocity, p the pressure, ρ the density, μ the dynamic viscosity, $\mathbf{g}$ the gravity and t the time. The solid phase is assumed as a fluid with a specific rheology and the generalized Navier–Stokes equations (1) for two-phase flows apply everywhere. The solid behavior is obtained by using an Implicit Tensorial Penalty Method (ITPM) [1, 9] that assumes $\mu \rightarrow +\infty$ in the solid zone, i.e. $C = 1$. Numerically, μ lies in the range 10 to 10000. Velocity-pressure coupling and incompressibility are treated with an augmented Lagrangian algorithm [4]. It has also been generalized to account for solid behavior [12, 13]. Finally, an Uzawa algorithm [11] is applied to solve the penalized Navier–Stokes equations by means of a minimization procedure.

The conservation equations are approximated with implicit finite volume and staggered grids. Centered schemes and second order Euler discretization are respectively used to discretize the space fluxes and the time derivatives. A Piecewise Linear Interface Construction VOF-PLIC method is adopted for the advection of C [15]. The viscosity and density evolutions are directly obtained thanks to numerical laws depending on the phase function as

$$\text{if } C \geq 0.5 \text{ then } \rho = \rho_s \text{ and } \mu = \mu_s$$
$$\text{else } \rho = \rho_f \text{ and } \mu = \mu_f. \tag{2}$$

An explicit collision model can be included in the ITPM approach following for example the work of Singh et al. [10]. Nevertheless, no explicit treatment of particle-particle or particle-wall interaction has been considered in the present formulation of the ITPM method. As it will be demonstrated in the next section, the implicit penalty technique is sufficient to manage the collision under the control of the size of the grid cells.

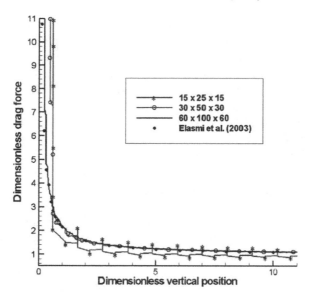

Fig. 1. Lubrication when a particle interacts with a solid wall – Comparison between analytical and numerical solutions for 2 ($15 \times 25 \times 15$), 4 ($30 \times 50 \times 30$) and 8 ($60 \times 100 \times 60$) points in particle diameter.

3 Physical validation

Two types of physical tests, representative of local flow structures in fluidized beds, have been considered [8]. The correct representation of interactions between two particles or a particle and a wall (lubrication phenomenon) are demonstrated first. Then, the group effect of several particles in a periodic domain is conditered in a Face Centered Cubic (CFC) array.

3.1 Lubrication effect and particle-wall interaction

In the existing Eulerian numerical approaches of the literature (for example, [6]), the interactions between particles or between particles and domain walls are explicitely added in the models through a repulsive body force depending on the distance between the particles and their obstacles [10]. The present physical test case aims at demonstrating that our ITPM method is able to implicitly account for lubrication effects without explicit modelling. A Stokes flow between a particle and a wall is considered as a reference case for comparison and validation. Analytical solutions are available [2] for the drag force of the particle when it comes near the wall.

We consider a sphere settling in a cylindrical tank of radius $R_t = 5.23$ cm and height 0.1046 m, filled with a liquid of viscosity 30 Pa.s and density 900 kg.m^{-3}. The density of the particle is 3847 kg.m^{-3} and its radius is $R_p = 9.198$ mm. Initially, the vertical position of the particle is $z_p = 10R_p$. The particle Reynolds number $Re =$

$\frac{2\rho_f R_p U_\infty}{\mu_f}$ is $2 \cdot 10^{-2}$ and the corresponding Stokes number $St = \frac{2\rho_s}{9\rho_f} Re$ is 10^{-2}. In Figure 1, we present the convergence of the numerical drag force compared to the one of the same sphere in an infinite medium with respect to the dimensionless vertical position $\frac{z_p}{R_p} - 1$. Three-dimensional simulations on three different grids demonstrate the convergence of the ITPM to the analytical solution of Elasmi. Even with only two Eulerian mesh points in a particle diameter ($15 \times 25 \times 15$ grids), the lubrication effect is well predicted as long as there is more than one cell between the particle and the wall. It should be mentionned that the lubrication effect is an important physical process occurring in fluidized beds.

3.2 Drag force of an array of ordered spheres

The drag force exerted on a fixed Cubic Face-Centered (CFC) array of particles by a viscous flow is finally investigated with the ITPM Direct Numerical Simulation approach. This particle framework is usually considered to mimick group effects in particulate flows. A parametric study is presented for two characteristic Reynolds numbers ($Re = 10$ and 100) for values of mean solid fractions between 0.05 and 0.6. The ratio between the drag force coefficient of the CFC array C_d over the drag force coefficient Cd_s of a single sphere is an important parameter for macroscopic models of fluidized beds.

$70 \times 70 \times 70$ grids are used. The domain is periodic in all Cartesian directions. Nylon particles of density $\rho = 2700 \, \text{kg.m}^{-3}$ are considered. The value of the particle radius R_p is varied according to the particle Reynolds number $Re = \frac{2\rho R_p U_f}{\mu_f}$. The particles are plunged into a viscous fluid whose properties are 10^{-3} Pa.s for the viscosity and $1000 \, \text{kg.m}^{-3}$ for the density. In this study, the particles are fixed and the fluid motion is obtained applying a pressure gradient in a normal direction to one face. We demonstrate the good behavior of the ITPM method by comparing our numerical results to empirical drag force laws [3, 14] and single phase DNS on body meshes adapted to the CFC array [7] (see Figure 2). The differences observed between the fixed CFC simulation, both with Eulerian and body fitted grids, and the correlations are due to the fixed character of the particles in the simulations while the empirical correlations were derived for moving fluidized beds. In the near future, numerical experiments will be carried out using the ITPM method with moving particle arrays, various fluid properties (inertial flows) and radii of the spheres varying between particles (polydisperse particulate flows).

4 Simulation of 2D fluidized beds

A two-dimensional liquid-solid fluidized bed is considered corresponding to the configuration proposed by Gevrin et al. [5]. In a rectangular tank full of water, 2860 glass particles are fluidized under a vertical pressure gradient. The solid fraction is 0.2, the particle Reynolds number is 555 and the Stokes number is 70. Typical DNS and averaged particulate flow configurations are presented in Figure 3. On the averaging

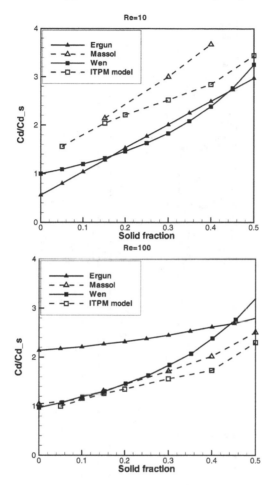

Fig. 2. Evolution of the dimensionless drag force coefficient according to solid fraction for $Re = 10$ and $Re = 100$. Comparisons witn the results of Ergun [3], Wen and Yu [14] and Massol et al. [7].

10×500 grids corresponding to Gevrin's macroscopic simulations, the mean velocities and solid fraction demonstrate the alternation of void fraction and concentrated bands in a characteristic planar like instability.

Based on the DNS of the fluidized bed (Figure 3 left), statistical and averaging procedures are implemented in order to extract macroscopic quantities from the small scale simulations. For example, the mean solid fractions, bed heigth, granular pressure, particle fluctuation velocities and fluidizations velocities are obtained after scale and time integration of the 100×3930 grid simulations on a corser 10×500 mesh (see Figure 4).

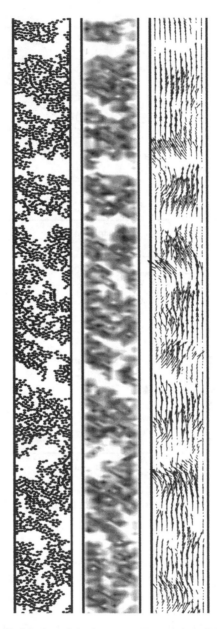

Fig. 3. Fluidization of cylindrical particles in water – From left to right, DNS solid concentration, solid fraction and particle velocities integrated on a 10×500 macroscopic grids corresponding to results of [5], granular pressure averaged over time and space.

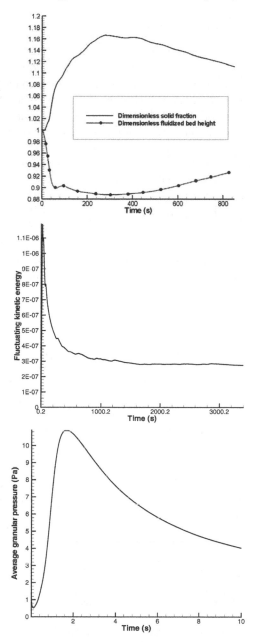

Fig. 4. Examples of macroscopic datas available after statistical treatment of the DNS simulations. Top: time evolution of the dimensionless solid fraction and fuidized bed heigth. Middle: Fluctuating kinetic energy integrated over the whole calculation domain. Bottom: averaged granular pressure calculated with the velocity fluctuations.

5 Future works

A new Implicit Tensorial Penalty Method has been presented and validated for the simulation of particulate flows. We proved the accuracy and the consistency of the ITPM approach for simulating the lubrication effect between a particle and a wall in Stokes flows and the drag force of a CFC array of fixed particle. We have presented the first results of 2D simulations of fluidized beds involving thousands of particles and we have provided some illustrations of macroscopic datas extracted from simulations at local lengthscale.

Three-dimensional simulations of typical fluidized bed configurations involving hundreds of particles will be carried out in the near future . In parallel, numerical experiments will be realized in periodic boxes containing 10 to 100 particles subjected to simple stresses. In both configurations, a statistical analysis of local particulate flows will be developed in order to extract interesting meso or macroscopic features such as velocity fluctuations, mean inter-particle distance and the mean drag force in an integration volume. Our objective is to understand the local hydrodynamics effects and to provide useful informations for macroscopic modelling.

References

1. Caltagirone, J.P. and Vincent, S., 2001, *C.R. Acad. Sci., Série IIb* **329**, 607–613.
2. Elasmi, L., Berzig, M. and Feuillebois, F., 2003, *Z. Angew. Math. Phys.* **54**, 304–327.
3. Ergun, S., 1952, *Chem. Eng. Prog.* **48**, 89–94.
4. Fortin, M. and Glowinski, R., 1982, *Collection Méthodes Mathématiques de l'Informatique*, Dunod, Paris.
5. Gevrin, F., Masbernat, O. and Simonin, O., 2001, Granular pressure and particle velocity fluctuation predictions in liquid-solid fluidized beds, in *Proceedings 4th International Conference on Multiphase Flows, ICFM'01*, New Orleans, USA, May 27–June 1.
6. Pan, T.W. and Glowinski, R., 2002, *J. Comput. Phys.* **181**, 260–279.
7. Massol, A., 2004, Numerical simulations of flows through networks of mono- and bidisperse spheres for moderate Reynolds numbers, PhD Thesis of Mechanical Engineering, Paul Sabatier University, Toulouse, France.
8. Randrianarivelo, T.N., Vincent, S., Pianet, G., Simonin, O. and Caltagirone, J.P., 2004, Towards direct numerical simulation of fluidized beds, in *Proceedings 5th International Conference on Multiphase Flows, ICFM'04*, Yokohama, Japan, May 30–June 4.
9. Randrianarivelo, T.N., Pianet, G., Vincent, S. and Caltagirone, J.P., 2005, *Int. J. Numer. Meth. Fluids* **47**, 1245—1251.
10. Singh, P., Joseph, D.D., Hesla, T.I., Glowinski, R. and Pan, T.W., 2000, *J. Non-Newtonian Fluid Mech.* **91**, 165–188.
11. Uzawa, H., 1958, in *Studies in Linear and Non-Linear Programming*, K.J. Arrow, L. Hurwicz and H. Uzawa (eds), Standford University Press.
12. Vincent, S. and Caltagirone, J.P., 1999, *Int. J. Numer. Meth. Fluids* **30**, 795–811.
13. Vincent, S. and Caltagirone, J.P., 2000, *J. Comput. Phys.* **163**, 172–215.
14. Wen, C. and Yu, Y., 1966, *Chem. Eng. Prog. Symp. Ser.* **62**, 100–111.
15. Youngs, D.L., 1982, *Numerical Methods for Fluid Dynamics*, Academic Press, New York.

Performance of Various Fluid-Solid Coupling Methods for DNS of Particulate Flow

Markus Uhlmann and Alfredo Pinelli

Departamento de Combustibles Fósiles, CIEMAT, Avenida Complutense 22, 28040 Madrid, Spain; e-mail: markus.uhlmann@ciemat.es

1 Introduction

In the present article we are concerned with efficient and accurate methods for the numerical simulation of the dynamics of rigid particles suspended in an incompressible fluid. We focus our attention on so-called fictitious domain methods, a framework in which the presence of suspended particles is accounted for by means of an artificial forcing term added to the Navier–Stokes equations. Thereby, simple fixed grids can be used and the additional cost of repeated adaptation of the computational mesh is avoided.

Existing fictitious domain methods fall into two main classes: those where the constraint force is explicitly formulated ("direct forcing") and those where some kind of feedback mechanism is employed ("indirect forcing"). The latter type of formulation is usually based upon the virtual spring-and-damper technique of Goldstein et al. [1], which has the drawback of introducing two additional free parameters into the problem. Also, since the characteristic time scale of the feedback system needs to be resolved, very small time steps are sometimes required for stability reasons.

Direct forcing methods, on the other hand, are in general free from the above mentioned problems. However, it has been observed [2] that a linear interpolation procedure (as used in [3, 4]) can lead to highly oscillatory hydrodynamic forces when a solid immersed body is in motion relative to the fixed grid.

In [5] the artificial force term is instead weighted by the solid fraction of the surrounding grid cell, providing some amount of smoothing. However, the resulting drag/lift variation obtained in our computations with this method still shows substantial grid-induced oscillations (see results in Section 3.1).

In Peskin's immersed boundary method [6] – which originally uses indirect forcing – quantities are transferred between arbitrary (Lagrangian) positions and the fixed (Eulerian) grid by means of a so-called "regularized delta function" with desirable smoothness properties and compact support.

Recently, a new direct forcing scheme which makes use of Peskin's delta function during the necessary interpolation steps was formulated in [7]. In the following

S. Balachandar and A. Prosperetti (eds), Proceedings of the IUTAM Symposium on Computational Multiphase Flow, 215–223.

we will discuss this method and its validation through a selection of test cases. Finally we will present preliminary results from the simulation of the sedimentation of several hundred spherical particles.

2 Numerical method

Let us write the time-discretized momentum equation in the following form

$$\frac{\mathbf{u}^{n+1} - \mathbf{u}^n}{\Delta t} = \mathbf{rhs}^{n+1/2} + \mathbf{f}^{n+1/2} , \tag{1}$$

where $\mathbf{rhs}^{n+1/2}$ regroups all usual forces (convective, pressure-related, viscous) and $\mathbf{f}^{n+1/2}$ is an artificial force term, both evaluated at some intermediate time level. Since Fadlun et al. [3] it is common to express the additional force term by simply rewriting the above equation as

$$\mathbf{f}^{n+1/2} = \frac{\mathbf{u}^{(d)} - \mathbf{u}^n}{\Delta t} - \mathbf{rhs}^{n+1/2} , \tag{2}$$

where $\mathbf{u}^{(d)}$ is the desired velocity at the point where forcing is to be applied. Formula (2) is characteristic for direct forcing methods. Problems arise from the fact that in general the locations where the desired velocity is known do not coincide with the Eulerian grid. In [7] the definition of the force term was instead formulated at Lagrangian positions attached to the surface of the particles, viz.

$$\mathbf{F}^{n+1/2} = \frac{\mathbf{U}^{(d)} - \mathbf{U}^n}{\Delta t} - \mathbf{RHS}^{n+1/2} , \tag{3}$$

where uppercase letters indicate quantities evaluated at Lagrangian coordinates. Obviously, the velocity in the particle domain $\mathscr{s}$ is simply given by the solid-body motion,

$$\mathbf{U}^{(d)}(\mathbf{X}) = \mathbf{u}_c + \boldsymbol{\omega}_c \times (\mathbf{X} - \mathbf{x}_c) \qquad \mathbf{X} \in \mathscr{s} , \tag{4}$$

as a function of the translational and rotational velocities of the particle, $\mathbf{u}_c, \boldsymbol{\omega}_c$.

The final element of the method of Uhlmann [7] is the transfer of the velocity (and r.h.s. forces) from Eulerian to Lagrangian positions as well as the inverse transfer of the forcing term to the Eulerian grid positions. For this purpose we define a Cartesian grid $\mathbf{x}_{ijk}$ with uniform mesh width h in all three directions and distribute so-called discrete Lagrangian force points $\mathbf{X}_l$ evenly on the particle surface. Using Peskin's [6] regularized delta function formalism, the transfer can be written as:

$$\mathbf{U}(\mathbf{X}_l) = \sum_{ijk} \mathbf{u}(\mathbf{x}_{ijk}) \, \delta_h(\mathbf{x}_{ijk} - \mathbf{X}_l) \, h^3 , \tag{5a}$$

$$\mathbf{f}(\mathbf{x}_{ijk}) = \sum_{l} \mathbf{F}(\mathbf{X}_l) \, \delta_h(\mathbf{x}_{ijk} - \mathbf{X}_l) \, \Delta V , \tag{5b}$$

where ΔV_l designates the forcing volume assigned to the lth force point. We use the particular function δ_h given in [8] which has the properties of continuous differentiability, second order accuracy, support of three grid nodes in each direction and consistency with basic properties of the continuous delta function.

The algorithm for each time-step can then be summed up as follows:

1. compute $\tilde{\mathbf{u}} = \mathbf{u}^n + \mathbf{rhs}^{n+1/2}$
2. transfer $\tilde{\mathbf{u}}$ to Lagrangian positions, using (5a)
3. compute $\mathbf{F}(\mathbf{X}_l)$ from (3)
4. transfer the force back to Eulerian positions, using (5b)
5. solve Navier–Stokes on the fixed grid with the added force term $\mathbf{f}(\mathbf{x}_{ijk})$.

The above method has been implemented in a staggered finite-difference context, involving central, second-order accurate spatial operators, an implicit treatment of the viscous terms and a three-step Runge–Kutta procedure for the non-linear part. Continuity in the entire domain is enforced by means of a projection method. The particle motion is determined by the Runge–Kutta-discretized Newton equations for rigid-body motion, which are weakly coupled to the fluid equations. In the present simulations direct particle interactions (collisions) are not considered.

3 Results

3.1 Uniform flow around an oscillating cylinder

In this first test case the particle motion is prescribed, i.e. one-way coupled. We consider the flow around a cylinder with diameter D located at the origin in a domain which measures $\Omega_1 = [-6.17, 20.5]D \times [-13.33, 13.33]D$. The uniform grid has 1024×1024 nodes, i.e. $D/h = 38.4$. The time step was set to $\Delta t = 0.003$, corresponding to a maximum CFL number of approximately 0.6. The cylinder follows a prescribed periodic motion perpendicular to the mean flow, i.e.:

$$y_c(t) = A \sin(2\pi f_f t), \qquad (6)$$

with the amplitude set to $A = 0.2D$ and the frequency $f_f/f_n = 0.8$, where f_n is the natural shedding frequency obtained from the value of the Strouhal number from the literature: $St = f_n D/u_\infty = 0.195$ (for $Re_D = 185$). This case corresponds to one of the cases simulated in [9]. The maximum velocity of the cylinder is $\max(|\mathbf{u}_c|)/u_\infty = 2\pi f_f A/u_\infty = 0.196$. The boundary conditions are: uniform velocity at the inflow and along the top and bottom boundaries; convective condition at the outflow.

Figure 1 shows that the temporal variation of the drag follows a reasonably smooth periodic curve when using the current scheme. The same goes for the lift force which has been omitted. On the other hand, the method of Kajishima and Takiguchi [5], implemented as shown in [2], yields significant oscillations on the time-scale of the mesh-width divided by the cylinder velocity. In other words, the smoothing provided by the present method proves more efficient in hiding the influence of the fixed grid.

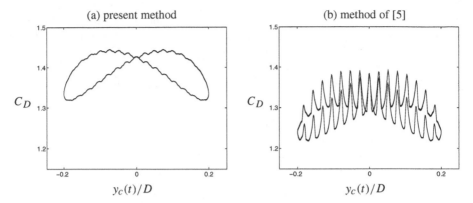

Fig. 1. Time-periodic variation of the drag coefficient in the case of a translationally oscillating cylinder in uniform cross-flow at $Re_D = 185$ with $D/h = 38.4$ and $CFL \approx 0.6$. Left graph: present method. Right graph: method of Kajishima and Takiguchi [5], implemented into the present solver as described in [2].

Table 1. Dimensionless coefficients obtained from the simulation of the flow around a cylinder at $Re_D = 185$ which oscillates near the natural shedding frequency and using $D/h = 38.4$ and $\Delta t = 0.003$. The domain Ω_1 has been used, except where otherwise stated.

	$\bar{C}_D$	C'_D	$(C_L)_{rms}$
present	1.380	±0.063	0.176
present, enlarged domain Ω_2	1.354	±0.065	0.166
Kajishima & Takiguchi's scheme [5]	1.282	±0.088	0.223
Lu and Dalton [9]	1.25		0.18

The mean values and fluctuations of drag and lift are given in Table 1. The mean drag is over-predicted by approximately 10% with the current scheme. A similar over-prediction was noted in [10] where the original immersed boundary method was used for the prediction of the flow around a stationary cylinder on the same grid. In the latter reference the over-prediction was attributed to an insufficient domain size. Here we verify this argument by repeating the simulation in an enlarged domain $\Omega_2 = 1.5\Omega_1$, while maintaining the mesh width and time step. The effect is that indeed both mean drag and lift fluctuations decrease, yielding an error of approximately 8% in the larger domain. The difference in mean drag between the result of the present method and the method of Kajishima and Takiguchi [5] is probably due to the different smoothing properties.

3.2 Sedimentation of a single sphere

We consider a single sphere which is released from rest at $t=0$ in a quiescent fluid. The physical parameters of the simulation are chosen in order to match cases 1, 2, 4 of the experiment of Mordant and Pinton [11], where the motion of spherical

Table 2. Parameters of the experiment of Mordant and Pinton [11] and resulting terminal particle Reynolds number Re_D in the case of a single sedimenting sphere.

case	$\dfrac{\rho_p}{\rho_f}$	$\nu \times 10^3$	Re_D exp. [11]	Re_D present
1	2.56	5.41637	41.17	41.12
2	2.56	1.04238	362.70	366.69
4	7.71	2.67626	280.42	282.45

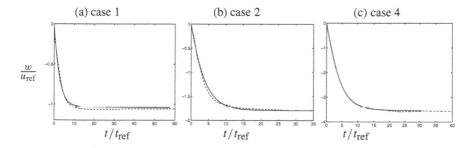

Fig. 2. Sedimentation of a single sphere corresponding to [11]. Vertical velocity: —— present, – – – – experimental data.

beads in water was investigated, while their material and diameter were varied from case to case. The experiment takes place in a large container, justifying the use of periodic conditions in the simulation. By similarity with the experiment (density ratio, Froude number, particle Reynolds number) we have selected the values for the particle diameter $D = 1/6$ and the gravitational acceleration $|\mathbf{g}| = 9.81$ alongside the parameters given in Table 2. The range of Reynolds numbers spans $40 \ldots 360$ and the density ratio is varied between 2.56 and 7.71. The values for the numerical parameters are: mesh width $h = 1/76.8$, i.e. $D/h = 12.8$; time step $\Delta t = 0.0025$, i.e. yielding a maximum CFL number of 0.3, 0.75, 0.5, respectively.

Figure 2 shows the vertical particle velocity as a function of the elapsed time. Gravitational scaling is used, i.e. $u_{\text{ref}} = \sqrt{|\mathbf{g}|D}$ and $t_{\text{ref}} = \sqrt{D/|\mathbf{g}|}$ are the reference velocity and time, respectively. The computational results are shown for times before the particle motion in the periodic domain is affected by the remnants of its own wake. A very good agreement with the experimental measurements can be observed. In Table 2 the terminal value of the Reynolds number is reported for all three cases. It can be seen that the maximum error is below 2% (case 2).

3.3 Many-particle sedimentation

Here we consider a similar case as the one studied in [5]. A large number of particles are sedimenting in a tri-periodic domain. The dynamic formation of particle agglomerations is the object of the investigation in [5]. In the present study, all particles have a density ratio $\rho_p/\rho_f = 2.56$ and a diameter $D = 1/6$; the fluid has a viscosity value

Table 3. Definitions for the two different configurations used in triply-periodic many-particle simulations in Section 3.3, listing the number of particles N_p, the volume fraction of solid ϵ_p and the domain size in the three coordinate directions L_i (gravity acts in the negative z-direction).

case	N_p	ϵ_p	L_x/D	L_y/D	L_z/D
A	512	0.4%	40	20	80
B	1000	0.8%	40	40	40

of $\nu = 10^{-3}$ and the gravitational acceleration measures $|\mathbf{g}| = 9.81$. This leads to a terminal Reynolds number of approximately 400 for a single sphere, similar to case 2 in Section 3.2. Our series of simulations have only been initialized recently and, therefore, the present results only show the behavior of the systems for early times.

Table 3 gives the details of the two configurations presently studied. The volume fraction is kept below 1%, meaning that the dilute regime is addressed. Figure 3 shows the initial particle positions and the configuration after 100 time units (gravitational scaling) for case A. At that time the particles have reached a seemingly disordered state with an inhomogeneous spatial distribution. Statistically, this means that the range of inter-particle distances changes. Most prominently, the global minimum of that distance rapidly approaches the limit of one particle diameter (cf. Figure 4). In fact, since we do not use any explicit collision strategy, the distance can drop below this limit and cause non-physical overlap. When this occurred we have stopped the simulation. Figure 4 also shows that the average distance to the nearest particle neighbor decreases significantly from the initial homogeneous state. From Figure 5 we can see that during the initial phase the average sedimentation velocity reaches a minimum (where $\bar{w}/u_{\text{ref}} \approx 2.6$) and then levels out to approximately $\bar{w}/u_{\text{ref}} = 2$. This is a manifestation of a strong wake-sheltering effect as already observed in [5]. Obviously, the minimum is not observed in the case of a single sedimenting sphere. Figure 5 also shows the r.m.s. values of the angular particle velocity. It is interesting to note that the values for rotation vectors in the horizontal plane are by a factor 6 higher than those in the vertical direction.

3.4 Efficiency of the method

Operation count

The following numbers refer to the operations carried out during one Runge–Kutta sub-step, of which there are three per full time step. The main work in the pure fluid part of the code is done while solving the Helmholtz problems during the prediction step and when solving the Poisson problem of the projection step. Using a multigrid method, the number of operations scales as $\mathcal{O}(N_x N_y N_z)$. On the other hand, the particle-related work scales as $\mathcal{O}(N_p \cdot (D/h)^2)$, i.e. linear with the number of particles (since we neglect collisions) and with the square of the number of grid points per diameter (since we only force the surface of the particle).

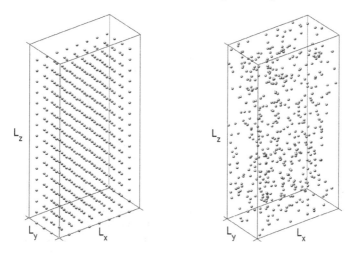

Fig. 3. Particle positions during the simulation of many-particle sedimentation, case A: $t = 0$ (left graph); $t/t_{ref} = 100$ (right graph).

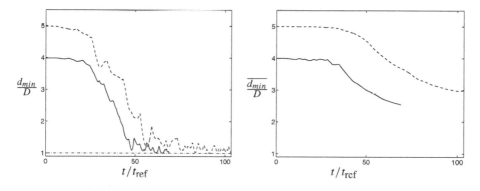

Fig. 4. Statistical particle-related quantities during the simulation of many-particle sediment-ation. The left graph shows the minimum inter-particle distance for case A (– – – –), case B (————). The graph on the right shows the average distance to the nearest neighbor for the two cases.

Time step

It was observed that the present method does not have a noticeable influence upon the theoretical temporal stability limit, $CFL < \sqrt{3}$.

Parallelization

Standard domain decomposition over a three-dimensional Cartesian processor grid was used for the fluid solver. The particle-related operations are performed by a

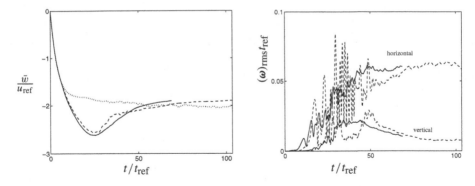

Fig. 5. Statistical particle-related quantities during the simulation of many-particle sediment-ation. The left graph shows the mean sedimentation velocity for case A (– – – –), case B (———) and a single sphere (········). The graph on the right shows the root-mean-square angular particle velocities in the horizontal plane and the vertical direction for the former two cases. In both graphs gravitational scaling is used.

Table 4. Execution times on an IBM Cluster with Power 4 processors at 1.1 GHz (64 bit arithmetic), using different grid sizes $N_x \times N_y \times N_z$, numbers of particles N_p and numbers of processors nproc. The resolution of the spherical particles was set to $D/h = 12.8$ in all cases.

$N_x \times N_y \times N_z$	N_p	nproc	$t_{\text{exec}}[s]$
$512 \times 512 \times 512$	1000	64	115.0
$512 \times 512 \times 1024$	1000	128	144.9
$512 \times 512 \times 1024$	2000	128	147.4

master processor who is responsible for all particles contained in its private sub-volume of the domain. Particles overlapping more than one sub-volume are handled by one or more slave processors.

Timing

Table 4 shows some execution times per full time step for the present scheme. Good scaling with the number of processors can be observed. Also, it becomes clear that the particle-related work makes up only a small fraction of the total execution time for the problems under consideration.

4 Conclusions

We have discussed the problems associated with fictitious domain methods of the direct and indirect type, presenting a recently proposed variant which uses the regu-larized delta function of Peskin and co-workers [6, 10, 8] for the association between arbitrary Lagrangian and discrete Eulerian positions. Thereby, the hydrodynamic forces acting upon the solid domains, which are at the same time driving the particle motion, are free from significant oscillations. This effect was demonstrated for the flow around an oscillating cylinder.

The comparison of the new scheme with well-established experimental results for the sedimentation of a spherical particle shows its accuracy over a significant range of Reynolds numbers while using a very realistic resolution of only 13 grid points per diameter.

Taking into account the fast execution speed of the current method we can conclude that it is indeed very competitive. Further work – which is currently underway – will include the performance of simulations with $\mathcal{O}(10000)$ particles in larger domains. The evaluation of numerical collision strategies will be an important issue which needs to be addressed in detail.

Acknowledgements

This work was supported by the Spanish Ministry of Education and Science under the Ramón y Cajal program (contract DPI-2002-040550-C07-04) and through grant DPI-2002-1314-C07-04.

References

1. D. Goldstein, R. Handler, and L. Sirovich, 1993, Modeling a no-slip boundary with an external force field, *J. Comput. Phys.* **105**, 354–366.
2. M. Uhlmann, 2003, First experiments with the simulation of particulate flows, Technical Report No. 1020, CIEMAT, Madrid, Spain, ISSN 1135-9420.
3. E.A. Fadlun, R. Verzicco, P. Orlandi and J. Mohd-Yusof, 2000, Combined immersed-boundary finite-difference methods for three-dimensional complex flow simulations, *J. Comput. Phys.* **161**, 35–60.
4. J. Kim, D. Kim and H. Choi, 2001, An immersed-boundary finite-volume method for simulations of flow in complex geometries, *J. Comput. Phys.* **171**, 132–150.
5. T. Kajishima and S. Takiguchi, 2002, Interaction between particle clusters and particle-induced turbulence, *Int. J. Heat Fluid Flow* **23**, 639–646.
6. C.S. Peskin, 2002, The immersed boundary method, *Acta Numerica* **11**, 479–517.
7. M. Uhlmann, 2005, An immersed boundary method with direct forcing for the simulation of particulate flows, *J. Comput. Phys.* **209**(2), 448–476.
8. A.M. Roma, C.S. Peskin and M.J. Berger, 1999, An adaptive version of the immersed boundary method, *J. Comput. Phys.* **153**, 509–534.
9. X.Y. Lu and C. Dalton, 1996, Calculation of the timing of vortex formation from an oscillating cylinder, *J. Fluids Structures* **10**, 527–541.
10. M.-C. Lai and C.S. Peskin, 2000, An immersed boundary method with formal second-order accuracy and reduced numerical viscosity, *J. Comput. Phys.* **160**, 705–719.
11. N. Mordant and J.-F. Pinton, 2000, Velocity measurement of a settling sphere, *Eur. Phys. J. B* **18**, 343–352.

Numerical Study of Particle Migration in Tube and Plane Poiseuille Flows

B.H. Yang[1], J. Wang[1], D.D. Joseph[1], H.H. Hu[2], T.-W. Pan[3] and R. Glowinski[3]

[1]Department of Aerospace Engineering and Mechanics, University of Minnesota, Minneapolis, MN, 5USA
[2]Department of Mechanical Engineering and Applied Mechanics, University of Pennsylvania, Philadelphia, PA, 19104USA
[3]Department of Mathematics, University of Huston, Huston, X7204USA

Abstract. The lateral migration of a single spherical particle in tube Poiseuille flow is simulated by ALE scheme, along with the study of the movement of a circular particle in plane Poiseuille flow with consistent dimensionless parameters. These particles are rigid and neutrally buoyant. A lift law $L = CU_s(\Omega_s - \Omega_{se})$ analogous to $L = \rho U \Gamma$ is validated in both two dimensions and three dimensions here; U_s and Ω_s are slip velocity and angular slip velocity, Ω_{se} is the angular slip velocity at equilibrium. A method of constrained simulation is used to generate data which is processed for correlation formulas for the lift force, slip velocity, and equilibrium position. Our formulas predict the change of sign of the lift force which is necessary in the Segré–Silberberg effect. Correlation formulas are compared between tube and plane Poiseuille flows by fixing the dimensionless size of particle and the Reynolds number. Our work provides a valuable reference for a better understanding of the migration of particle in Poiseuille flows and the Segré–Silberberg effect.

1 Introduction

The literature on the migration of rigid particles in shear flow has been reviewed by Yang et al. [6] and else where and will not be reviewed here. Yang et al. [6] used the ALE scheme to study the lift force on a neutrally buoyant sphere in tube Poiseuille flow. They validated the lift law in three dimensions and established a general procedure for obtaining correlation formulas from numerical experiments. Their correlation formulas and predictions obtained good agreement with the literature.

The main goal of this work is to correlate the lift laws in two dimensions and three dimensions simultaneously by fixing some important dimensionless parameters such as Reynolds number and the dimensionless size of particle. Another goal is to study the analogy and difference between the migration of a spherical particle in tube Poiseuille flow and that of a circular particle in plane Poiseuille flow by analyzing the results obtained from the same procedure of data interrogation.

S. Balachandar and A. Prosperetti (eds), Proceedings of the IUTAM Symposium on Computational Multiphase Flow, 225–235.

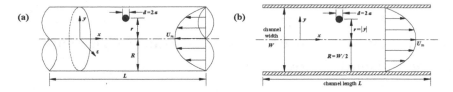

Fig. 1. Sketches for the problem of (a) a rigid spherical particle in tube Poiseuille flow and (b) a rigid circular particle in plane Poiseuille flow.

2 Governing equations and dimensionless parameters

The fluid-particle system is governed by the Navier–Stokes equations for the fluid and Newton's equations for rigid body motions. The dimensionless governing equations in a general three-dimensional case are (see [6])

$$
R_e \left(\frac{\partial \mathbf{u}}{\partial t} + (\mathbf{u} \cdot \nabla)\mathbf{u} \right) = -\nabla p + \nabla^2 \mathbf{u}, \tag{1}
$$

$$
\frac{\rho_p}{\rho_f} R_e \frac{d\mathbf{U}_p}{dt} = G\mathbf{e}_g + \frac{6}{\pi} \int [-p\mathbf{1} + \boldsymbol{\tau}] \cdot \mathbf{n} \, d\Gamma,
$$

$$
\frac{\rho_p}{\rho_f} R_e \frac{d\boldsymbol{\Omega}_p}{dt} = \frac{60}{\pi} \int (\mathbf{x} - \mathbf{X}_p) \times ([-p\mathbf{1} + \boldsymbol{\tau}] \cdot \mathbf{n}) \, d\Gamma. \tag{2}
$$

The dimensionless parameters are

$$
R_e = \frac{\rho_f V(2a)}{\mu} = \frac{\rho_f \dot{\gamma}_w (2a)^2}{\mu} = \frac{8a^2 \rho_f U_m}{\mu R}, \quad \text{the Reynolds number;} \tag{3}
$$

where $\dot{\gamma}_w$ is the wall shear rate and $V = 2a\dot{\gamma}_w$.

$$
G = \frac{(\rho_p - \rho_f)g(2a)^2}{\mu V}, \quad \text{the gravity number;} \tag{4}
$$

$$
\rho_p/\rho_f, \quad \text{the density ratio.} \tag{5}
$$

It is convenient to carry out the analysis of correlations in terms of dimensionless forms of correlating parameters. The ratio of the particle radius a to tube radius R and the dimensionless radial position $\bar{r}$ are defined by

$$
\bar{a} = a/R, \quad \bar{r} = r/R. \tag{6}
$$

The dimensionless lift is given by

$$
\bar{L} = \frac{6\rho_f}{\pi \mu^2} L. \tag{7}
$$

Relative motions between the fluid and the particle, which may be characterized by slip velocities, are essential to understand the lift force on the particle. We use U_p and Ω_p to denote the translational and angular velocities of the particle at steady state. The slip velocities are defined as:

$$U_s = U_f - U_p, \quad \text{the slip velocity;} \tag{8}$$

$$\Omega_s = \Omega_p - \Omega_f = \Omega_p + \dot{\gamma}/2, \quad \text{the slip angular velocity,} \tag{9}$$

where U_f and $\dot{\gamma}$ are the fluid velocity and the local shear rate evaluated at the location of the particle center in the undisturbed flow.

We express the flow quantities $U_m, U_s, \Omega_s, \Omega_{se}$ in the form of Reynolds numbers. A flow Reynolds number is given by

$$\bar{U}_m = \frac{\rho_f U_m R}{\mu} = \frac{R_e}{8\bar{a}^2}. \tag{10}$$

Slip velocity Reynolds numbers are defined as

$$\bar{U}_s = \rho_f U_s (2a)/\mu, \quad \bar{\Omega}_s = \rho_f \Omega_s (2a)^2/\mu, \quad \bar{\Omega}_{se} = \rho_f \Omega_{se} (2a)^2/\mu. \tag{11}$$

A dimensionless form of the product $U_s (\Omega_s - \Omega_{se})$ which enters into our lift law is given as the product $\bar{U}_s (\bar{\Omega}_s - \bar{\Omega}_{se})$.

We draw the reader's attention to the fact that the flow is in the negative x direction in our three-dimensional simulation (see Figure 1(a)). The symbol U_m in (3) and (10) should be understood as the magnitude of the fluid velocity at the tube centerline. Similarly, we use the magnitude of U_f and U_p to calculate the slip velocity U_s defined in (8). We shall focus on the steady state flow of a neutrally buoyant spherical particle, in which the left-hand side of (2) and the term Ge_g in (2) vanish. Thus, R_e and $\bar{a}$ are the two parameters at play.

Here, we do not describe again the equations and parameters in two dimensions. Interested readers are referred to [1, 2, 5] for details. The only change is that in this paper the coordinate is at the centerline of channel.

In the plane Poiseuille flow, the Reynolds number is

$$R_e = \frac{\rho_f V (2a)}{\mu} = \frac{2\rho_f \dot{\gamma}_w (2a)^2}{\mu} = \frac{\rho_f W (2a)^2 \bar{p}}{\mu^2}, \tag{12}$$

where $\bar{p}$ is the constant pressure gradient. We also introduce the dimensionless parameter $\bar{r}$ to the two-dimensional cases,

$$\bar{r} = \frac{|y|}{R} = \frac{|y|}{W/2}, \tag{13}$$

where W is the width of channel and R is half of the channel width. The dimensionless lift is given by

$$\bar{L} = \frac{6\rho_f (2a)}{\pi \mu^2} L. \tag{14}$$

Other dimensionless quantities, such as $\bar{U}_s, \bar{\Omega}_s$ and $\bar{\Omega}_{se}$, result in the same expressions as those in the three dimensions.

3 Correlations from the numerical simulation

In our numerical simulations, we perform both unconstrained and constrained simulations. In the unconstrained simulation, the particle moves freely until it reaches its equilibrium position. In the constrained one, the particle is only allowed to move along a line parallel to the axis of tube and rotate freely; its lateral migration is suppressed. Numerical experiments using constrained simulation provide us with the distribution of the lift force and particle velocity in the tube and the position and velocity of the particle at equilibrium. We develop correlations for these quantities in this section. The key correlation is for the lift force, which shows the dependence of the lift force on the slip angular velocity discrepancy $\Omega_s - \Omega_{se}$. The lift force correlation predicts the change of sign of the lift force, which is necessary to explain the two-way migration in the Segré–Silberberg effect. The correlations for the equilibrium state of the particle are also of interest, because they may be used to predict the position and the velocities of the particle at equilibrium.

3.1 Correlation for the lift force

The steady state values of the lift forces on a particle at different radial positions computed in constrained simulation are plotted in Figure 2 for a spherical particle with the radius ratio $\bar{a} = 0.15$ in three-dimensional tube Poiseuille flow. The same correlations for the migration of a circular particle in two-dimensional plane Poiseuille flow are also given in this figure. The positive direction of the lift force is in the negative $\mathbf{e}_r$ direction. In other words, $\bar{L}$ is positive when pointing to the centerline and negative when pointing away from the centerline.

The equilibrium positions of a neutrally buoyant particle are the points where $\bar{L} = 0$. The stability of the equilibrium at a zero-lift point can be determined from the slope of the $\bar{L}$ vs. $\bar{r}$ curve. The centerline is on a negative-slope branch of the $\bar{L}$ vs. $\bar{r}$ curve. When a particle is disturbed away from the centerline, the lift force is negative and drives the particle further away from the centerline. Therefore the centerline is an unstable equilibrium position. The other zero-lift point is between the centerline and the wall and it is on a positive-slope branch of the curve. When the particle is disturbed away from this point, the lift force tends to push the particle back. Thus the zero-lift point between the centerline and the wall is a stable equilibrium position. It is a surprise to see that the stable equilibrium position $\bar{r}_e$ moves towards the wall as the Reynolds numbers increases for the three-dimensional cases but away from the wall for the two-dimensional cases.

We discuss the three-dimensional cases with the radius ratio $\bar{a} = 0.15$. When the Reynolds number is small ($R_e = 1, 2, 9$ or 18), only one stable branch and one unstable branch can be observed in the $\bar{L}$ vs. $\bar{r}$ curves (Figures 2(b), 2(d)). For higher Reynolds numbers, the distributions of the lift force as a function of the radial position become more complicated (Figure 2(f)). A refined mesh was necessary to obtain converged results at high R_e.

We seek expressions for the lift force in terms of the slip velocities. The slip velocity Reynolds numbers have been defined in (11). We plot $\bar{\Omega}_s - \bar{\Omega}_{se}$ at different radial

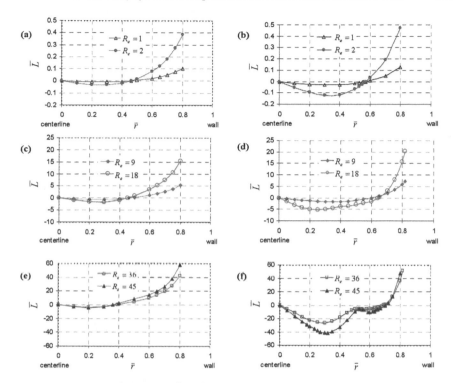

Fig. 2. The dimensionless lift force $\bar{L}$ at different radial positions for a particle with the radius ratio $\bar{a} = 0.15$. The curves in (a), (c) and (e) are for the migration of a circular particle in plane Poiseuille flow with different Reynolds numbers; the curves in (b), (d) and (f) are for the migration of a spherical particle in tube Poiseuille flow.

positions in Figure 3 for a particle with $\bar{a} = 0.15$ for both the two-dimensional and the three-dimensional cases. Comparison of Figures 2 and 3 shows that the quantity $\bar{\Omega}_s - \bar{\Omega}_{se}$ always changes sign above and below the stable equilibrium position for either the two dimensions or the three dimensions.

The lift $\bar{L}$ changes sign with the discrepancy $\bar{\Omega}_s - \bar{\Omega}_{se}$ near the stable equilibrium position at all the Reynolds numbers. The lift correlation is developed in the region near the stable equilibrium position.

We seek correlations between the lift force $\bar{L}$ and the product

$$F = \bar{U}_s(\bar{\Omega}_s - \bar{\Omega}_{se}). \tag{15}$$

From our data, we noted that in the vicinity of the stable equilibrium position, the relation between $\bar{L}$ and F may be represented by a linear correlation:

$$\bar{L}(\bar{r}, R_e, \bar{a}) = k(R_e, \bar{a})F(\bar{r}, R_e, \bar{a}), \tag{16}$$

where k is the proportionality coefficient which depends on the Reynolds number and the radius ratio $\bar{a}$. Some examples of the linear correlation between $\bar{L}$ and F

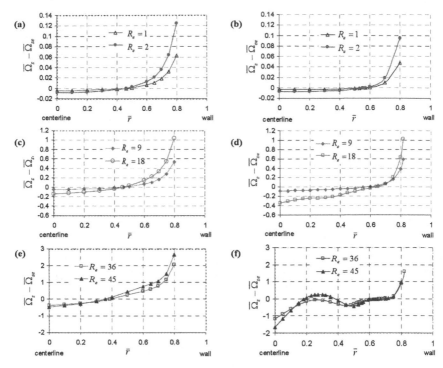

Fig. 3. The dimensionless slip angular velocity discrepancy at different radial positions for a particle with $\bar{a} = 0.15$. The curves in (a), (c) and (e) are for the migration of a circular particle in plane Poiseuille flow with different Reynolds numbers; the curves in (b), (d) and (f) are for the migration of a spherical particle in tube Poiseuille flow.

are plotted in Figure 4. The linear correlation (16) is not valid far away from the equilibrium position.

We use power laws to fit the expressions for k in terms of the Reynolds number and then obtain the linear correlations between $\bar{L}$ and F by Equation (16). To reveal the dependence of the lift force on the slip velocities explicitly, we substitute the definitions of $\bar{L}$ and F into these correlations and then obtain the lift laws in Table 2.

The lift force in our correlation is on a freely rotating particle translating at steady velocity. Thus correlations in Table 1 apply to particles with zero acceleration. For a migrating particle with substantial acceleration, these correlations may not be valid.

3.2 Correlations for slip velocity U_s and slip angular velocity Ω_s

Besides the lift force on the particle, the translational and the angular velocities of the particle at steady state are also of interest. We use power laws to fit the correlations between the slip velocities and the Reynolds number. All of coefficients in the power law correlations can be explicitly expressed in terms of $\bar{r}$. Details about the construction of the correlations for the slip velocity U_s and the slip angular velocity

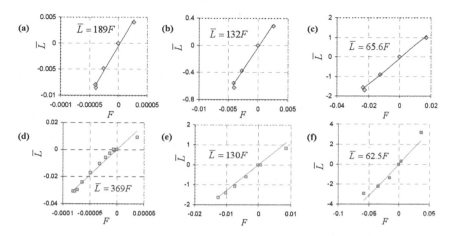

Fig. 4. The linear correlation between $\bar{L}$ and F in the vicinity of the stable equilibrium position of a neutrally buoyant particle. (a) $R_e = 1$, $\bar{a} = 0.15$, plane Poiseuille flow; (b) $R_e = 9$, $\bar{a} = 0.15$, plane Poiseuille flow; (c) $R_e = 18$, $\bar{a} = 0.15$, plane Poiseuille flow; (d) $R_e = 1$, $\bar{a} = 0.15$, tube Poiseuille flow; (e) $R_e = 9$, $\bar{a} = 0.15$, tube Poiseuille flow; (f) $R_e = 18$, $\bar{a} = 0.15$, tube Poiseuille flow.

Table 1. Lift laws for the migration of a single neutrally-buoyant particle with $\bar{a} = 0.1$ and $\bar{a} = 0.15$ in plane and tube Poiseuille flows.

	Circular particle migrates in plane Poiseuille flow	
	$\bar{a} = 0.1$	$\bar{a} = 0.15$
range of R_e	$1 \le R_e \le 18$	$1 \le R_e \le 45$
$k = \bar{L}/F$	$k = 619 R_e^{-0.757}$	$k = 272 R_e^{-0.609}$
$L = \dfrac{\pi}{4}\dfrac{\mu^2}{\rho_f (2a)}\bar{L}$	$L = 486 R_e^{-0.757} \rho_f U_s (\Omega_s - \Omega_{se})(2a)^2$	$L = 214 R_e^{-0.609} \rho_f U_s (\Omega_s - \Omega_{se})(2a)^2$
	Spherical particle migrates in tube Poiseuille flow	
	$\bar{a} = 0.1$	$\bar{a} = 0.15$
range of R_e	$1 \le R_e \le 12$	$1 \le R_e \le 22.5$
$k = \bar{L}/F$	$k = 809 R_e^{-0.604}$	$k = 450 R_e^{-0.658}$
$L = \dfrac{\pi}{6}\dfrac{\mu^2}{\rho_f}\bar{L}$	$L = 424 R_e^{-0.604} \rho_f U_s (\Omega_s - \Omega_{se})(2a)^3$	$L = 236 R_e^{-0.658} \rho_f U_s (\Omega_s - \Omega_{se})(2a)^3$

Ω_s can be found in [6] and will not be shown here. The final correlations for U_s and Ω_s and the corresponding applicable ranges are listed in Table 2.

Table 2. Correlations of slip velocity and slip angular velocity for the migration of a single neutrally-buoyant particle with $\bar{a} = 0.1$ and $\bar{a} = 0.15$ in plane Poiseuille flow and tube Poiseuille flow.

	Circular particle migrates in plane Poiseuille flow	
	$\bar{a} = 0.1$	$\bar{a} = 0.15$
range of $\bar{r}$	$0.10 \leq \bar{r} \leq 0.85$	$0.10 \leq \bar{r} \leq 0.80$
$\bar{U}_s$	$\bar{U}_s = 4.3 \times 10^{-3} \exp(3.2\bar{r})R_e^{-0.45\bar{r}+1.1}$	$\bar{U}_s = 6.8 \times 10^{-3} \exp(3.0\bar{r})R_e^{-0.27\bar{r}+1.1}$
$\bar{\Omega}_s$	$\bar{\Omega}_s = 6.4 \times 10^{-5} \exp(7.3\bar{r})R_e^{-0.95\bar{r}+1.9}$	$\bar{\Omega}_s = 1.3 \times 10^{-4} \exp(7.3\bar{r})R_e^{-0.69\bar{r}+1.6}$
	Spherical particle migrates in tube Poiseuille flow	
	$\bar{a} = 0.1$	$\bar{a} = 0.15$
range of $\bar{r}$	$0.05 \leq \bar{r} \leq 0.85$	$0.10 \leq \bar{r} \leq 0.80$
$\bar{U}_s$	$\bar{U}_s = 7.4 \times 10^{-3} \exp(2.1\bar{r})R_e^{-1.4\bar{r}+1.9}$	$\bar{U}_s = 1.1 \times 10^{-2} \exp(2.2\bar{r})R_e^{-1.1\bar{r}+1.7}$
$\bar{\Omega}_s$	$\bar{\Omega}_s = 6.8 \times 10^{-6} \exp(9.6\bar{r})R_e^{-3.3\bar{r}+3.9}$	$\bar{\Omega}_s = 2.1 \times 10^{-5} \exp(9.2\bar{r})R_e^{-2.1\bar{r}+2.8}$

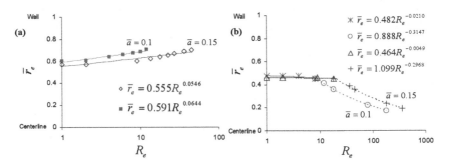

Fig. 5. The stable equilibrium position $\bar{r}_e$ of a neutrally buoyant particle as a function of the Reynolds number in (a) tube Poiseuille flow and (b) plane Poiseuille flow.

3.3 Correlations for parameters at equilibrium

The equilibrium state of a particle is always the focus of the study of particle migration. We obtain the particle parameters at stable equilibrium, such as the equilibrium position $\bar{r}_e$, the slip velocity U_{se} and the slip angular velocity Ω_{se} by unconstrained simulation and find that they may be correlated to the Reynolds number. We summarize the particle parameters at stable equilibrium in Table 3.

The correlations for the equilibrium position $\bar{r}_e$ are shown in Figure 5. In two dimensions, multiple power law fittings are used in different ranges of Reynolds numbers (Figure 5(b)). As mentioned before, $\bar{r}_e$ moves closer to the wall as the Reynolds number increases for the three-dimensional cases but moves to the centerline for the two-dimensional cases.

Figure 6 shows that power law correlations also exist between the dimensionless slip angular velocity at equilibrium $\bar{\Omega}_{se}$ and the Reynolds number R_e for either the two dimensions or the three dimensions. These correlations are important because they give explicitly the slip angular velocity when the particle is at stable equilibrium.

Table 3. Particle parameters at stable equilibrium: the equilibrium position $\bar{r}_e$, the dimensionless slip angular velocity $\bar{\Omega}_{se} = \rho_f \Omega_{se}(2a)^2/\mu$ and the dimensionless slip velocity $\bar{U}_{se} = \rho_f U_{se}(2a)/\mu$.

$\bar{a}$	R_e	Circular particle migrates in plane Poiseuille flow				Spherical particle migrates in tube Poiseuille flow			
		$\bar{U}_m$	$\bar{r}_e$	$\bar{\Omega}_{se}$	$\bar{U}_{se}$	$\bar{U}_m$	$\bar{r}_e$	$\bar{\Omega}_{se}$	$\bar{U}_{se}$
0.05	2	-	-	-	-	100	0.731	0.00710	0.0247
0.1	1	1.25	0.478	0.00210	0.0155	12.5	0.603	0.00188	0.0219
	2	2.5	0.478	0.00460	0.0317	25	0.608	0.00509	0.0444
	4	5	0.476	0.0124	0.0661	50	0.638	0.0209	0.0901
	6	-	-	-	-	75	0.661	0.0498	0.152
	8	10	0.456	0.0406	0.134	100	0.674	0.0901	0.470
	10	-	-	-	-	125	0.684	0.139	0.712
	12	15	0.413	0.0724	0.194	150	0.708	0.202	0.296
	18	22.5	0.357	0.110	0.276	-	-	-	-
	80	100	0.222	0.499	1.01	-	-	-	-
	180	225	0.174	1.09	2.12	-	-	-	-
0.15	1	0.56	0.463	0.00419	0.0225	5.56	0.573	0.00354	0.0338
	2	1.11	0.463	0.00849	0.0452	11.1	0.573	0.00765	0.0675
	9	5	0.464	0.0491	0.214	50	0.601	0.0861	0.306
	13.5	-	-	-	-	75	0.623	0.197	0.482
	18	10	0.454	0.145	0.439	100	0.642	0.342	0.730
	22.5	-	-	-	-	125	0.657	0.513	0.785
	27	-	-	-	-	150	0.670	0.705	1.07
	36	20	0.388	0.368	0.799	200	0.691	1.16	1.18
	45	25	0.359	0.455	0.967	250	0.700	1.67	1.74
	180	100	0.234	1.70	3.14	-	-	-	-
	360	200	0.190	3.21	5.62	-	-	-	-
0.2	32	-	-	-	-	100	0.598	0.793	1.74
0.25	50	-	-	-	-	100	0.567	1.49	2.84

The correlations for parameters at equilibrium are summarized in Table 4.

4 Conclusion

- A lift law $L = CU_s(\Omega_s - \Omega_{se})$ analogous to $L = \rho U \Gamma$ of the classical aerodynamics is valid in both two dimensions and three dimensions.
- Equilibrium may be identified at the Segré–Silberberg radius at which the lift vanishes (for a neutrally buoyant particle).

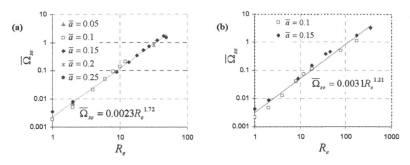

Fig. 6. The correlations between $\bar{\Omega}_{se}$ and the Reynolds number R_e for the migration of (a) a spherical particle in tube Poiseuille flow and (b) a circular particle in plane Poiseuille flow.

Table 4. Correlations of particle parameters at equilibrium for the migration of a single neutrally-buoyant particle with $\bar{a} = 0.1$ and $\bar{a} = 0.15$ in plane Poiseuille flow and tube Poiseuille flow.

	Circular particle migrates in plane Poiseuille flow	Spherical particle migrates in tube Poiseuille flow
$\bar{\Omega}_{se}$	$\bar{\Omega}_{se} = 0.0031R_e^{1.21}$ (for $0.10 \le \bar{a} \le 0.15$)	$\bar{\Omega}_{se} = 0.0023R_e^{1.72}$ (for $0.05 \le \bar{a} \le 0.25$)
Ω_{se}	$\Omega_{se} = 0.0031R_e^{1.21}\mu/\left(\rho_f 4a^2\right)$ (for $0.10 \le \bar{a} \le 0.15$)	$\Omega_{se} = 0.0023R_e^{1.72}\mu/\left(\rho_f 4a^2\right)$ (for $0.05 \le \bar{a} \le 0.25$)
$\bar{r}_e$	$\bar{r}_e = 0.482R_e^{-0.0210}$ (for $\bar{a} = 0.1$ and $1 \le R_e \le 8$); $\bar{r}_e = 0.888R_e^{-0.3147}$ (for $\bar{a} = 0.1$ and $8 \le R_e \le 180$); $\bar{r}_e = 0.464R_e^{-0.0049}$ (for $\bar{a} = 0.15$ and $1 \le R_e \le 18$); $\bar{r}_e = 1.099R_e^{-0.0869}$ (for $\bar{a} = 0.15$ and $18 \le R_e \le 360$).	$\bar{r}_e = 0.591R_e^{0.0644}$ (for $\bar{a} = 0.1$ and $1 \le R_e \le 12$); $\bar{r}_e = 0.555R_e^{0.0546}$ (for $\bar{a} = 0.15$ and $1 \le R_e \le 45$).

- The slip angular velocity discrepancy $\Omega_s - \Omega_{se}$ is the circulation for the free particle and it is shown to change sign at the equilibrium position where the lift reaches zero on its stable branch. The behaviors of L and $\Omega_s - \Omega_{se}$ are very similar between the two dimensions and the three dimensions at comparatively low Reynolds numbers.
- The equilibrium position (the Segré–Silberberg radius) moves towards the wall as R_e increases at each fixed $\bar{a}$ for the migration of a spherical particle in tube Poiseuille flow but moves away from the wall for the migration of a circular particle in plane Poiseuille flow.

References

1. Joseph, D.D. and Ocando, D., 2002, Slip velocity and lift, *J. Fluid Mech.* **454**, 263–286.
2. Patankar, N.A., Huang, P.Y., Ko, T. and Joseph, D.D., 2001, Lift-off of a single particle in Newtonian and viscoelastic fluids by direct numerical simulation, *J. Fluid Mech.* **438**, 67–100.
3. Segré, G. and Silberberg, A., 1961, Radial Poiseuille flow of suspensions, *Nature* **189**, 209.
4. Segré, G. and Silberberg, A., 1962, Behavior of macroscopic rigid spheres in Poiseuille flow: Part I, *J. Fluid Mech.* **14**, 136–157.
5. Wang, J. and Joseph, D.D., 2003, Lift forces on a cylindrical particle in plane Poiseuille flow of shear thinning fluids, *Phys. Fluids* **15**, 2267–2278.
6. Yang, B.H., Wang, J., Joseph, D.D., Hu, H.H., Pan, T.-W. and Glowinski, R., 2005, Migration of a sphere in tube flow, *J. Fluid Mech.* **540**, 109–131.

New Advances in Force-Coupling Method: From Micro to Macro

Martin R. Maxey[1], Don Liu[2], Suchuan Dong[1] and George E. Karniadakis[1]

[1]*Division of Applied Mathematics, Brown University, Providence, RI 02912, USA;*
e-mail: maxey@cfm.brown.edu, sdong@dam.brown.edu, gk@dam.brown.edu
[2]*Space Geodesy Laboratory, Code 697, NASA Goddard Space Flight Center,*
Greenbelt, MD 20771, USA; e-mail: dliu@bowie.gsfc.nasa.gov

Abstract. The force-coupling method (FCM) provides an efficient tool for computing particle motion and the flow in the surrounding fluid both in confined microflow systems and in larger scale suspensions. Here we present results for the interaction of individual particles in a shear flow showing that FCM captures reliably the changes in lift and drag forces. We note too the extension from spherical to non-spherical particles and comment on the use of FCM to analyze flow systems, bridging the gap between simulation data and macroscopic descriptions of dispersed two-phase flows.

1 Introduction

In order to investigate microflow systems that involve a small number of particles, or larger scale suspensions it is important to have a reliable way to compute the motion of the particles and the surrounding fluid. Several techniques for full numerical simulation have been developed such as the Distributed Lagrange Multiplier (DLM) method, see [1, 2], or the Lattice Boltzmann method (LBM) [3]. Both schemes are adapted for solid particles and use a static computational grid but require significant numerical resolution per particle. The Force-Coupling Method (FCM) was initially proposed [4] as a self-consistent approximation for finite-sized particles in a turbulent flow and to avoid the limitations of particle-tracking models [5]. It has proven to be a robust tool for studying both low Reynolds number, Stokes flows for which an analytic theory is available, and particle motion at finite Reynolds numbers. The spatial resolution required for each particle with the FCM scheme is modest and using spectral methods reliable results have been obtained using 3–4 grid points per particle diameter, though 4–5 points is preferable. LBM and DLM simulations usually require a minimum of 8 grid points and typically 12 or more points for accuracy. The scheme is also computationally efficient, requiring orders of magnitude less effort as compared to direct numerical simulations [6]. All of the FCM simulations presented here were performed on a single CPU Linux PC workstation, while the corresponding DNS results were obtained with 32 CPUs on an IBM SP3 parallel computer or comparable system.

S. Balachandar and A. Prosperetti (eds), Proceedings of the IUTAM Symposium on Computational Multiphase Flow, 237–246.

A further advantage of FCM is that all the flow variables vary continuously so that it is straightforward to go from simulation data to continuum descriptions and averaged flow quantities. It is possible to see immediately the significance of averaging procedures. In the following sections we briefly summarize the FCM scheme, present some examples of particle motion in shear flows including nonspherical particles and discuss how the results for large systems of particles may be analyzed as a continuous system.

2 Force-Coupling Method (FCM)

The Force-Coupling Method (FCM), developed in [7, 8], uses a set of finite force multipoles to represent the presence of each particle in the flow. Fluid is assumed to fill the whole flow domain, including the volume occupied by the particles. The flow is specified in terms of a 'volumetric' velocity field $\mathbf{u}(\mathbf{x}, t)$ that is incompressible and satisfies

$$\rho D\mathbf{u}/Dt = -\nabla p + \mu \nabla^2 \mathbf{u} + \mathbf{f}(\mathbf{x}, t), \tag{1}$$

where the fluid density and pressure are ρ and p respectively, μ is the viscosity. The body force density $\mathbf{f}$ is made up of the contributions from the individual spherical particles centered at $\mathbf{Y}^{(n)}(t)$ and is given by

$$f_i(\mathbf{x}, t) = \sum_{n=1}^{N} F_i^{(n)} \Delta(\mathbf{x} - \mathbf{Y}^{(n)}(t)) + G_{ij}^{(n)} \partial \Delta'(\mathbf{x} - \mathbf{Y}^{(n)}(t))/\partial x_j. \tag{2}$$

The local density distributions $\Delta(\mathbf{x})$, $\Delta'(\mathbf{x})$ are Gaussian functions

$$\Delta(\mathbf{x}) = (2\pi\sigma^2)^{-3/2} \exp(-\mathbf{x}^2/2\sigma^2), \tag{3}$$

with length scales σ and σ' set in terms of the particle radius a as $\sigma/a = \sqrt{\pi}$ and $(\sigma'/a)^3 = 6\sqrt{\pi}$.

The first term in (2) represents a finite force monopole of strength $\mathbf{F}$ while the second term is a force dipole G_{ij}, that combines the effect of an external torque on the particle and a symmetric stresslet. The symmetric part of $G_{ij}^{(n)}$ is adaptively set to eliminate any net rate of strain on the rigid particles and for each particle

$$\int \left(\frac{\partial u_i}{\partial x_j} + \frac{\partial u_j}{\partial x_i} \right) \Delta'(\mathbf{x} - \mathbf{Y}^{(n)}(t))d^3\mathbf{x} = 0. \tag{4}$$

The strength of the force monopole is set by the external force $\mathbf{F}^{\text{ext}}$acting on the particle and the inertia of the particle. Under conditions of steady motion $\mathbf{F}^{(n)}$ is equal to the force exerted by the particle on the fluid. The velocity of the particle $\mathbf{V}^{(n)}(t)$ is found by forming a local average of the fluid velocity over the region occupied by the particle as

$$\mathbf{V}^{(n)}(t) = \int \mathbf{u}(\mathbf{x}, t)\Delta(\mathbf{x} - \mathbf{Y}^{(n)}(t))d^3\mathbf{x}. \tag{5}$$

A spectral/hp element method [9] has been used to solve for the primitive variables **u**, p in the Navier–Stokes equations.

The force-coupling method was initially derived so as to reproduce standard results for Stokes flow. Detailed comparisons for individual particles and groups of particles at low Reynolds numbers are given in [7, 8, 10]. Further comparisons against full direct numerical simulations for finite Reynolds number flows are given in [6, 11, 12]. All of these show that there is either exact or good agreement, within 1–3%, for the fluid forces in a range of situations. A comparison of experiments [13] involving particles in a channel flow with corresponding simulations with FCM also showed good agreement. The flow representation at distances of $0.25 - 0.5a$ from a particle surface agrees generally with DNS, while nearer the surface the representation is smoothed out over the interior region of the particle.

3 Particles in shear flows

3.1 Spherical particles

An illustration of how FCM can predict particle motion is given by a calculation of the forces acting on particles held fixed, without rotation, in a Poiseuille flow. We compare results from a full direct numerical simulation (DNS) with FCM results for both a single particle and two particles held fixed near a wall. The flow configuration is shown in Figure 1. The interaction between the two particles and the presence of the wall illustrates too the limitations of particle-tracking models for such situations.

In terms of the particle radius a, the channel length is $20a$ and the distance between the two planar, no-slip walls is $7a$. In the spanwise direction, $-3.5 < x_3/a < 3.5$, periodic boundary conditions are applied. The Poiseuille flow is driven by a fixed pressure gradient such that in the absence of any particles the parabolic velocity profile has a value $U_0 = 1.225$ at the centerline and the approach velocity for the center of the particle is 1.0. Otherwise a periodic boundary condition is applied in the streamwise direction. For both the case of a single sphere and the case of two spheres, the particles are positioned at $Y_2 = 1.5$ so that the gap between a particle and the adjacent wall is $1.0a$. The fluid viscosity and density are both equal to one and the particle Reynolds number is 2.0, based on the local approach velocity of the flow. The corresponding shear Reynolds number, based on the local velocity gradient, is 4.2.

The DNS for a single particle uses a structured mesh of 1728 hexahedral spectral elements that fully resolves the flow about the spherical surface and adequately resolves the flow elsewhere in the channel. Each element has an eighth-order Jacobi polynomial representation. For the two particle case, where $Y_1^{(1)} = 6a$ and $Y_1^{(2)} = 10a$, a new nonuniform structured mesh of 768 ninth-order spectral elements is used. For all the computations with FCM a much simpler mesh of 480 rectangular elements is used, distributed uniformly in the streamwise and spanwise directions as a 20×4 array. In the x_2 direction there are 6 elements distributed to provide greater resolution of the region $0 < x_2 < 3.5a$. Each element has an sixth-order Jacobi

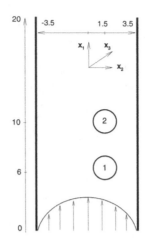

Fig. 1. Positions of spheres in channel flow.

Table 1. Drag, lift and torque on particles held fixed in a channel flow.

	Drag $F_1/\mu a U_0$	Lift $F_2/\mu a U_0$	Torque $T_3/\mu a^2 U_0$
FCM Particle 1	18.14	−1.36	2.60
DNS Particle 1	17.48	−1.33	2.69
FCM Particle 2	17.39	0.41	2.51
DNS Particle 2	16.68	0.43	2.59
Single particle DNS	21.82	−0.98	3.22
Single particle FCM	22.58	−1.00	3.11

polynomial representation. Further details and illustrations of the meshes used are given by Liu [12].

A comparison of the drag and lift forces, and the torque acting on the single particle is given in Table 1. The FCM results agree well with those from DNS especially for the sensitive features of lift and torque. As FCM is a mobility, as opposed to resistance, formulation a penalty scheme [14] is employed to maintain the particle at its fixed position. The drag and other forces are determined from the resulting FCM parameters in (2). The results show that the lift force is directed away from the wall, consistent with general expectations from the theory of Saffman [15] and McLaughlin [16]. These theories apply to lower Reynolds number flows and isolated particles in a uniform shear flow but are often the basis for determining lift forces in particle-tracking models. The drag is substantially larger than for an isolated particle due to the influence of the nearby wall.

For the two particles, there is again good agreement between the forces computed from FCM and the DNS results. The drag on both particles is significantly lower due to their mutual interaction effects on the flow. Most noticeably, the lift on the upstream particle is increased and directed away from the wall whereas the lift force

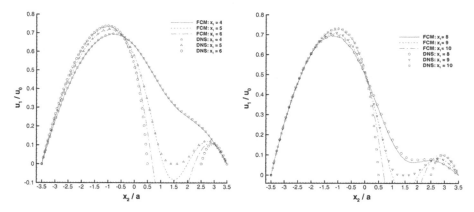

Fig. 2. Velocity profiles of u_1 versus x_2 at selected streamwise locations for the two spheres: sphere 1, left; sphere 2, right.

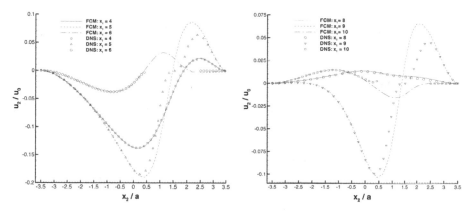

Fig. 3. Velocity profiles of u_2 versus x_2 at selected streamwise locations for the two spheres: sphere 1, left; sphere 2, right.

on the downstream particle is directed towards the wall. Standard methods based on particle tracking models would fail to capture this effect yet such interactions are quite likely at even modest void fractions.

There is also good comparison of the flow fields computed by the two methods. In Figures 2 and 3 we compare the profiles for the streamwise and wall-normal velocity components at selected streamwise locations within the symmetry plane, $x_3 = 0$. These show the flow variation just upstream of the particles and in the region between them. As expected there is a good correspondence in the region half a radius or further from the particle surface. FCM does not resolve the boundary conditions on the particles and simply matches lower-order integral moments. The usually strong surface vorticity is distributed in a smoother variation extending into the volume nominally occupied by the particle. For the calculation of viscous dissipation and

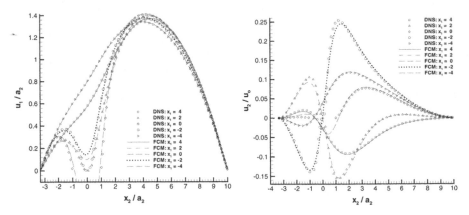

Fig. 4. Flow past a fixed spheroid in a channel flow: velocity profiles of u_1 (left) and u_2 (right) at different streamwise locations.

other physical properties, the particle volume is an active part of the flow domain [7, 8].

3.2 Spheroidal particles

Originally, FCM was developed for the motion of rigid spherical particles in a fluid at low to moderate particle Reynolds numbers. The procedures have recently been extended to rigid ellipsoidal particles [12]. The extension requires a specification of the Gaussian distribution (3) in terms of principal body axes of each particle and scaling the length scale σ to account for the different values of the semi-axes. The ratios of σ and σ' to the semi-axes are the same as for a spherical particle.

As an illustration, the results for flow past a fixed spheroid in a channel are shown in Figure 4. The particle, with semi-axes equal to 2.0, 1.0, 1.0 is placed at $\mathbf{x} = 0$, with the channel walls at $x_2 = -3.33, 10$. The major axis is parallel to the wall. A parabolic velocity profile $u_1 = (1 + 0.3x_2)(1 - 0.1x_2)$ is prescribed at the inflow at $x_1 = -14$ such that the approach velocity $u_0 = 1.0$ at $x_2 = 0$. A simple outflow condition is set at $x_1 = 6$ while periodic conditions are specified in the spanwise direction at $x_3 = \pm 5$. The fluid density and viscosity are again both set equal to one. The FCM results are compared to corresponding results for a full direct numerical simulation resolving all the boundary conditions. The agreement is generally good except in the immediate region of the particle. The results for the drag force and torque agree to within 0.5%, while the lift force is 3.08 (FCM) versus 2.90 (DNS) and is directed away from the adjacent wall.

4 Macroscopic flow analysis

Although the FCM scheme follows the motion of individual finite-sized particles it may be expressed in terms of the more familiar continuum field variables of two-phase flow. The instantaneous particle concentration for any realization of a system of N particles, each of volume $\Omega_P^{(n)}$ is

$$c(\mathbf{x}, t) = \sum_{n=1}^{N} \Omega_P^{(n)} \Delta(\mathbf{x} - \mathbf{Y}^{(n)}), \qquad (6)$$

where $\Omega_P \Delta$ plays the role of an indicator function. We can form the conditional, particle-phase velocity field $\mathbf{v}(\mathbf{x}, t)$ as

$$c\mathbf{v}(\mathbf{x}, t) = \sum_{n=1}^{N} \Omega_P^{(n)} \mathbf{V}^{(n)} \Delta(\mathbf{x} - \mathbf{Y}^{(n)}), \qquad (7)$$

defined where $c \neq 0$. It is straightforward to verify from these definitions the particle-phase conservation law

$$\frac{\partial c}{\partial t} + \nabla \cdot c\mathbf{v} = 0. \qquad (8)$$

The liquid-phase density is $\rho(1 - c(\mathbf{x}, t))$ and a conditional, liquid-phase velocity field $\mathbf{w}(\mathbf{x}, t)$ is similarly given as

$$(1 - c(\mathbf{x}, t))\mathbf{w}(\mathbf{x}, t) = \mathbf{u}(\mathbf{x}, t) - c\mathbf{v}(\mathbf{x}, t). \qquad (9)$$

Again it is straightforward to verify the usual liquid-phase mass conservation law.

In contrast to the volumetric velocity field $\mathbf{u}(\mathbf{x}, t)$, both $\mathbf{v}(\mathbf{x}, t)$ and $\mathbf{w}(\mathbf{x}, t)$ are compressible. FCM ensures that the volume of each particle is constant and that the underlying liquid phase is incompressible. The particle-phase and liquid-phase velocities are only defined in their respective phase while $\mathbf{u}$ is defined at all locations and varies continuously.

The volumetric velocity field $\mathbf{u}$ is easier to interpret physically, especially when the flow variables are averaged either for a random suspension or for a turbulent flow. For example in a turbulent shear flow, the conditional, mean particle velocity is $\langle c\mathbf{v}(\mathbf{x}, t) \rangle / \langle c(\mathbf{x}, t) \rangle$ and is commonly used to determine whether on average the particles are moving faster or slower than the ambient flow. If the particles are nonuniformly distributed and possibly dispersing away from a near-wall region then this particle velocity is not defined where $\langle c \rangle = 0$ and has poor statistical value where the particles are sparsely distributed. A more robust procedure is to compare $\langle c(\mathbf{x}, t) \rangle \langle \mathbf{u}(\mathbf{x}, t) \rangle$ and $\langle c\mathbf{v}(\mathbf{x}, t) \rangle$. These quantities are defined everywhere and are most significant in the regions of higher particle concentration. These concepts may be applied to other variables such as the turbulent fluctuating velocities. If a simulation of dispersed two-phase flow has been computed using conditional phase variables then it is still possible to construct the volumetric velocity from (9). A useful check would then be that this velocity field is indeed incompressible.

This discussion is not limited to FCM and may equally be used to interpret simulation data obtained by other methods such as LBM or DLM. Such direct numerical simulations provide the instantaneous volumetric flow field directly.

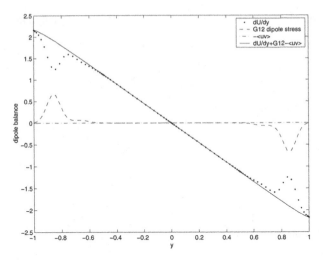

Fig. 5. Profiles of viscous stresses and dipole stress in laminar channel flow.

FCM is also helpful in understanding features such as the enhanced viscosity of a particle suspension. The usual Stokes–Einstein estimate for a dilute suspension of neutrally buoyant, spherical particles is that the effective viscosity of the suspension is $\mu_{\text{eff}} = \mu(1 + 2.5\langle c \rangle)$. This assumes that the particles are small compared to the scale over which the flow is varying and that there are sufficiently many particles to make such an average meaningful. In many applications, especially at finite Reynolds numbers, these assumptions may not be appropriate. The enhanced viscosity is directly linked to the stresslet induced in the flow by each particle in response to the external velocity gradient. The FCM equations (1–2) show that the stresslet distribution gives rise to a stress tensor $\sum_{n=1}^{N} G_{ij}^{(n)} \Delta'(\mathbf{x} - \mathbf{Y}^{(n)}(t))$.

In a Poiseuille flow the mean shear stress is then

$$\tau_{12} = \mu \frac{\partial U_1}{\partial x_2} + \langle -\rho u_1' u_2' \rangle + \left\langle \sum_{n=1}^{N} G_{12}^{(n)} \Delta'(\mathbf{x} - \mathbf{Y}^{(n)}(t)) \right\rangle. \tag{10}$$

In equilibrium, this stress profile will vary linearly across the channel regardless of the respective contributions. In Figure 5 results are shown for a laminar Poiseuille flow at $Re = 1$, based on the channel half-width $h = 1$ and centerline velocity $U_0 = 1$. Particles of radius $a = 0.1$ have been seeded near each wall at $Y_2 = \pm 0.85$ at an average void fraction of 4.2%, or a peak concentration $\langle c \rangle = 0.0955$ at $x_2 = \pm 0.85$. Without the particles, the viscous stress at the wall is 2.0 and we see here that there is about a 10% increase in the drag. The Reynolds stresses are absent in this regular arrangement of the particles, while the dipole stress term counterbalances the variations in the regular viscous shear stress term.

5 Conclusion

Results have been given to illustrate the effectiveness of FCM in calculating the motion of individual particles at finite Reynolds numbers. The method is applicable to both spherical and nonspherical particles. This approach is valuable for investigating microflow systems such as peristaltic pumps created from the forced motion of microspheres [14]. The FCM scheme is also valuable in guiding the analysis of dispersed two-phase flows and provides a physically realizable connection between direct numerical simulations and continuum descriptions. In particular, the volumetric velocity field provides an unambiguous representation of the flow dynamics and preserves the properties of incompressible flow and mass conservation.

Acknowledgments

This work was supported by the National Science Foundation under an award CTS-0326702 and by DARPA-ATO.

References

1. N.A. Patankar, P. Singh, D.D. Joseph, R. Glowinski and T.W. Pan, 2000, A new formulation of the distributed Lagrange multiplier/fictitious domain method for particulate flows, *Int. J. Multiphase Flow* **26**, 1509–1524.
2. R. Glowinski, T.W. Pan, T.I. Hesla, D.D. Joseph and J. Periaux, 2001, A fictitious domain approach to the direct numerical simulation of incompressible viscous flow past moving rigid bodies: Application to particulate flow, *J. Comput. Phys.* **169**, 363–426.
3. A.J.C. Ladd and R. Verberg, 2001, Lattice-Boltzmann simulations of particle-fluid suspensions, *J. Stat. Phys.* **104**, 1191–1251.
4. M.R. Maxey, B.K. Patel, E.J. Chang and L.-P. Wang, 1997, Simulations of dispersed turbulent multiphase flow, *Fluid Dyn. Res.* **20**, 143–156.
5. M.R. Maxey and J.J. Riley, 1983, Equation of motion for a small rigid sphere in a nonuniform flow, *Phys. Fluids* **26**, 883–889.
6. S. Dong, D. Liu, M.R. Maxey and G.E. Karniadakis, 2004, Spectral distributed lagrange multiplier method: Algorithm and benchmark test, *J. Comput. Phys.* **195**, 695–717.
7. M.R. Maxey and B.K. Patel, 2001, Localized force representations for particles sedimenting in stokes flow, *Int. J. Multiphase Flow* **9**, 1603–1626.
8. S. Lomholt and M.R. Maxey, 2003, Force Coupling Method for particulate two-phase flow: Stokes flow, *J. Comput. Phys.* **184**, 381–405.
9. G.E. Karniadakis and S.J. Sherwin, 1999, *Spectral/hp Element Methods for CFD*, Oxford University Press.
10. S.L. Dance and M.R. Maxey, 2003, Incorporation of lubrication effects into the force-coupling model for particulate two-phase flow, *J. Comput. Phys.* **189**, 212–238.
11. D. Liu, M.R. Maxey and G.E. Karniadakis, 2002, A fast method for particulate microflows, *J. Microelectromech. Syst.* **11**, 691–702.
12. D. Liu, 2004, *Spectral Element/Force Coupling Method: Application to Colloidal Micro-Devices and Self-Assembly Structures in 3D Complex-Geometry Domains*, Ph.D. Thesis, Brown University.

13. S. Lomholt, B. Stenum and M.R. Maxey, 2002, Experimental verification of the force coupling method for particulate flows, *Int. J. Multiphase Flow* **28**, 225–246.
14. D. Liu, M.R. Maxey and G.E. Karniadakis, 2004, Modeling and optimization of colloidal micro-pumps, *J. Micromech. Microeng.* **14**, 567–575.
15. P.G. Saffman, 1965, The lift on a small sphere in a slow shear flow, *J. Fluid Mech.* **22**, 385–400.
16. J.B. McLaughlin, 1991, Inertial migration of a small sphere in linear shear flows, *J. Fluid Mech.* **224**, 261–274.

Treatment of Particle Collisions in Direct Numerical Simulations of High Speed Compressible Flows

Robert Nourgaliev, Nam Dinh, Loc Nguyen and Theo Theofanous

Center for Risk Studies and Safety, UC Santa Barbara, Santa Barbara, CA 93106-5130, USA;
e-mail: robert@engr.ucsb.edu, nam@engr.ucsb.edu, loc@engr.ucsb.edu, theo@engr.ucsb.edu

1 Introduction

This work is motivated by needs of modeling high-speed particulate flows, such as encountered in explosive dispersal of solid materials, shock-induced powder compaction and fluidization, protection of structures against explosions using particle layers or foams, etc. [13]. It is well known, that such flows are dominated by particle-to-particle collisions, and our aim here is to address this need at the DNS level.

Particle-to-particle collision models can be classified into two major groups: *"hard-sphere"* (integral-form) and *"soft-sphere"* (differential-form) [4]. Upon a collision in the "hard-sphere" models, the particles instantaneously change their velocities based on momentum conservation principles. The loss of energy during a collision can be represented by introducing *restitution coefficients*. An example of "hard-sphere" model used in the DNS of suspensions can be found in a study by Johnson and Tezduyar [8]. In the "soft-sphere" models, the dynamics of colliding particles is represented by solving the equations of motion, in which the force due to collision is computed as a function of deformation ς, i.e. $\mathbf{F}(\varsigma)$. In DNS of particulate flows, two major classes of the "soft collision" models have been used: the *"short-range repulsive"* model of Glowinski et al. [7] and the *"lubrication theory-based collision"* model of Nguyen and Ladd [10]. The collision force in the "short-range repulsive" model mimics the force of the classic Hertz impact theory and is expressed as $F_{\mathrm{col}} \sim \kappa \varsigma^{n}$, where κ is a stiffness coefficient, and $1 \leq n \leq 2$. This model is elastic and unable to represent the loss of energy, which is significant in high-speed collisions. Furthermore, the stiffness coefficient is an empirical constant, which is chosen to smear-out the collision in order to avoid particle overlaps and numerical stability problems. The "lubrication theory-based collision" model on the other hand does not treat solid-solid interactions at all. As such it is limited to low speed collisions, where all energy is dissipated in the hydrodynamics.

Here we are interested in the other extreme of behavior, where the principal effects are due to the solid-sold interaction itself. This involves both elastic, and dissipative effects, as well as fluid-dissipative effects to a degree however that is likely to

S. Balachandar and A. Prosperetti (eds), Proceedings of the IUTAM Symposium on Computational Multiphase Flow, 247–259.

depend on the flow conditions. Our aim is the creation of a computational framework appropriate for representing these phenomena in high speed, compressible flows.

For the gas dynamics we utilize the computational framework which is based on explicit high-order-accurate Godunov-type schemes, implemented in the adaptive mesh refinement (AMR) environment [1, 14], and capable of effectively capturing the motion of fluid-solid boundaries using an Eulerian formulation [11, 12]. Our interface boundary-condition-capturing approach, denoted here as the "Level-Set-based Cartesian Grid (LSCG)/Adaptive Characteristics-Based Matching (aCBM)", is the crucial element of the approach.

Particle collisions can be possibly implemented into this framework using two approaches. The first one is the *"restitution coefficient-based hard sphere (RBHS)"* model. In this approach, upon a collision, the particles instantaneously change their velocities, as in the above-discussed "hard-sphere" low-speed models, with the restitution coefficient ε provided from an *apriori*-developed inelastic collision database. This would be basically a multi-dimensional map, constructed using a physics-based collision model for the particular material involved and all possible collision configurations, including different impact velocities, collision angles and offsets. The second alternative would be a direct implementation of a *physics-based collision (PBC)* model into the algorithm for time advancement of gas dynamics and particle logics, using a "soft-sphere" formulation.

We have found that (i) provided a sufficient grid resolution is feasible (in our case, for example, due to AMR), both RBHS and PBC approaches are numerically stable; (ii) due to specifics of high-speed collision, the time step of gas dynamics ($\Delta t_{gd} \sim$ns) is much less than the collision time ($t_c \sim \mu$ s), which means that the PBC model does not introduce any significant burden to the computation. Therefore, we believe that the PBC collision approach is preferable in high-speed flow simulations, while the RBHS is inefficient (especially when contact geometries and surfaces are taken into account in construction of collision maps) and inelegant. Here, we demonstrate the implementation of PBC collision approach, using Brilliantov et al.'s model [2] for viscoelastic collisions, and test the approach on numerical examples of shock-induced head-on and offset two-particle collisions.

2 Overview of the numerical method

Level-Set-Based Cartesian Grid Method (LSCG). For description of moving fluid-solid boundaries in compressible flows, we employ the Level-Set-based Cartesian Grid (LSCG) method. In this approach, the physical time-space is discretized using structured adaptively refined meshes [1, 12]. The fluid-solid boundaries are represented by the level set function, defined in all computational nodes as a signed distance to the boundary φ. Zero-level of this function represents an interface which separates fluid ($\varphi > 0$) and solid ($\varphi < 0$). We do not solve the level set equation as in the case of fluid-fluid interface [12], but analytically reconstruct the distance function based on currently available particle's positions, updated as discussed below, after each time step of the numerical solution for fluid. More detail description of the LSCG concept is presented in [11].

Gas dynamics. As our basic numerical method for gas dynamics, we employ the Godunov-type scheme, which is based on the third-order-accurate Runge-Kutta TVD time advancement, and the third-order-accurate MUSCL$_3$ space discretization, supplemented by van Albada's limiter. Flux terms are treated with either Local-Lax-Friedrichs (LLF) or AUSM$^{+,up}$ schemes. Heat transfer and viscous effects are neglected for purpose of this illustration.

Characteristics-Based Matching (CBM). The purpose of the CBM is to "infuse" the desirable set of boundary conditions in the numerical solution for computational cells near zero level set. The key features of the CBM [12] are (i) a Riemann-solver-based treatment, which is essential for accuracy in the case of very strong shock waves; and (ii) elimination of the need for ghost fields and corresponding ghost cells, which we found to be necessary for compatibility with AMR (see below). The CBM is based on the generation/tracking/disposal of the subcell-interface-markers (denoted as CBM points), which exist only during one time step Δt. Applying the one-sided Riemann solver at CBM points, the wave structure and gas dynamics solutions at the interface are computed and applied for direct modification of numerical fluxes in the fluid ($\varphi > 0$) Eulerian cells near the interface, using the subcell position of the interface and a flux inter-/extrapolation algorithm. The concept of subcell markers and Riemann solutions are borrowed from front-tracking methods. The natural-neighbor interpolation (NNI) procedure is employed to correct the numerical solutions at Eulerian computational cells whose material occupancy has changed during the time step. These are denoted here as "degenerate" cells. The modification of numerical fluxes and treatment of "degenerate" cells are the substitutes for GFMs [5, 6] ghost fields/cells and the related to it PDE- or FM-based extrapolation techniques. Algorithmic details of the CBM are given in [12].

Structured Adaptive Mesh Refinement (SAMR). In the present study, we use Berger's and Colella's SAMR algorithm [1], as implemented in LLNL's SAMRAI package/(infrastructure) [14] for SAMR applications. SAMR is based on a sequence of nested, logically rectangular meshes, organized in a hierarchy of L grid levels with the coarsest grid covering the entire computational domain. Grids are refined in both time Δt_k and space Δh_k, $k=0,...,L-1$, using the same ratio $r = \frac{h_k}{h_{k+1}}$ for refinement, i.e. $\frac{\Delta t_0}{h_0} = \frac{\Delta t_1}{h_1} = ... = \frac{\Delta t_{L-1}}{h_{L-1}}$. Each level consists of a union of logically rectangular regions, or patches at the same grid resolution h_k. The utilities for dynamic management of AMR patches require tagging criteria for refinement. In the present paper, the grid is refined near fluid-solid boundaries, shocks and contact discontinuities, as discussed in [12]. An example of the SAMR mesh is shown in Figure 1.

Particle logics. The solid phase is assumed to consist of two-dimensional rigid (at the gas dynamic level) circular particles. The governing equations for motion of the

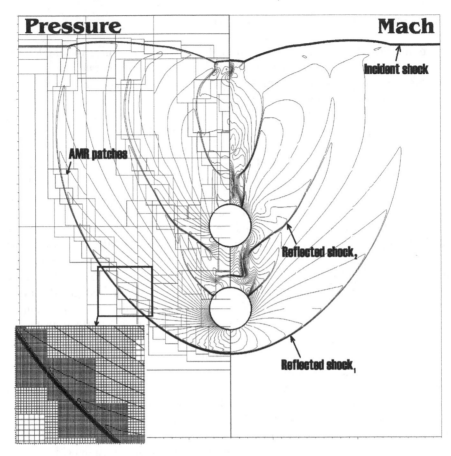

Fig. 1. Example of AMR grid (4 levels of adaptation with refinement ratio 2), outline of patches, pressure (left) and Mach number (right) fields for the head-on collision test. Effective grid resolution is 80 nodes/Dia. The apparent discontinuities in contours are due to the post-processor limitations at junctions of AMR patches.

ath particle are [9]:

$$
\left[\begin{array}{c} \frac{d\mathbf{V}_a}{dt} = \frac{\mathbf{F}_a}{m_a} \\[2mm] \frac{d\mathbf{r}_a}{dt} = \mathbf{V}_a \end{array}\right] \quad \text{and} \quad \left[\begin{array}{c} \frac{d\Omega_{z_a}}{dt} = \frac{K_{z_a}}{I_{z_a}} \\[2mm] \frac{d\theta_{z_a}}{dt} = \Omega_{z_a} \end{array}\right], \tag{1}
$$

where $\mathbf{V}_a$ and $\mathbf{F}_a$ are the velocity of the particle's center of inertia and the total force acting on the ath particle of mass m_a and radius R_a; $\mathbf{r}_a$ and θ_{z_a} are the position of the center of inertia and orientation angle of the particle; Ω_{z_a}, K_{z_a} and $I_{z_a} = \frac{\pi}{2} R_a^4$ are the angular velocity, torque and principle moment of inertia, respectively. The total force acting on the ath particle is composed of the hydrodynamic and "collision" forces, $\mathbf{F}_a = \mathbf{F}_a^{(HD)} + \mathbf{F}_a^{(Col)}$.

Equations (1) are discretized using the third-order-accurate Runge–Kutta differencing scheme. Due to the complex parallel structure of the AMR grid, communication of local forces and torques as needed for integration over each particle is non-trivial, as the particle's surfaces in general reside on different patches and processors. Thus, we have developed an algorithm, in which (i) the surface of each particle is sub-divided on the patch-scale segments, using current arrangement of patches on the finest level; (ii) each segment is uniformly discretized on grid-size segments; (iii) the hydrodynamic forces are interpolated into the center of each grid-size segment, using the natural-neighbor- and bi-cubic-spline- interpolation techniques; and, finally, (iv) the forces for each particle are integrated over all grid-size and patch-size segments, to determine the total force and torques acting on each particle. Item (iv) is implemented by using an *apriorily* constructed map reflecting the residence of patch-scale segments in the patch (processor) network. The above map is dynamically created during each time step on the finest AMR level, as the patches are dynamically created/removed according to the flow conditions.

3 Description of particle-particle collisions

Consider a collision between two circular particles i and j. The relative velocity of the colliding particle's surfaces at the point of contact is $\mathbf{g}_{ij} = \mathbf{V}_i - \mathbf{V}_j$. Defining the normal vector as $\mathbf{n} = \frac{\mathbf{r}_i - \mathbf{r}_j}{|\mathbf{r}_i - \mathbf{r}_j|}$, the normal and tangential components of relative velocity are $\mathbf{g}_N = \mathbf{n}(\mathbf{n} \cdot \mathbf{g}_{ij})$ and $\mathbf{g}_T = \mathbf{g}_{ij} - \mathbf{g}_N$, respectively. Next, we define the deformation as $\varsigma(t) = \tilde{R}_i + \tilde{R}_j - |\mathbf{r}_i - \mathbf{r}_j|$, where $\tilde{R}_a = R_a + \delta_{\text{s.z.}}$, $a=i,j$ and the "safety zone", used to prevent particle-particle overlap, is typically set to $\delta_{\text{s.z.}} = h_{L-1}$.

"Hard-sphere (RBHS) model". In the "hard-sphere" model, upon a collision event $\varsigma > 0$, the particle's velocities are instantaneously changed as

$$
\begin{aligned}
\mathbf{V}_i^{(\text{after})} &= \mathbf{V}_i^{(\text{before})} - \frac{m_{\text{eff}}}{m_i}\left\{(1+\varepsilon)\mathbf{g}_N + \frac{2}{7}|\mathbf{g}_T|\mathbf{t}\right\}, \\
\mathbf{V}_j^{(\text{after})} &= \mathbf{V}_j^{(\text{before})} + \frac{m_{\text{eff}}}{m_j}\left\{(1+\varepsilon)\mathbf{g}_N + \frac{2}{7}|\mathbf{g}_T|\mathbf{t}\right\},
\end{aligned}
\tag{2}
$$

where ε, $m_{\text{eff}} = \frac{m_i m_j}{m_i + m_j}$ and $\mathbf{t}$ are the restitution coefficient, effective mass and tangential vector, respectively.

"Soft-sphere (PBC) model". We employ a *"viscoelastic"* model introduced by Brilliantov et al. in [2]. In this model, the collision force is computed as

$$
\mathbf{F}_a^{(\text{Col})} = \begin{cases} 0 & \text{if } \varsigma < 0, \\ F_N \mathbf{n} + F_T \mathbf{t} & \text{otherwise}, \end{cases}
\tag{3}
$$

where the normal force consists of an elastic (conservative) part due to the deformation ς of the particles and a viscous part due to the dissipation of energy

Table 1. On grid convergence for hydrodynamic forces. $\mathcal{L}_1$-norms of error are computed as $\sum_{N_t} \frac{|C_{D_{g_1}}(t) - C_{D_{g_2}}(t)|}{N_t}$, where N_t is the total number of time steps, and g_i is the grid resolution expressed in terms of number of nodes per diameter.

Test-case	Particle	$\mathcal{L}_1^{10-20/\mathrm{Dia}}$	*Rate*	$\mathcal{L}_1^{20-40/\mathrm{Dia}}$
Head-on elastic (normal) collision	1	$1.328 \cdot 10^{-1}$	2.037	$3.2345 \cdot 10^{-2}$
	2	$7.92 \cdot 10^{-2}$	1.454	$2.89 \cdot 10^{-2}$
Offset viscoelastic (normal) collision	1	$1.247 \cdot 10^{-1}$	1.18	$5.5 \cdot 10^{-2}$
	2	$1.373 \cdot 10^{-1}$	2.035	$3.348 \cdot 10^{-2}$

in the bulk of the particle material, which depends on the deformation rate, i.e. $F_N = C \varsigma^{1/2} (\varsigma + A\dot{\varsigma})$. The conservative part is computed using Hertz's theory of elastic contact, with the elasticity coefficient computed as

$$C = \frac{2Y}{3(1 - v^2)} \sqrt{\frac{R_i R_j}{R_i + R_j}},$$

where Y and v are the Young modulus and the Poisson ratio. The coefficient of viscoelasticity A in addition depends on the particle shape. For tangential force, we use $F_T = \mathrm{sgn}(-|\mathbf{g}_T|)\frac{1}{2}|F_N|$, which is the simplified version of the model employed by Campbell and Brennen in their Lagrangian, particle-only simulations [3].

4 Numerical results and discussion

Formulation. Sample calculations will be presented on examples of two-particle collisions. In a two-dimensional computational domain of size 10×10 mm, two circular particles of radius $R_p = 1$ mm and density $\rho_p = 2,000$ kg/m^3 are suspended in initially motionless air ($\gamma = 1.4$) under atmospheric conditions, i.e. $P = 10^5$ Pa and $\rho = 1.19$ kg/m^3. Two configurations are considered: *"head-on"* ($\mathbf{r}_1 = [5; 3]$ mm and $\mathbf{r}_2 = [5; 5]$ mm) and *"offset"* ($\mathbf{r}_1 = [4.5; 3]$ mm and $\mathbf{r}_2 = [5; 5]$ mm). A planar incident $M_{\mathrm{sh}} = 30$ shock wave hits the particles from below, causing their motion and collision. For the "soft-sphere" model, the following material parameters are used: $Y = 10^{10}$ N/m^2, $v = 0.3$ and $A = 5 \cdot 10^{-7}$ s.

Results and discussion. Figures 2 and 3a,b present the results for the case of head-on elastic collision. As it can be seen, the durations of the collision due to elastic (Hertz-theory)/viscoelastic (Brilliantov's) model are $\approx 0.5/1.5$ μs, whereas the time

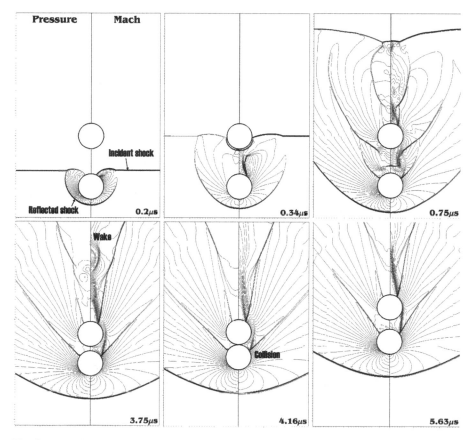

Fig. 2. Dynamics of the pressure (left) and Mach number (right) fields for the head-on elastic collision test. Effective grid resolution is 80 nodes/Dia.

step for gas dynamics is ~ns. Such a small time step is characteristic of the CFL controlled explicit flow solver used for the simulation of highly compressible systems. Consequently, the collision dynamics can be fully resolved by the same time step as used for the flow solver. The longer duration of the viscoelastic collision is due to dissipative term ζ, which makes the collision force less repulsive. Comparison between hydrodynamic and collision forces during the collision event shows that the collision force is about an order of magnitude larger than the hydrodynamic force, Figures 3a vs. 3b. Other practical situations with weaker incident shock would further decrease the significance of the hydrodynamic force during the collision. The above observation is true for both elastic (Hertz's) and viscoelastic (Brilliantov's) models.

As shown in Table 1, our numerical method is high-order-accurate, with the Richardson extrapolation-based convergence rate for hydrodynamic drag coefficient varied between 1 and 2 (assuming the grids at play are already in the asymptotic convergence regime).

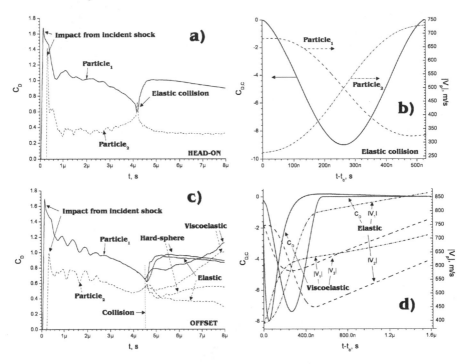

Fig. 3. History of the "hydrodynamic" drag coefficient, (a) and (c); "collision" drag coefficient and particle's velocity during collision, (b) and (d). Head-on collision (a) and (b) vs offset collision (c) and (d). $C_D \equiv \dfrac{|\mathbf{F}|}{\rho_{ps} U_{ps}^2 / 2}$, where "ps" is "post-shock".

Computational results for shocked particles in an offset configuration are given in Figures 3c,d, 4, 5 and 6, which include Hertz's elastic, Brilliantov's viscoelastic, and elastic hard-sphere ($\varepsilon = 1$) collisions. Simulation with the inelastic hard-sphere model, using the same restitution coefficient as in Brilliantov's viscoelastic collision ($\varepsilon = 0.008$), is found to be unstable, as the particles tend to cluster, which requires a special treatment of the collision logics to prevent spurious collisions due close proximity of particles.

As can be seen from Figure 4, the particles appear to get aligned due to the effect of the wake behind the first particle. In a general collision situation, both normal and tangential collision forces are important. The tangential force is often represented as linearly-dependent on the normal force. However, a collision model, which provides a consistent basis for simultaneous treatment of the normal and tangential collision forces, is yet to be developed. More importantly, there are fundamental issues in applying Hertz theory and its extended versions for situations with a high collision velocity. First, the Hertz theory requires that the collision velocity is much smaller than the speed of sound in the particle material. Second, Hertz's treatment assumes that the collision area and deformation is several orders of magnitude smaller than particle's diameter. Both conditions may not be satisfied when the particles collide

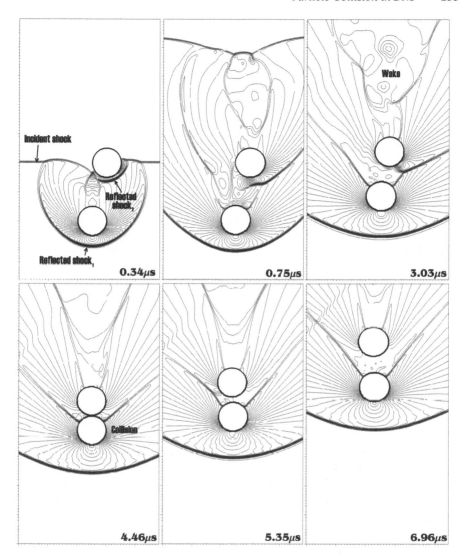

Fig. 4. Dynamics of the pressure field for the offset elastic collision test. Effective grid resolution is 40 nodes/Dia.

at high speed. In the generalized Hertz theory, such as Brilliantov's viscoelastic collision model, although accounting for energy dissipation during the collision, the treatment remains elastic. Elasticity requires the particle to return to its initial shape by the end of collision. Physically, such a model leads to a prolongation of the collision in order for the material in the collision zone to fully relax. We found that the elasticity condition was manifested in Brilliantov's viscoelastic model by inducing a prolonged period of unphysical attractive force (to keep particles together) during the collision run-apart phase. Obviously, a better inelastic collision model should

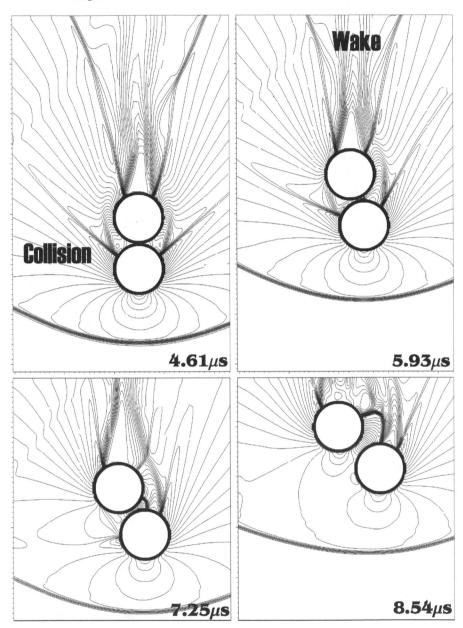

Fig. 5. Dynamics of the Mach number field for the offset viscoelastic collision test. Effective grid resolution is 40 nodes/Dia.

take into account the realistic change of the particle's shape under large, irreversible deformations (plasticity effects).

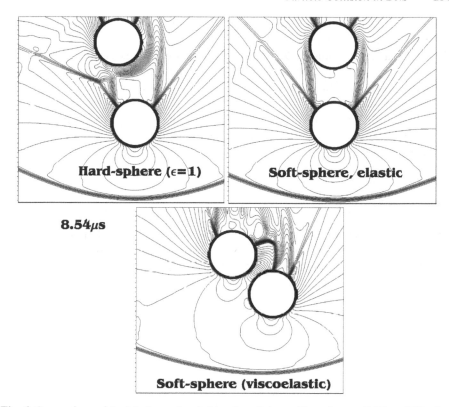

Fig. 6. Comparison of the Mach number field and particle positions for $t = 8.54$ μs. Effective grid resolution is 40 nodes/Dia.

It is interesting to note a significant difference between hard-sphere, and elastic soft-sphere simulations, Figures 3c and 6. The drag coefficient for the second (wake) particle is nearly 50% of that in the soft-sphere collision. This is associated with a finite duration of the collision time (≈ 1 μs) for the soft-sphere model, which is comparable to acoustic time scales of gas dynamics and sufficient to alter the particle trajectory, keeping it in the wake of the first particle.

5 Conclusions and recommendations

To our knowledge, this is the first time direct numerical simulations of collisions in particulate systems, under highly compressible gas flow conditions (shock waves), were performed. The computations were made possible by the Characteristics-Based Matching method implemented on an Adaptive Mesh Refinement platform. The following findings are drawn from these first computations:

1. With a sufficient grid resolution (via AMR), calculations are stable and do not require smearing over time-space.

2. The effects of hydrodynamic forces on particle-particle collision are insignificant compared to collision forces.
3. Viscoelastic collision models, such as Brilliantov's, do not correctly reflect the nature of high-speed collision. Plasticity is a key feature needed for the numerical simulation of high speed flows.
4. For large-deformation collisions, the simplified particle-particle interactions models utilized here must be substituted by the full description of structural mechanics inside the particles. Methods employed here provide readily for such an extension.

Acknowledgments

This work was supported by the Lawrence Livermore National Laboratory (ALPHA and MIX projects), the National Ground Intelligence Center (NGIC), the US Army (DURIP Project), and the JSTO/DTRA (ASOS Project). Support of Drs. Frank Handler and Glen Nakafuji at LLNL, Dr. Rick Babarsky at NGIC, Mr. Ed Conley, JBCCOM at RDEC, and Mr. John Pace and Mr. Charles Fromer of the JSTO at DTRA is gratefully acknowledged.

References

1. Berger, M.J. and Colella, P., 1989, Local adaptive mesh refinement for shock hydrodynamics, *Journal of Computational Physics* **82**, 67–84.
2. Brilliantov, N.V., Spahn, F., Hertzsch, J.-M. and Poschel, T., 1996, Model for collisions in granular gases, *Physical Review E* **53**(5), 5382–5392.
3. Campbell, C.S. and Brennen, C.E., 1985, Computer simulation of granular shear flows, *Journal of Fluid Mechanics* **151**, 167–188.
4. Crowe, C., Sommerfeld, M. and Tsuji, Y., 1998, *Multiphase Flows with Droplets and Particles*, CRC Press, Boca Raton, FL, 1998.
5. Fedkiw, R.P., Aslam, T., Merriman, B. and Osher, S., 1999, A non-oscillatory Eulerian approach to interfaces in multimaterial flows (the ghost fluid method), *Journal of Computational Physics* **152**, 457–492.
6. Fedkiw, R.P., 2002, Coupling an Eulerian fluid calculation to a Lagrangian solid calculation with the ghost fluid method, *Journal of Computational Physics* **175**, 200–224.
7. Glowinski, R., Pan, T.-W., Hesla, T.I. and Joseph, D.D., 1999, A Distributed Lagrange multiplier/fictitious domain method for particulate flows, *International Journal of Multiphase Flow* **25**, 755–794.
8. Johnson, A.A. and Tezduaer, T.E., 1997, 3D simulation of fluid-particle interactions with the number of particles reaching 100, *Comput. Methods. Appl. Mech. Engrg.* **145**, 301–321.
9. Landau, L.D. and Lifschitz, E.M., 1988, *Theoretical Physics, Vol. 1: Mechanics*, fourth edition, Nauka, Moscow.
10. Nguyen, N.Q. and Ladd, A.J.C., 2002, Lubrication corrections for lattice-Boltzmann simulations of particle suspensions, *Physical Review E* **66**, 046708.

11. Nourgaliev, R.R., Dinh, T.N. and Theofanous, T.G., 2004, The 'Characteristics-Based Matching' (CBM) method for compressible flow with moving boundaries and interfaces, *ASME Journal of Fluids Engineering* **126**, 586–604.
12. Nourgaliev, R.R., Dinh, T.N. and Theofanous, T.G., 2006, Adaptive characteristics-based matching for compressible multifluid dynamics, *Journal of Computational Physics*, in press.
13. Theofanous, T., Nourgaliev, R., Li, G. and Dinh, N., 2006, Compressible multi-hydrodynamics (CMH): Breakup, mixing and dispersal of liquids/solids in high speed flows, in *Proceedings of the IUTAM Symposium on Computational Multiphase Flow*, S. Balachandar and A. Prosperetti (eds), Springer, Dordrecht, pp. 353–369.
14. Wissink, A.M., Hornung, R., Kohn, S., Smith, S. and Elliott, N., Large scale parallel structured AMR calculations using the SAMRAI framework, in *Proc. SC01 Conf. High Perf. Network. & Comput.*, Denver, CO, November 10–16, 2001. Also available as LLNL Technical Report UCRL-JC-144755.

PART IV
FREE SURFACE FLOWS, DROPS AND BUBBLES

Struggling with Boundary Layers and Wakes of High-Reynolds-Number Bubbles

Jacques Magnaudet, Dominique Legendre and Guillaume Mougin

Institut de Mécanique des Fluides de Toulouse, UMR 5502, Allée Camille Soula, 31400 Toulouse, France

Abstract. We discuss two sets of non-trivial effects affecting the motion of high-Reynolds number gas bubbles rising in still liquid which have been significantly clarified thanks to direct numerical simulations making use of a suitable boundary-fitted technique. We first summarize some features of the interaction between two spherical bubbles rising side by side in a viscous liquid. Then we briefly discuss some aspects of the path instability of a spheroidal bubble rising in a low-viscosity liquid.

1 Introduction

Computational techniques for multiphase flows have undergone a tremendous development over the last fifteen years. Thanks to the spectacular increase of computer capabilities and to the improvements of methods such as Front-Tracking [1], ALE [2] and others, direct numerical simulations of dispersed flows involving some hundreds of deformable bubbles or rigid particles is now possible, provided the particle Reynolds number is of $O(10)$ or less. For instance, in [1] the authors carried out a DNS of 256 bubbles rising with a Reynolds number in the range 10–30 using four months of CPU time on eight processors. On a different scale, the point-force approximation and its extensions allow global effects of a large number of particles on the carrying flow to be investigated at a reasonable cost, provided the expression of the various forces acting on each particle is known to a good approximation. However, for different reasons, none of these two streams of methods is currently capable of resolving properly the thin boundary layer and wake that develop around particles moving at Reynolds numbers of $O(10^2)$ or more, or the extremely thin boundary layers associated with the diffusion of high-Schmidt-number contaminants, like surfactants. For such problems the boundary-fitted technique remains by far the most appropriate approach, even though it can only deal with simple topologies. We illustrate the above point of view by considering briefly two problems that we recently addressed. Other physical situations considered by groups using the same gneral approach may be found in, for instance, [3, 4].

S. Balachandar and A. Prosperetti (eds), Proceedings of the IUTAM Symposium on Computational Multiphase Flow, 263–271.

2 Outline of the numerical technique

The computations discussed below were carried out with the JADIM code developed in our group. This code has been extensively described in previous publications (e.g. [5–7]). It solves the three-dimensional unsteady Navier–Stokes equations written in velocity-pressure variables in a general system of orthogonal curvilinear coordinates. The discretisation makes use of a staggered grid and the equations are integrated in space using a finite-volume method with second-order accuracy, all spatial derivatives being approximated with second-order centered schemes. Time advancement is achieved through a Runge–Kutta/Crank–Nicolson algorithm which is second-order accurate in time. Incompressibility is satisfied at the end of each time step by solving a Poisson equation for an auxiliary potential. The points we wish to stress here concern the curvilinear grids used in the applications discussed below. While it is quite easy to design such grids (in two dimensions) by using either conformal or quasiconformal mappings, the treatment of highly curved regions, frequently located very close to the body, requires special care. This is because extra source terms due to the rotation of the local grid axes arise in the momentum equations and may have a crucial influence on the overall results. A specific procedure allowing us to ensure that these extra source terms do not create artificial sinks or sources of momentum in the discrete equations is used. In brief this procedure enforces the obvious property that, once discretized on the orthogonal grid, any constant vector must have a zero curl, while any constant second-order tensor must have a zero divergence [7]. Another technical point deserves some comments. When dealing with high-Re bubbles or particles moving in an unbounded domain, two opposite requirements are encountered. First, the outer boundary must be located at a sufficient distance from the body to avoid artificial confinement effects, even though non-reflecting boundary conditions are used in the wake. This distance is generally of some tens of equivalent radii of the body for the Reynolds numbers considered below, which, in order to reduce the cost of the computations, suggests to use large cells near the outer boundary. Second, it is crucial to describe properly the boundary layer around the body, which frequently means that the thickness of the first row of cells surrounding it has to be only a few percents of the body equivalent radius. As a consequence of these two requirements, the grids used in such computations are highly stretched in the radial direction, with a ratio between the thickest and the thinnest cells frequently of $O(10^2)$. Maintaining the spatial accuracy on such stretched grids is not obvious and requires the staggered velocity and pressure nodes to be properly located with respect to each other. An example of the influence of this relative location on the accuracy of the overall results is discussed in [6]. Note that in the applicatons discussed below, at least 5 grid points lie within the boundary layer whatever the Reynolds number, which allows effects of the vorticity generated on the body to be fully resolved, as shown in [8].

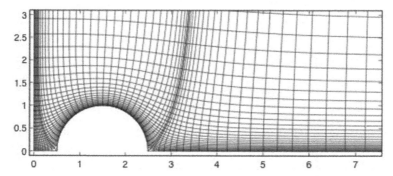

Fig. 1. Detail of the grid used in the computation of the interaction between two spherical bubbles ($S = 3$). The vertical plane on the left is the symmetry plane of the flow. The horizontal line at the bottom is the symmetry axis of the grid.

3 Interaction of two spherical bubbles rising side by side

This problem is of particular relevance in the understanding and determination of the properties of bubbly suspensions in which the bubbles move at moderate-to-large Reynolds number. The reason for this is that potential flow theory predicts the stable position of two neighboring bubbles to be reached when they come in contact with their line of centres perpendicular to gravity [9]. This finding was confirmed a decade ago in "direct" numerical simulations of the potential flow induced by the rise of a cloud of bubbles, which revealed that the bubbles eventually agglomerate in thin horizontal clusters [10, 11]. This conclusion is clearly not realistic, since the bubble distribution observed in many laboratory experiments is close to homogeneity. A possible explanation for this discrepancy is the role of the vorticity produced at the surface of the bubbles. A thorough investigation of these effects was carried out in [12], using grids such as those shown in Figure 1. Here we only summarize the most significant conclusions of this study.

Figure 2 shows the evolution of the transverse force between the two bubbles as a function of the rise Reynolds number $Re = 2UR/\nu$ and for various separations $S = d/R$ (U is the rise velocity, R the bubble radius, d the distance between the two bubble centers, and ν the kinematic viscosity of the fluid). It is clear that the force is always positive (i.e. repulsive) for low enough Reynolds numbers, whereas it becomes negative (i.e. attractive) when the Reynolds number exceeds a critical value. Obviously this critical value depends on the separation and lies roughly in the range 30-80; the smaller the separation, the larger the critical Re.

Figure 3 allows us to understand the origin of the evolution depicted in Figure 2. The vorticity contours corresponding to $Re = 300$ are seen to be almost symmetrical with respect to the vertical plane passing through the bubble center, which indicates that there is nearly no direct interaction between the two vorticity fields. Hence the dominant interaction results from the two potential dipoles associated with the bubbles. This irrotational interaction is known to accelerate the flow in the gap (Venturi effect), resulting in a pressure gradient directed away from the symmetry plane.

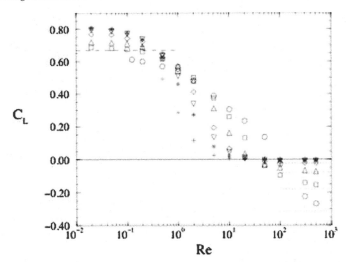

Fig. 2. The interaction force (normalized by $\pi \rho U^2 R^2 / 2$) vs. the Reynolds number for various separations. ○ $S = 2.25$, □ $S = 2.5$, △ $S = 3$, ◊ $S = 4$, ∇ $S = 5$, ∗ $S = 6$, + $S = 10$; - - - irrotational prediction.

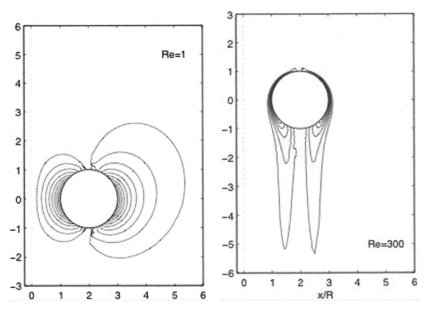

Fig. 3. Iso-contours of the vorticity field for two widely different Reynolds numbers ($S = 4$); the symmetry plane of the flow is on the left.

This is of course in line with the existence of the attractive force observed in Figure 2 in this range of Re. The situation is drastically different at a moderate-to-low Reynolds number. Now, vorticity tends to spread quite far around each bubble but this diffusion is prevented in the gap, owing to the presence of the second bubble. This blocking effect results in an asymmetry of the vorticity distribution. It is an easy matter to show that, in the horizontal plane where the two bubbles lie, the vorticity induces an upward contribution in the velocity field. As the vortex lines are tightened in the gap, this upward contribution is maximum there, thus tending to make the vertical velocity minimum in between the two bubbles. This velocity minimum is of course associated with a pressure maximum and the transverse force is now repulsive. From this analysis one can infer that the change of sign of the interaction force occurs when the irrotational and the vortical mechanisms balance each other.

4 Zigzagging/spiraling bubbles

Millimetric spheroidal bubbles rising in water (or more generally in low-viscosity liquids) frequently exhibit zigzagging or spiraling trajectories. While this intriguing phenomenon has been described in many experiments, its physical origin remained unclear until recently. This is due to the many different mechanisms that can be involved in the phenomenon, among which deformation and oscillations of the bubble, effects of possible contamination by surfactants, and wake instability [13]. Direct numerical simulation is particularly useful to help elucidating such complicated situations because it allows us to separate or even suppress arbitrarily each of the various possible mechanisms. This was the line of thinking we decided to follow. Based on various considerations, we guessed that the basic mechanism responsible for the path instability lies in the wake dynamics. As we showed in the past [8] that the strength of the vorticity produced on a clean spheroidal bubble increases tremendously with its aspect ratio (the aspect ratio χ is defined as the ratio of the lengths of the major and minor axes), we decided to study the model problem of the free rise of a fixed-shape spheroidal bubble submitted to a shear-free condition at its surface. This model problem is in some sort a fluid-structure interaction problem in which a moving body induces a disturbance in the surrounding fluid, and this disturbance in turn generates forces and torques that drive the body motion. From a technical point of view, solving numerically such problems requires either the grid to deform according to the body displacements (ALE technique such as the one used in [2]), or (when the flow domain is unbounded externally) the governing equations to be transformed in such a way that the effects of the body translation and rotation are directly incorporated in the formulation. Albeit it cannot be easily extended to multibody problems, we chose this second possibility which allows high-accuracy results to be obtained at a moderate cost. Hence we transformed properly the Navier–Stokes equations and the boundary conditions so as to avoid to regenerate the grid during the computation. We supplemented these equations with the Newton's equations so as to ensure that the total force and torque acting on the bubble remain zero at all time, as the bubble may be considered as an inertialess body (note that here as well as in the example above, the flow inside the bubble is not solved). All technical details concerning this formulation may be found in [14]. It is just worth pointing out here that the computations discussed below require about $2 \cdot 10^5$ grid points and several hundreds thousand time steps, which represents about two months CPU time on a standard Linux PC.

The model problem depends on two control parameters, namely the aspect ratio χ and the so-called Galileo number $Ga = g^{1/2} R_{eq}^{3/2} / \nu$ which may be thought of as a Reynolds number based on a gravitational scaling of the velocity (R_{eq} is the equivalent radius of the bubble and g denotes gravity). The computational protocol consists in releasing the bubble from rest with its minor axis aligned with gravity. A small ($10^{-4} g$) sinusoidal perturbation with a random frequency is added to gravity to trigger the path instability, but we checked that it just shortens the time required for the transition to occur without influencing the final result.

In what follows we shall describe the observations made in the case $Ga = 138$ which corresponds to a bubble with $R_{eq} = 1.25$ mm rising in water under standard conditions [15]. Nothing special happens for $\chi < 2.25$, approximately: the bubble rises in straight line and its wake remains axisymmetric. In contrast, when the aspect ratio is slightly increased beyond $\chi = 2.25$, the bubble first rises in straight line and then quickly bifurcates toward a planar zigzag path (the plane of which is selected by the artificial perturbation). A detailed analysis reveals that this transition occurs through a supercritical Hopf bifurcation. When χ is further increased up to $\chi = 2.5$, the crest-to-crest amplitude and the Strouhal number $St = 2f(R_{eq}/g)^{1/2}$ of the zigzag motion saturate at a value of $4.8 R_{eq}$ and 0.09, respectively (Figure 4).

The path remains in a plane for a very long time. However a careful observation indicates that the horizontal component of the bubble velocity perpendicular to the plane of the zigzag slowly grows in time. At a certain moment, this growth becomes larger, resulting in a significant motion of the bubble out of the plane of the zigzag; simultaneously, a slight reduction of the amplitude of the zigzag is observed. The two horizontal components of the bubble velocity eventually reach the same amplitude, yielding a perfectly circular helical motion of the bubble. This second transition from a planar zigzag to a circular helix was observed in all cases where the rectilinear/zigzag bifurcation occurred. This suggests that the zigzag is only a very long transient, i.e. the zigzag/helical transition is not a secondary bifurcation. However this has still to be confirmed since, owing to the cost of the computations, our coverage of the (χ, Ga) plane was quite coarse and we may have missed intermediate values of the control parameters for which the zigzag could be stable at long time.

The connection between the path of the bubble and the structure of the wake can be established by recording some characteristic features of the latter. An excellent indicator is provided by the streamwise vorticity shown in Figure 5. While this quantity is obviously zero during the straight part of the path (since the wake is then axisymmetric), it becomes nonzero as soon as the zigzag path sets in. The streamwise vorticity is then concentrated within two counter-rotating vortex tubes. By examining simultaneous records of the path, it is found that these vortex tubes disappear when the bubble crosses the inflexion point of its trajectory, and then reappear with an interchange in the sign of the vorticity (Figures 5b and 5b'). Finally,

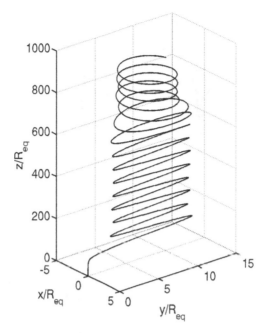

Fig. 4. Complete path of a zigzagging/spiraling bubble corresponding to $\chi = 2.5$ and $Ga = 138$.

when the bubble enters the helical stage of its trajectory, the two vortex tubes wrap up around one another and the wake becomes steady in a system of axes rotating with the bubble. Figures 4 and 5 clearly indicate that there is a one-to-one correspondence between the path and the wake structure; in other terms the "devil" responsible for the non-rectilinear path of the spheroidal bubble lies in the wake.

Further insight into the physical mechanisms at work may be obtained by examining the lateral force balance during the zigzag stage. Figure 6 shows the evolution of the various contributions to this force. The largest of them is the wake-induced force F_ω resulting from the two counter-rotating vortices. This force is balanced by the combination of the lateral component of the buoyancy force (which is nonzero since the bubble inclines itself so as to maintain its minor axis essentially parallel to its instantaneous velocity), and the added-mass force due to the acceleration of the liquid displaced by the bubble. Note that in Figure 6 all forces are normalized by the buoyancy force, which reveals that the maximum of F_ω is of the same order as the Archimedes force that drives the whole system.

Combining all the ingredients contained in Figures 4 to 6, we obtain the following scenario. The instability that breaks the initial axial symmetry of the wake generates a new wake topology in which streamwise vorticity is concentrated within two vortex threads of opposite sign. The flow field due to this vorticity distribution results in a lateral lift force which generates horizontal displacements of the bubble and makes the rectilinear path unstable. A complete analysis of the force and torque

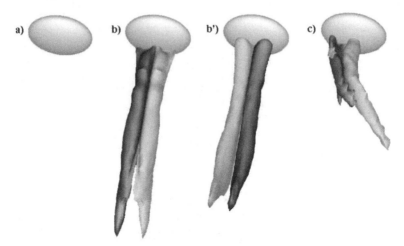

Fig. 5. The four successive stages of the wake corresponding to the path of Figure 5. (a) No streamwise vorticity (rectilinear path); (b) and (b′) the streamwise vorticity is concentratted within two vortex tubes and changes sign twice during a period of the zigzag; (c) the wake is steady in a system of axes rotating with the bubble (helical path).

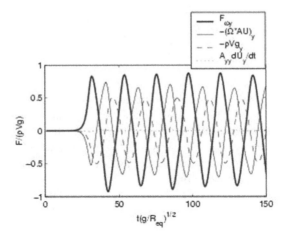

Fig. 6. The various contributions to the lateral force balance during the zigzag stage ($\chi = 2.5$, $Ga = 138$).

balances (not detailed here) then allows us to understand why the zigzag path is selected among all possible nonrectilinear paths.

5 Conclusions

We discussed two fundamental problems of bubble hydrodynamics for which the use of direct numerical simulation based on boundary-fitted grids proved to be extremely useful. This technique, while limited in terms of the geometrical complexity of the flows it can handle, is highly accurate. It probably remains the most efficient tool to investigate problems involving thin gaps or thin boundary layers, as in the two cases considered here, and allows subtle physical mechanisms to be elucidated.

References

1. B. Bunner and G. Tryggvason, 2002, Dynamics of homogeneous bubbly flows. Part 1. Rise velocity and microstructure of the bubbles, *J. Fluid Mech.* **466**, 17–52.
2. J. Feng, H.H. Hu and D.D. Joseph, 1994, Direct simulation of initial value problems for the motion of solid bodies in a Newtonian fluid. Part 1. Sedimentation, *J. Fluid Mech.* **261**, 95–134.
3. E.J. Chang and M.R. Maxey, 1994, Unsteady flow about a sphere at low to moderate Reynolds number. Part 1. Oscillatory motion, *J. Fluid Mech.* **277**, 347–379.
4. P. Bagchi and S. Balachandar, 2002, Steady planar straining flow past a rigid sphere at moderate Reynolds number, *J. Fluid Mech.* **466**, 365–407.
5. J. Magnaudet, M. Rivero and J. Fabre, 1995, Accelerated flows around a rigid sphere or a spherical bubble. Part I: Steady straining flow, *J. Fluid Mech.* **284**, 97–135.
6. I. Calmet and J. Magnaudet, 1997, Large-Eddy Simulation of high-Schmidt number mass transfer in a turbulent channel flow, *Phys. Fluids* **9**, 438–455.
7. D. Legendre and J. Magnaudet, 1998, The lift force on a spherical bubble in a viscous linear shear flow, *J. Fluid Mech.* **368**, 81–126.
8. A. Blanco and J. Magnaudet, 1995, The structure of the high-Reynolds number flow around an ellipsoidal bubble of fixed shape, *Phys. Fluids* **7**, 1265–1274.
9. L. Van Wijngaarden, 1993, The mean rise velocity of pairwise-interacting bubbles in liquid, *J. Fluid Mech.* **251**, 55–78.
10. A.S. Sangani and A.K. Didwania, 1993, Dynamic simulations of flows of bubby liquids at large Reynolds numbers, *J. Fluid Mech.* **250**, 307–437.
11. P. Smereka, 1993, On the motion of bubbles in a periodic box, *J. Fluid Mech.* **254**, 79–112.
12. D. Legendre, J. Magnaudet and G. Mougin, 2003, The interaction between two spherical bubbles rising side by side in a viscous liquid, *J. Fluid Mech.* **497**, 133–166.
13. J. Magnaudet and I. Eames, 2000, The motion of high-Reynolds-number bubbles in inhomogeneous flows, *Annu. Rev. Fluid Mech.* **32**, 659–708.
14. G. Mougin and J. Magnaudet, 2002, The generalized Kirchhoff equations and their application to the interaction of a rigid body with an arbitrary time-dependent viscous flow, *Int. J. Multiphase Flow* **28**, 1837–1851.
15. G. Mougin and J. Magnaudet, 2002, Path instability of a rising bubble, *Phys. Rev. Lett.* **88**, 014502.

Direct Numerical Simulations of Bubbly Flows

Gretar Tryggvason, Jiacai Lu, Souvik Biswas and Asghar Esmaeeli

Worcester Polytechnic Institute, Worcester, MA 01609, USA

Abstract. The status of direct numerical simulations of bubbly flows is reviewed and a few recent results are presented. The development of numerical methods based on the one-field formulation has made it possible to follow the evolution of a large number of bubbles for a sufficiently long time so that converged statistics for the averaged properties of the flow can be obtained. In addition to extensive studies of homogeneous bubbly flows, recent investigations have helped give insight into drag reduction due to the injection of bubbles into turbulent flows and two-fluid modeling of laminar multiphase flows in channels.

1 Introduction

Boiling heat transfer, cloud cavitation, aeration and stirring of reactors in water purification and waste water treatment plants, bubble columns and centrifuges in the petrochemical industry, cooling circuits of nuclear reactors, propagation of sound in the ocean, the exchange of gases and heat between the oceans and the atmosphere, and explosive volcanic eruptions, are just a few examples of multiphase bubbly flows occurring in both industrial and natural processes. As these examples show, understanding the evolution and properties of bubbly flows is therefore of major technological as well as scientific interest.

Although Direct Numerical Simulations (DNS) of bubbly flows have come into their own only in the last few years, computational studies of multiphase flow date back to the beginning of computational fluid dynamics, when the MAC method of Harlow and collaborators was used for simulations of the Rayleigh–Taylor instability, splats due to impacting droplets, and other problems involving a free surface or a fluid interface. Although the MAC method, and its successor the VOF method, slowly gained popularity, in the late seventies and the early eighties, serious computational studies relied mostly on boundary integral methods and body fitted grids for intermediate Reynolds numbers [23]. The current surge of activities in multiphase flow simulations goes back to the beginning of the nineties, when significant improvements in methods that use fixed grids took place. Fixed grids offer great flexibility in the geometric complexity of the multiphase flow under investigation, com-

S. Balachandar and A. Prosperetti (eds) Proceedings of the IUTAM Symposium on Computational Multiphase Flow, 273–281.
© 2006 *Springer. Printed in the Netherlands.*

bined with the efficiency inherent in the use of regular structured grids. The continuous surface force (CSF) method [4] to compute surface tension in VOF methods, the level set [24], the phase field, and the CIP [27] methods were all introduced at that time, along with the front tracking method of Unverdi and Tryggvason [30]. By now, a large number of refinement and new methods have been introduced and the development of numerical methods for multiphase flow is currently a "hot" topic. There is, for example, hardly an issue of the *Journal of Computational Physics* that does not have at least one paper in some way related to multiphase flow simulations.

The development of more efficient, accurate, and robust methods continues to be of considerable interest, as well as the extension of the various methods to handle more complex physics. It is, however, the use of numerical methods to conduct direct numerical simulations of complex multiphase flows that is sure to have the greatest impact in the future. Such simulations are already yielding unprecedented insight, even though DNS have only been used to examine a tiny fraction of the systems that can be explored with current capabilities. Those studies that have been done have focused mostly on suspensions of solid particles and bubbly flows. Here we will discuss the current status of DNS of bubbly flows.

2 Numerical method

For non-dilute disperse multiphase flows at intermediate Reynolds numbers, it is necessary to solve the full unsteady Navier–Stokes equations. Most methods currently in use for DNS of multiphase flows are based on writing one set of the governing equations for the whole flow field by allowing the density and viscosity fields to be discontinuous across the phase boundary and by including a singular term representing the surface forces. The momentum equation is:

$$\rho \frac{\partial \mathbf{u}}{\partial t} + \rho \nabla \cdot \mathbf{u}\mathbf{u} = -\nabla P + \nabla \cdot \mu (\nabla \mathbf{u} + \nabla \mathbf{u}^T) + \sigma \int_F \kappa_f \mathbf{n}_f \delta(\mathbf{x} - \mathbf{x}_f) \, dA_f. \quad (1)$$

Usually both fluids are assumed to be incompressible:

$$\nabla \cdot \mathbf{u} = 0. \quad (2)$$

Here, $\mathbf{u}$ is the velocity, P is the pressure, and ρ and μ are the discontinuous density and viscosity fields, respectively. δ is a three-dimensional delta-function constructed by repeated multiplication of one-dimensional delta-functions. κ is twice the mean curvature. $\mathbf{n}$ is a unit vector normal to the front. Formally, the integral is over the entire front, thereby adding the delta-functions together to create a force that is concentrated at the interface, but smooth along the front. $\mathbf{x}$ is the point at which the equation is evaluated and $\mathbf{x}_f$ is the position of the front.

Numerical implementations of Equations (1) and (2) include the Volume of Fluid, the Level Set, and the CIP methods as well as the Front-Tracking/Finite Volume method of Unverdi and Tryggvason [30]. The front-tracking method is used for the simulations presented here. In all these methods, the governing equations are solved

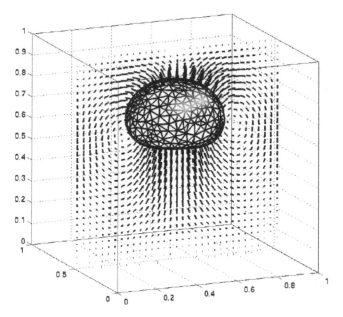

Fig. 1. The computational setup. The conservation equations are solved on a regular structured grid, but the phase boundary is tracked by a moving unstructured triangular grid.

by a projection method on a fixed grid. The way the phase boundary is tracked is, however, different. In most cases a marker function that identifies the different fluids is advected by the flow. In the method of Unverdi and Tryggvason, however, the phase boundary is tracked by connected marker points (the "front") and the marker function reconstructed from the location of the "front". This keeps the boundary between the phases sharp, and allows the accurate computation of the surface tension. The front points are advected by the flow velocity, interpolated from the fixed grid. As the front deforms, surface markers are dynamically added and deleted. The surface tension is represented by a distribution of singularities (delta-functions) located at the front. The gradients of the density and viscosity become delta functions when the change is abrupt across the boundary. To transfer the front singularities to the fixed grid, the delta functions are approximated by smoother functions with a compact support on the fixed grid. At each time step, after the front has been advected, the density and the viscosity fields are reconstructed by integration of the smooth grid-delta function. The surface tension is then added to the nodal values of the discrete Navier–Stokes equations. Finally, an elliptic pressure equation is solved by a multigrid method to impose a divergence-free velocity field. For a detailed description of the original method, including various validation studies, see [29, 30]. Figure 1, where the velocity field computed on a fixed grid is shown along with the tracked surface for a single buoyant bubble, summarizes the computational approach.

The original method of Unverdi and Tryggvason [30] has been extremely successful for relatively complex multiphase flows at modest Reynolds numbers. As it has been applied to more challenging problems, such as higher Reynolds numbers and large properties ratios, a number of improvements have been implemented. The development include improved regridding procedures for the tracked front, conservative techniques to compute the surface tension, high order upwind methods for the advection terms, the use of non-conservative form of the advection terms to eliminate spurious oscillations for high density ratios, and non-uniform grids. The basic method has, however, remained the same, confirming that robustness of direct tracking of the interface.

3 Results

Below we review briefly the status of our studies of bubbly flows, using direct numerical simulations. We have examined a large number of cases, but until recently we have mostly focused on homogeneous flows, modeled by fully periodic domains.

3.1 Homogeneous bubbly flows

The interactions of two bubbles in a periodic domain was examined briefly in [30]. The motion of many nearly spherical bubbles at moderate Reynolds numbers was studied by Esmaeeli and Tryggvason [10] for a case where the average rise Reynolds number of the bubbles remained relatively small (1–2) and Esmaeeli and Tryggvason [11] looked at another case where the Reynolds number was 20–30. Bunner and Tryggvason [5, 6] simulated a much larger number of three-dimensional bubbles using a parallel version of the method used by Esmaeeli and Tryggvason. Their largest simulations followed the motion of 216 three-dimensional buoyant bubbles per periodic domain for a relatively long time. The simulations showed, among other things, that modest Reynolds bubbles generally interact through "drafting, kissing, and tumbling" collision [13] and that the probability of finding horizontal bubble pairs increases with the Reynolds numbers. While "low order" statistical quantities like the rise velocity of the bubbles converged rapidly as the size of the system increases, other quantities like the dispersion coefficients converge slower. Esmaeeli et al. [12] briefly examined bubbles that settle down into periodic wobbling and showed that the bubbles slow down significantly once they start to wobble. Göz et al. [14] examined higher Reynolds number bubbles and found what looked like chaotic motion at high enough Reynolds numbers. The effect of deformability was studies by Bunner and Tryggvason [7] who found that relatively modest deformability could lead to a streaming state where bubbles gathered in a stream or a chimney. Other studies of the motion and interactions of many bubbles have been done by several Japanese authors. Early work, using the VOF method to compute the motion of a single two-dimensional bubble can be found in [28] and more recent work on bubble interactions, using both VOF and the Lattice Boltzman Method, is presented in [25, 26]. The various simulations that have been done for bubbly flows suggest that we are well on our way to understand elementary behavior of homogeneous bubbly flow when the Reynolds number is relatively low.

3.2 Using DNS to validate and develop models

For industrial systems DNS are generally not practical and it is necessary to use models that predict the average behavior of the system. Reynolds Averaged Navier–Stokes (RANS) computations of homogeneous flows have a long history, starting with the pioneering work of Launder and Spalding [20] and Harlow and collaborators (e.g. [2]). For multiphase flows, several averaged models have been developed, ranging from simple mixture models to more sophisticated two-fluid models. Considerable effort has, in particular, been devoted to the development of two-fluid models for disperse flows (see, for example, [8, 9, 31]. The averaging leads to an equation for the void fraction and separate momentum equations for each phase. It also results in the usual Reynolds stresses and the force between the phases as the terms that must be modeled. The force is usually split into several parts, including the steady-state drag and lift, added mass, Basset force, wall drag and wall-repulsion, and dispersion force. These terms are modeled using a combination of analytical solutions for Stokes flow and empirical correlations/corrections to account for higher Reynolds numbers. For a spherical isolated particle the forces are reasonably well understood, with the exception of lift, but for higher concentrations and deformable bubbles the situation is more uncertain. The momentum equation for the continuous phase is always solved using an Eularian approach, where the averaged equations are solved on a fixed grid, but the dispersed phase can be treated either using a Lagrangian or an Eularian approach. In the Eularian approach the momentum equation for the averaged particle velocity is solved in the same way as for the continuous phase, but in the Lagrangian approach the dispersed phase is represented by point particles that are tracked through the flow domain. As the particles move, they generate velocity disturbances in the continuous phase, even if it is initially quiescent. These velocity fluctuations show up as Reynolds stresses in the averaged equations and are usually modeled using potential flow solutions for flow over a sphere. In turbulent flows they are simply added to the Reynolds stresses generated by the fluid turbulence. For well-behaved flows, such as flows in pipes and ducts, current two-fluid models generally do well and capture the main flow features.

We have recently started to look at two-fluid models by comparing results from direct numerical simulations with the predictions of the model of Antal et al. [1] for laminar bubbly flow. The primary goal of this study is simply to find out what kind of domain sizes are needed to produce results with well-converged averages. The simulations were done assuming a two-dimensional flow so we had to adjust the model parameters slightly. We did, however, find that once we adjusted the parameters for one flow, other situations were well predicted by the same model parameters. For steady-state flow, where the slip velocity between the bubbles and the continuous phase is given by an algebraic relation, there is no question of ill-posedness and the model converged rapidly when the grid is refined. We are currently in the progress of doing fully three-dimensional simulations that will allow us to do a more thorough term-by-term assessment of the closure models and to explore their sensitivity to a distribution of bubble sizes, bubble deformability, and so on. We have also done a low resolution preliminary simulation of bubbles in a turbulent channel flow (similar

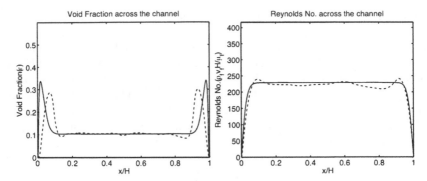

Fig. 2. Comparison of the average velocity and void fraction profile from a simulation of 64 bubbles and the two-flluid model of Antal et al. [1]. While the model does not capture completely the location of the wall peaks in void fraction, the overall agreement is good.

to our study of drag reduction described below, but with bubbles rising due to buoyancy) and we plan to use the results to shed some light on the interaction of bubble induced velocity fluctuations and the already existing turbulence. Figure 2 shows a comparison of the simulated void fraction and velocity profiles, averaged over the entire computational domain, and profiles predicted using the two-fluid model of Antal et al. [1]. Except for the location of the wall-peak, the model captures the simulated results reasonably well. The results also show that the bubbles are pushed to the wall until the flow in the center of the channel is in hydrostatic equilibrium and essentially homogeneous. For details, see [3].

3.3 Drag reduction due to bubble injection

In addition to providing data and insight for modeling, DNS studies can help explain complex interactions between the bubbles and the flow. Sometimes such interactions involve very subtle effects. Figure 3 shows one frame from a simulation of bubbles in a turbulent channel flow with a Reynolds number of 4000. In addition to the bubbles, isocontours of spanwise vorticity are shown, with different shading indicating positive and negative vorticity. The wall shear on the bottom wall is also shown. The goal of this investigation is to cast some light on the mechanisms underlying drag reduction due to bubble injection and to provide data to help with the modeling of such flows. Experimental studies (see [17, 19, 22], for a review) show that the injection of a relatively small amount of bubbles into a turbulent boundary layer can result in a significant drag reduction. While the general belief seems to be that the bubbles should be as small as possible (a few wall units in diameter), drag reduction is found experimentally in situations where the bubbles are considerably larger (order of 100 wall units).

We have examined the effect of bubbles on turbulent channel flow, mostly using simulations with sixteen bubbles in the so-called "minimum turbulent channel" of Jimenez and Moin [15]. The results, discussed in detail in [21], show that slightly

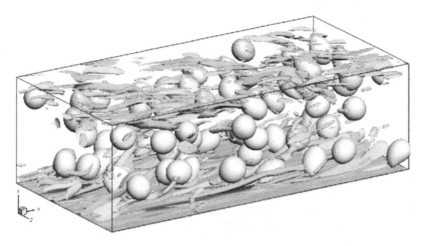

Fig. 3. One frame from a simulation of the effect of bubbles on the wall shear in a turbulent flow, at a relatively early time. The computational domain is $2\pi \times \pi \times 2$ in the streamwise, spanwise and wall-normal direction, respectively.

deformable bubbles can lead to significant reduction of the wall drag (up to 20%) by sliding over streamwise vortices and forcing them toward the wall where they are cancelled by the wall bound vorticity of the opposite sign. Spherical bubbles, on the other hand, often reach into the viscous sublayer where they are slowed down and lead to a increase in drag. This study has demonstrated powerfully the ability of DNS to explain very subtle effects that could probably not be understood in any other way. While the minimum turbulent channel flow is, admittedly, a somewhat special situation, preliminary simulations using larger channels (Figure 3), suggest that the evolution does not depend sensitively on the channel size. Kanai and Miyata [16] and Kawamura and Kodama [18] have also examined the motion of bubbles in turbulent channel flows, but did not see drag reduction.

4 Conclusions

The goal of numerical studies of multiphase flows is to obtain insight into the dynamics of the flow as well as quantitative data. Such data is essential for the modeling of industrial flows. Major progress has been made in using DNS to understand bubbly flows in the last few years, but much remains to be done. Very little has been done to examine the effect of different bubbles sizes and essentially noting for bubble breakup and coalescence. Similarly, the application of the results to help with the improvements of models is only beginning.

References

1. S.P. Antal, R.T. Lahey and J.E. Flaherty, 1991, Analysis of phase distribution in fully developed laminar bubbly two-phase flow, *Int. J. Multiphase Flow* **17**, 635–652.
2. A.A. Amsden and F.H. Harlow, 1968, Transport of turbulence in numerical fluid dynamics, *J. Comput. Phys.* **3**, 94–110.
3. S. Biswas, A. Esmaeeli and G. Tryggvason, 2005, Comparison of results from DNS of bubbly flows with a two-fluid model for two-dimensional laminar flows, *Int. J. Multiphase Flows* **31**, 1036–1048.
4. J.U. Brackbill, D.B. Kothe and C. Zemach, 1992, A continuum method for modeling surface tension, *J. Comput. Phys.* **100**, 335–354.
5. B. Bunner and G. Tryggvason, 2002, Dynamics of homogeneous bubbly flows: Part 1, Rise velocity and microstructure of the bubbles, *J. Fluid Mech.* **466**, 17–52.
6. B. Bunner and G. Tryggvason, 2002, Dynamics of homogeneous bubbly flows. Part 2, Fluctuations of the bubbles and the liquid, *J. Fluid Mech.* **466**, 53–84.
7. B. Bunner and G. Tryggvason, 2003, Effect of bubble deformation on the stability and properties of bubbly flows, *J. Fluid Mech.* **495**, 77–118.
8. C. Crowe, M. Sommerfeld and Y. Tsuji, 1998, *Multiphase Flows with Droplets and Particles*, CRC Press.
9. D.A. Drew and S.L. Passman, 1999, *Theory of Multicomponent Fluids*, Springer.
10. A. Esmaeeli and G. Tryggvason, 1998, Direct numerical simulations of ubbly flows. Part I – Low Reynolds number arrays, *J. Fluid Mech.* **377**, 313–345.
11. A. Esmaeeli and G. Tryggvason, 1999, Direct numerical simulations of bubbly flows. Part II – Moderate Reynolds number arrays, *J. Fluid Mech.* **385**, 325–358.
12. A. Esmaeeli, E.A. Ervin and G. Tryggvason, 1994, Numerical simulations of rising bubbles, in *Proceedings of the IUTAM Conference on Bubble Dynamics and Interfacial Phenomens*, Birmingham, UK, 6–9 September 1993, J.R. Blake, J.M. Boulton-Stone and N.H. Thomas (eds), pp. 247–255.
13. A. Fortes and D.D. Joseph and T. Lundgren, 1987, Nonlinear mechanics of fluidization of beds of spherical particles, *J. Fluid Mech.* **177**, 467–483.
14. M.F. Göz, B. Bunner, M. Sommerfeld and G. Tryggvason, 2000, EDSM2000-11151: The unsteady dynamics of two-dimensional bubbles in a regular array, in *Proceedings of the ASME FEDSM'00 ASME 2000 Fluids Engineering Division Summer Meeting*, Boston, MA, June 11–15.
15. J. Jimenez and P. Moin, 1991, The minimal flow unit in near-wall turbulence, *J. Fluid Mech.* **225**, 213–240.
16. A. Kanai and H. Miyata, 2001, Direct numerical simulation of wall turbulent flows with microbubbles, *Int. J. Num. Meth. Fluids* **35**,, 593–615.
17. H. Kato, M. Miyanaga, H. Yamaguchi and M.M. Guin, 1995, Frictional drag reduction by injecting bubbly water into turbulent boundary layer and the effect of plate orientation, in *Proceedings of the 2nd International Conference on Multiphase Flow '95, ICMF95*, Kyoto, pp. 31–38.
18. T. Kawamura and Y. Kodama, 2002, Numerical simulation method to resolve interactions between bubbles and turbulence, *Int. J. of Heat and Fluid Flow* **23**, 627–638.
19. Y. Kodama, A. Kakugawa, T. Takahashi, S. Nagaya and K. Sugiyama, 2003, Microbubbles: Drag reduction and applicability to ships, in *Twenty-Fourth Symposium on Naval Hydrodynamics*, 2003, Naval Studies Board (NSB). Available at: http://books.nap.edu/books/NI000511/html/
20. B.E. Launder and D.B. Spalding, 1972, *Mathematical Models of Turbulence*, Academic Press, New York.

21. J. Lu, A. Fernandez and G. Tryggvason, 2005, The effect of bubbles on the wall shear in a turbulent channel flow, *Phys. Fluids* **17**, 095102.
22. C.L. Merkle and S. Deutsch, 1990, Drag reduction in liquid boundary layers by gas injection, *Progr. Astronautics Aeronautics* **123**), 351–412.
23. G. Ryskin and L.G. Leal, 1984, Numerical solution of free-boundary problems in fluid mechanics. Part 2. Buoyancy-driven motion of a gas bubble through a quiescent liquid, *J. Fluid Mech.* **148**, 19–35.
24. M. Sussman, P. Smereka and S. Osher, 1994, A level set approach for computing solutions to incompressible two-phase flows, *J. Comput. Phys.* **114**, 146–159.
25. N. Takada, M. Misawa, A. Tomiyama and S. Fujiwara, 2000, Numerical simulation of two- and three-dimensional two-phase fluid motion by lattice Boltzmann method, *Comput. Phys. Commun.* **129**, 233–246.
26. N. Takada, M. Misawa, A. Tomiyama and S. Hosokawaî, 2001, Simulation of bubble motion under gravity by lattice Boltzmann method, *J. Nucl. Sci. Technol.* **38**, 330–341.
27. H. Takewaki, A. Nishiguchi and T. Yabe, 1985, Cubic interpolated pseudoparticle method (CIP) for solving hyperbolic-type equations, *J. Comput. Phys.* **61**, 261–268.
28. A. Tomiyama, I. Zun, A. Sou and T. Sakaguchi, 1993, Numerical analysis of bubble motion with the VOF method, *Nucl. Engr. Design* **141**, 69–82.
29. G. Tryggvason, B. Bunner, A. Esmaeeli, D. Juric, N. Al-Rawahi, W. Tauber, J. Han, S. Nas and Y.-J. Jan, 2001, A front tracking method for the computations of multiphase flow, *J. Comput. Phys.* **169**, 708–759.
30. S.O. Unverdi and G. Tryggvason, 1992, A front tracking method for viscous incompressible flows, *J. Comput. Phys.* **100**, 25–37.
31. D.Z. Zhang and A. Prosperetti, 1994, Ensemble phase-averaged equations for bubbly flows, *Phys. Fluids* **6**, 2956–2970.

Direct Numerical Simulation of Droplet Formation and Breakup

Stéphane Zaleski

Laboratoire de Modélisation en Mécanique, CNRS and Université Pierre et Marie Curie (Paris VI), 4 Place Jussieu, 75005 Paris, France; e-mail: zaleski@lmm.jussieu.fr

1 Introduction

Droplet formation is a fascinating process. In industrial atomizers high-speed liquid jets are deformed and broken into drops by a sequence two-phase flow instabilities. In many others processes, such as droplet impact and splashing, volcanic eruptions, or turbulent liquid-gas flows droplet and bubbles are formed through diverse and complex mechanisms. The direct numerical simulation of flows with interfaces has brought significant insight on these processes. Here we review one of the most active aspects, the destabilizing of liquid gas mixing layers.

Atomizing devices exist in many different types, but of particular interest are those which involve parallel liquid and gas jets, especially coaxial atomizers (Figure 1). The parallel jets form mixing layers that create small droplets close to the nozzle exit. A liquid-gas mixing layer involves parallel liquid and gas streams that mix behind a splitter plate (Figure 2).

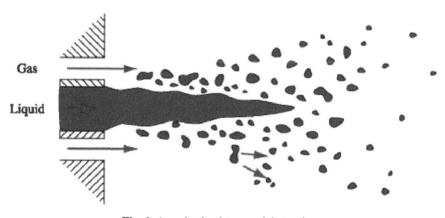

Gas

Liquid

Fig. 1. Atomization in a coaxial atomizer.

S. Balachandar and A. Prosperetti (eds), Proceedings of the IUTAM Symposium on Computational Multiphase Flow, 283–292.

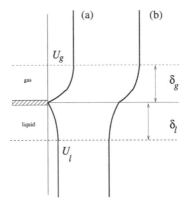

Fig. 2. Typical velocity profiles (a) behind a splitter plate and (b) some distance downstream. In a typical flow the Reynolds number is large and as a result the gas boundary layer will change little while the liquid boundary layer is reversed. The size of the liquid boundary layer is actually growing downstream.

2 Basic equations and methods

We assume two Newtonian, viscous fluids with constant surface tension. We use standard notations with σ the surface tension, κ the interface curvature, $\mathbf{n}$ the normal to the interface, μ the viscosity. The Navier–Stokes equations are

$$\partial_t \mathbf{u} + \mathbf{u} \cdot \nabla \mathbf{u} = -\frac{1}{\rho} \nabla p + \frac{1}{\rho} \nabla \cdot (2\mu \mathbf{D}) + \sigma \kappa \mathbf{n} \delta_S, \tag{1}$$

where δ_S is a distribution concentrated on the interface and $\mathbf{D}$ is the rate-of-strain tensor

$$D_{ij} = \frac{1}{2} \left(\frac{\partial u_j}{\partial x_i} + \frac{\partial u_i}{\partial x_j} \right). \tag{2}$$

Incompressible flow is assumed

$$\nabla \cdot \mathbf{u} = 0. \tag{3}$$

The interface follows the flow, or in other words, the normal velocity of the interface equals the normal flow velocity $\mathbf{u} \cdot \mathbf{n}$. To follow the interface, several methods have been described in the literature: the immersed boundary method [17, 18] or the Volume of Fluid method [15]. In the Volume of Fluid (VOF) method the interface is tracked by the volume fraction in each cell: C_{ij} being the volume fraction of the liquid phase in cell ij (Figure 3). In the Piecewise Linear Interface Calculation method (PLIC) the interface is reconstructed by linear segments in each cell, leading to errors of the order of κh^2. Thus the method is second order accurate and leads to exact reconstructions for straight lines. One of the critical components for reconstruction is the computation of the normal. The elementary finite-difference estimate of the normal is

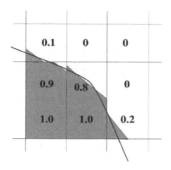

Fig. 3. The basic principle of the VOF-PLIC method: the interface is reconstructed by linear unconnected segments in each cell.

$$\mathbf{n}_{ij} = \frac{\nabla_h C}{||\nabla_h C||}, \tag{4}$$

with properly-centered finite differences. More advanced method for normal calculation have been known for some time [15].

Beyond reconstruction, the *propagation* of the interface is another essential component of the method. Several "advanced" propagation methods have been discussed in [1, 16]. A remarkable property of the recently developped methods, such as EI-LE or the geometrical method of Aulisa et al. [1] is that it conserves mass to machine accuracy. It may seem strange to the reader that the VOF methods do not *always* conserve mass to machine accuracy. This comes from small errors occuring during the propagation step. However, methods that do not conserve mass to machine accuracy still give good results: many of our results have been obtained with the method of Li [11] which does not conserve mass at machine accuracy, but has other advantages such as simplicity and the fact that it maintains the volume fraction between 0 and 1, consistent with its definition. It must be noticed that when the methods are used with relatively poor accuracy, for instance with unsufficient grid resolution, errors may become large, and the loss of mass becomes obvious. This may happen for instance when an entire atomizer is simulated, for instance the entire field shown on Figure 1.

The interface tracking is not the only important component of a numerical method. Of at least equal importance is the surface tension algorithm. The surface tension force added to the Navier–Stokes equations is

$$\mathbf{f}_\sigma \delta_S = -\sigma \kappa \mathbf{n} \delta_S. \tag{5}$$

In the Continuous Surface Force (CSF) method of Brackbill et al. [5], the δ_S distribution is approximated by $||\nabla_h C||$ (the subscript h indicates a finite difference). This approximation seems natural since the discrete color or marker function C_{ij} approximates the Heaviside function H and we have $\nabla H = -\delta_S \mathbf{n}$. Thus we add the following force $\mathbf{f}_{ij}$ to the velocity nodes,

$$\mathbf{f}_{ij} = \sigma \kappa^h ||\nabla_h C|| \mathbf{n}_{ij}^h, \tag{6}$$

where κ^h is an approximation of the curvature and $\mathbf{n}_{ij}^h$ is an approximation of the normal.

Using (4) for the normal we get $\mathbf{f}_{ij} = \sigma \kappa^h \nabla_h C$. In practice use of this (and other methods for surface tension) leads to problems when the surface tension is large. The dimensionless surface tension is the Laplace number La $= \rho d\sigma/\mu^2$ where d is a characteristic length scale, for instance the droplet diameter. For the discussion of surface tension the liquid ρ and μ should be used, as they allow the force on the interface to be compared with the smallest viscous dissipation.

The Laplace number La is of the order of 10^6 for a 1 cm air bubble in water. At such large La and for air-water density ratios, computation of surface tension becomes particularly difficult. To improve things, there are two main options: either smooth the color fraction C_{ij} or improve the computation of the curvature. A smoothed color function may be obtained by convolution

$$\tilde{H}(\mathbf{x}) = (H * K)(\mathbf{x}) = \int_V H(\mathbf{x}')K(\mathbf{x} - \mathbf{x}'; \epsilon)d\mathbf{x}', \tag{7}$$

where $K(\mathbf{x}; \epsilon)$ is an integration kernel of width ϵ. A discrete approximation of the convolution is

$$\tilde{C}_{ij} = A(\epsilon) \sum_m \sum_l C_{lm} K \left(1 - \frac{x_{il}^2 + y_{jm}^2}{\epsilon^2} \right) h^2, \tag{8}$$

where $x_{il} = x_i - x_l$, $y_{jm} = y_j - y_m$ and x_i is the abcissa of the ith column in a rectangular grid. The sum is over all l, m such that (x_l, y_m) is in the disk Ω_ϵ of radius ϵ, and $A(\epsilon)$ is a normalization constant (Figure 4). The other way to improve surface tension is to improve curvature estimates. This is the basis of the PROST method of Renardy and Renardy [14], which exactly fits a quadratic curve or a circle. Other methods such as the method of Popinet and Zaleski [13] will improve surface tension calculations in combination with marker methods instead of Volume of Fluid.

The marker methods are somewhat less complex to use, however their behavior will be very different in one respect: a marker method typically does not reconnect interfaces. The difficulty is then to design an algorithm or a realistic physical rule for reconnecting interfaces.

Another issue when solving the above equations is the viscosity interpolation in mixed cells. The arithmetic mean is most often used

$$\mu_{ij} = \mu_l C_{ij} + \mu_g(1 - C_{ij}), \tag{9}$$

where μ_{ij} the viscosity at node ij. The other method is the harmonic mean

$$\mu_{ij}^{-1} = \mu_1^{-1} C_{ij} + \mu_2^{-1}(1 - C_{ij}). \tag{10}$$

3 Theory of mixing layer instability

The theory of mixing layer instability is based on linearized perturbations of the viscous Navier Stokes or inviscid Euler equations. Results based on the full Orr–Sommerfeld viscous stability equations were obtained recently by Boeck and Zaleski

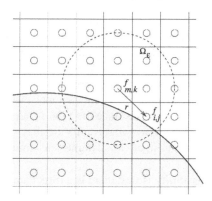

Fig. 4. The smoothing of the color function by a kernel to improve the surface tension calculation. The kernel averages the color function over a region of size ϵ.

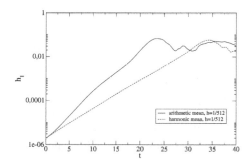

Fig. 5. Amplitude growth for the arithmetic and harmonic mean of the viscosity with small initial amplitude. The harmonic mean result is very close to linear theory even at small amplitudes while the arithmetic mean is close to linear theory only when the interface height is larger than one grid size (from [3].

[4]. They are based on smooth velocity profiles of the type of the boundary layer profile (b) shown in Figure 2. These profiles are naturally evolving from the parallel flow just behind the splitter plate. Recent agreement with laboratory experiments was obtained (Alain Cartellier, personal communication), and agreement is also obtained with the numerical code described above (Figure 5) . This agreement has been extremely difficult to get, because among other difficulties the oft-used results of the *inviscid* stability theory (as for instance in [12]) differ from the more realistic viscous theory by a large factor.

The dimensionless wavelength $\lambda_* = \lambda/\delta_g$ of the two-dimensional instability is predicted using simple scaling arguments [3] to increase like $\mathrm{Re}_*^{1/2}$ where $\mathrm{Re}_* = U_g \delta_g / \nu_g$ is the Reynolds number based on the gas boundary layer size δ_g (Figure 2). A similar analysis has been performed by Gordillo and Perez-Saborid [8].

Further into the flow the two-dimensional instability is supposed to lead to a three-dimensional instability, but there are several different mechanisms discussed in the literature. The simplest idea is to assume that the tip of ligaments form quasi-cylinders that break by the Rayleigh instability. A numerical demonstration of this mechanism was attempted by Fullana and Zaleski [7] but it failed, probably because the simulated ligaments were not thin enough. This issue of the thinness of the ligaments that can be simulated without breakup in a VOF method arises repeatedly. Other mechanisms involve three-dimensional instabilities that arise before ligaments are fully formed, for instance the two different types of "Rayleigh–Taylor" instabilites discussed by Marmottant and Villermaux [12] and Cartellier and Hopfinger (unpublished). Three-dimensional numerical simulations have not yet resolved the issue.

Another tantalizing possibility is that there is a direct amplification of the three-dimensional flow, through transient growth mechanisms [19]. This possibility awaits both experimental and numerical scrutiny.

4 Numerical simulation results

The numerical simulation of the full Navier–Stokes equations for the high speed jet atomization problem has been seldom performed. Work based on the Surfer code is now relatively old (beginning with Keller et al. [9], Tauber and Tryggvason [17], and Tauber et al. [18]ãin 2D and Zaleski et al. [20] in 3D), but recent results [3] have allowed to investigate the mixing layer instability [10] and the formation of filaments in much greater detail. In Figure 6 we show results for the amplification of the instability in a temporal setup: the domain is spatially periodic. It turns out that the old low-resolution results were in a way misleading: at low resolution the ligaments form and then break relatively early (as in Figure 7 on a 128×128 grid) while with finer resolution the ligaments are stretched much longer (Figure 6, on a 512×512 grid).)

When one attempts to perform 3D simulations with present day facilities, the resolution is even worse. Moreover, 3D temporal simulations such as those of Zaleski et al. [20] will still not be very realistic. The ultimate goal in direct numerical simulation of atoimization is to perform spatial 3D simulations. For instance, in Figure 8 we show results obtained on a $128 \times 128 \times 256$ grid for a spatially developing instability. Similar 3D calculations may be found in [2]. This type of simulation actually compounds two difficulties: one is the move to 3D, and the other is the spatial character: we have two simulate several wavelengths of instability in the same domain instead of one or two. To obtain in 3D the same kind of accuracy as in the simulation of Figure 6 we would need to have many boxes of size 512^3.

5 Conclusion and perspectives

One feeling that this short review of methods and physical problems hopefully conveys is the sheer complexity of the numerical simulation approach. Although experts generally agree that Volume Of Fluid methods have attained a satisfactory degree of

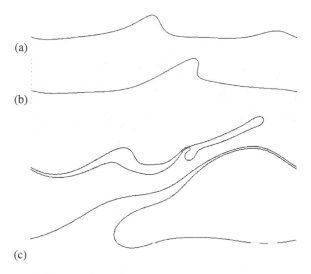

Fig. 6. Three stages of ligament formation and elongation at higher Reynods number, 512×512 grid: $\text{Re}_g = 4000$ and $\text{We}_g = 500$, (a) $t = 4$, (b) $t = 4.8$, (c) $t = 9.76$.

Fig. 7. Same simulation at lower (128×128) resolution: the breakup occurs much earlier than in the high resolution case. $t = 8.16$. Here the ligament is shown at $t = 8.4$ just before breakup.

accuracy and efficiency, this does not mean that efficient codes are available or easy to construct. For the physicist using Volume of Fluid methods, (or other methods for interface tracking) the sequence of algorithms to be used is complex and several important issues must be mastered. One is the resolution: Volume of Fluid methods tend to reconnect interfaces whenever they come within a grid cell of another interface. Thus thin ligaments can be followed only for a while: as they become thinner than one grid cell, the simulation becomes unrealistic.

Surface tension is also a source of difficulty. While most simulations show arbitrary Re, We and density ratios, the specific parameters of laboratory air/water experiments are difficult to attain. Surface tension calculations may be improved by smoothing using Equation (7), However smoothing introduces another difficulty: because the surface tension force is spread over a domain of width ϵ (see Figure 4), it acts as an effective interaction between interfaces, with a range of ϵ. Thus the minimum thickness of filaments before reconnection and breakup occur is not the grid

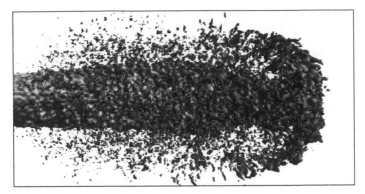

Fig. 8. The spatial development of an atomizing jet simulated in 3D (obtained by A. Lebois-setier). The domain shown is $128 \times 128 \times 256$.

size h but ϵ. The cost of the simulation is multiplied by $(\epsilon/h)^3$. Another possibility is the PROST method or other recent methods for improving the surface tension. However none of them has been tested yet on atomization problems, and PROST is expensive in CPU.

Another source of uncertainty is the averaging of viscosity inside mixed cells. Typically an arithmetic mean is used. However, the harmonic mean is superior for a parallel flow and interfaces aligned with it [6]. Changing the type of mean used may have a dramatic effect on the growth rate of the instability when the interface perturbation amplitude is smaller than a grid cell (Figure 5). It is not clear whether the small amplitude at which this phenomenon occurs makes it irrelevant or whether it would have a significant effect on full simulations.

The above difficulties make it hard to make predictions through numerical simulations that may be compared directly to experimental air-water results. This is both a question of sheer computer power (the objective being to increase the grid size), and of algorithm development (how to design even better algorithms that correct the principal defects *without* costing too much to implement and use). Because of the complexity of a numerical simulation code a third difficulty is numerical analysis and programming: the whole task of developing a code often overwhelms the abilities of isolated researchers. It is not enough to have pieces of code, corresponding to published algorithms, available on the internet (as on the author's web site http://www.lmm.jussieu.fr/~zaleski and links therein). In order to write good quality, easy to use codes from these pieces, an in-depth knowledge of the issues is needed which may be achieved only with sufficient expertise.

The question thus remains relatively open as to which type of code, with what degree of complexity will eventually be developed to simulate a fully 3D, realistic, experimentally relevant atomisation problem. It may involve other methods than Volume Of Fluid, such as marker methods for immersed boundaries which are less prone to reconnection, or involve adaptive mesh refinement. Once such a code is available and efficient, it will function as a kind of "virtual atomizer": a device allowing to design nozzle shapes in order to optimize characteristics of the outgoing spray. Similar conclusions may be drawn in other areas of two-phase flow simulation research, with in some cases even larger computer power perhaps necessary, such as splash calculations or boiling.

References

1. Aulisa, E., Manservisi, S., Scardovelli, R. and Zaleski, S., 2003, A geometrical area-preserving volume of fluid method, *J. Comput. Phys* **192**, 355–364.
2. Bianchi, G. M., Pelloni, P., Toninel, S., Scardovelli, R., Leboissetier, A. and Zaleski, S., 2005, A quasi-direct 3D simulation of the atomization of high-speed liquid jets, in *Proceedings of ICES05, 2005 ASME ICE Division Spring Technical Conference*, Chicago, IL, April 5–7, 2005.
3. Boeck, T., Li, J., Lopez-Pagès, E., Yecko, P. and Zaleski, S., 2006, Ligament formation in sheared liquid-gas layers, *Theoret. Comput. Fluid Dynam.*, to appear.
4. Boeck, T. and Zaleski, S., 2005, Viscous versus inviscid instability of two-phase mixing layers with continuous velocity profile, *Phys. Fluids* **17**, 032106.
5. Brackbill, J., Kothe, D.B. and Zemach, C., 1992, A continuum method for modeling surface tension, *J. Comput. Phys.* **100**, 335–354.
6. Coward, A.V., Renardy, Y.Y., Renardy, M. and Richards, J.R., 1997, Temporal evolution of periodic disturbances in two-layer Couette flow, *J. Comput. Phys.* **132**, 346–361.
7. Fullana, J.M. and Zaleski, S., 1999, Stability of a growing end-rim in a liquid sheet of uniform thickness, *Phys. Fluids* **11**, 952–954.
8. Gordillo, J.M. and Perez-Saborid, M., 2005, On the first wind atomization regime, *J. Fluid Mech.* **541**, 1–20.
9. Keller, F.X., Li, J., Vallet, A., Vandromme, D. and Zaleski, S., 1994, Direct numerical simulation of interface breakup and atomization, in *Proceedings of ICLASS94*, A.J. Yule (ed.), Begell House, New York, pp. 56–62.
10. Klein, M., 2005, Direct numerical simulation of a spatially developing water sheet at moderate reynolds number, *Int. J. Heat Fluid Flow* **26**, 722–731.
11. Li, J., 1995, Calcul d'interface affine par morceaux (piecewise linear interface calculation), *C. R. Acad. Sci. Paris, série IIb, (Paris)* **320**, 391–396.
12. Marmottant, P. and Villermaux, E., 2002, Atomisation primaire dans les jets coaxiaux, *Combustion (Revue des Sciences et Techniques de Combustion)* **2**, 89–126.
13. Popinet, S. and Zaleski, S., 1999, A front tracking algorithm for the accurate representation of surface tension, *Int. J. Numer. Meth. Fluids* **30**, 775–793.
14. Renardy, Y. and Renardy, M., 2002, PROST: A parabolic reconstruction of surface tension for the volume-of-fluid method, *J. Comput. Phys.* **183**, 400–421.
15. Scardovelli, R. and Zaleski, S., 1999, Direct numerical simulation of free-surface and interfacial flow, *Annu. Rev. Fluid Mech.* **31**, 567–603.
16. Scardovelli, R. and Zaleski, S., 2003, Interface reconstruction with least-square fit and split Lagrangian–Eulerian advection, *Int. J. Numer. Meth. Fluids* **41**, 251–274.
17. Tauber, W. and Tryggvason, G., 2000, Direct numerical simulation of primary breakup, *Comput. Fluid Dynam. J.* **9**, 158.
18. Tauber, W., Unverdi, S.O. and Tryggvason, G., 2002, The non-linear behavior of a sheared immiscible fluid interface, *Phys. Fluids* **14**, 2871.

19. Yecko, P. and Zaleski, S., 2005, Transient growth in two-phase mixing layers, *J. Fluid Mech.* **528**, 43–52.

20. Zaleski, S., Li, J., Scardovelli, R. and Zanetti, G., 1997, Direct simulation of multiphase flows with density variations, in *IUTAM Symposium on Variable Density Low Speed Turbulent Flows*, Marseille, France, July 7–10, 1996, L. Fulachier et al. (eds), Kluwer Academic Publishers, Dordrecht.

A Sharp-Interface Cartesian Grid Method for Computations of Droplet Impact and Spreading on Surfaces of Arbitrary Shape

S. Krishnan, H. Liu, S. Marella and H.S. Udaykumar

Department of Mechanical and Industrial Engineering, The University of Iowa, 2408 Seamans Center, Iowa City, IA 52242, USA; e-mail: ush@engineering.uiowa.edu

1 Introduction

A sharp-interface treatment of the solid-fluid boundaries [1] is combined with the Ghost Fluid Method (GFM) [2–4] to simulate interactions between droplets and solid surfaces [5]. All interfaces are represented by level sets [6, 7]. Application of contact angle conditions at the junction between the solid-fluid and fluid-fluid boundaries is fairly challenging in the level-set approach. In the VOF approach [8] the contact angle can be imposed by reconstructing the partial volume in the fluid-fluid interface cell that lies adjacent to the solid surface such that the reconstructed surface assumes the specified contact angle with respect to the solid surface [8]. In the Lagrangian moving mesh approach [9, 10] the mesh node that lies on the solid surface can be moved to apply the desired angle. An alternative approach based on a local level-set reconstruction has been modified and advanced in the present work. Additionally, the method is designed to enable simulations of droplet spreading on arbitrarily shaped solid surfaces. The results are compared with experimental as well as numerical results.

2 The current method

In the present work, the incompressible Navier–Stokes equations are solved in two-dimensional planar as well as axi-symmetric situations. For a solid-fluid boundary, a no-slip condition for the velocity and a Neumann condition for pressure [11] are applied at the interface. However, in the simulation of droplet impact on a solid surface, a slip boundary condition is applied in the immediate vicinity of the moving contact line. At a fluid-fluid boundary, jump conditions in velocity, velocity gradients and pressure [2] are applied.

S. Balachandar and A. Prosperetti (eds), Proceedings of the IUTAM Symposium on Computational Multiphase Flow, 293–300.

3 Flow solver

A cell-centered collocated arrangement of the flow variables is used to discretize the Navier–Stokes equations with a two-step fractional step method [11] to advance the solution in time. Interfaces are represented by level sets. Computational cells fall into two categories: (1) bulk cells with all neighbors in the same phase, (2) interfacial cells neighboring solid and other fluid phase cells, or both. For simplicity the discretization is shown below for a 1-dimensional case (x-direction) only. The discretization for multi-dimensions proceeds in similar fashion independently in each coordinate direction.

3.1 A general form of the discretization for the operators

A general discrete form for the diffusion operator $(\beta \psi_x)_x$ can be obtained as follows when multiple (say L_{max}) embedded boundaries are present in the flow.

$$
(\beta \psi_x)_x = \hat{\beta}_{+x} \alpha_{+x} \frac{(\psi_{+x} - \psi_{i,j})}{\gamma_x \Delta x^2} - \hat{\beta}_{-x} \alpha_{-x} \frac{(\psi_{i,j} - \psi_{-x})}{\gamma_x \Delta x^2} + \frac{\hat{\beta}_{+x} \alpha_{+x}}{\gamma_x \Delta x^2}
$$
$$
+ \frac{\hat{\beta}_{-x} \alpha_{-x}}{\gamma_x \Delta x^2} + \frac{\hat{\beta}_{+x}(1 - \chi_{+x})b_{+x}}{\beta_{i+1} \gamma_x \Delta x^2} + \frac{\hat{\beta}_{-x}(1 - \chi_{-x})b_{-x}}{\beta_{i-1} \gamma_x \Delta x^2}, \tag{1}
$$

where the coefficients $\hat{\beta}_{\pm x}$, $\alpha_{\pm x}$ and γ_x are obtained as described below (see [1, 5] for details). Using the level-set information switch functions (Equations (2) and (3)) that provide appropriate coefficients for mesh points adjoining fluid-fluid, fluid-solid and solid-solid interfaces are obtained in a generalized framework:

$$
(s_l)_{\pm x} = \left\{ \frac{(\phi_l)_{i,j}(\phi_l)_{i\pm 1,j}}{|(\phi_l)_{i,j}(\phi_l)_{i\pm 1,j}|} \right\}, \quad s_{\pm x} = \min_{l=1, L_{max}} \{(s_l)_{\pm x}\}, \tag{2}
$$

$$
\chi_{\pm x} = \min_{l=1, L_{max}} \left\{ |\max((s_l)_{\pm x}, 0)| + \frac{|(\phi_l)_{i,j}|}{|(\phi_l)_{i,j}| + |(\phi_l)_{i\pm 1,j}|} |\min((s_l)_{\pm x}, 0)| \right\}, \tag{3}
$$

$$
\delta_{\pm x} = \begin{cases} 1 & \text{if solid-fluid interface between } (i, j) \text{ and } (i \pm 1, j), \\ 0 & \text{otherwise,} \end{cases} \tag{4}
$$

$$
\psi_{\pm x} = \delta_{\pm x} \psi_{I_{\pm x}} + (1 - \delta_{\pm x}) \psi_{i\pm 1,j}, \tag{5}
$$

$$
\alpha_{\pm x} = \delta_{\pm x} \frac{1}{\chi_{\pm x}} + (1 - \delta_{\pm x}), \tag{6}
$$

$$
\hat{\beta}_{\pm x} = \frac{\beta_{i,j} \beta_{i\pm 1,j}}{\beta_{i,j} \chi_{\pm x} + \beta_{i\pm 1,j}(1 - \chi_{\pm x})}, \tag{7}
$$

$$
\gamma_x = \delta_{+x} \frac{\chi_{+x}}{2} + \delta_{-x} \frac{\chi_{-x}}{2} + (1 - \delta_{+x}) \left\{ \frac{1}{2} + \frac{1}{2} \chi_{-x} \delta_{-x} |\min(s_{+x}, 0)| \right\}
$$
$$
+ (1 - \delta_{-x}) \left\{ \frac{1}{2} + \frac{1}{2} \chi_{+x} \delta_{+x} |\min(s_{-x}, 0)| \right\}, \tag{8}
$$

$$a_{\pm x} = \frac{(\phi_{l_1})_{i,j}}{|(\phi_{l_1})_{i,j}|} a_{I_{\pm x}} |\min(s_{\pm x}, 0)|(1 - \delta_{\pm x}),\tag{9}$$

$$b_{\pm x} = \pm \frac{(\phi_{l_2})_{i,j}}{|(\phi_{l_2})_{i,j}|} b_{I_{\pm x}} |\min(s_{\pm x}, 0)|(1 - \delta_{\pm x}).\tag{10}$$

Note that the above equations reduce, in the appropriate cases by the use of simple level set based switch functions, to the discrete form for a solid-fluid interface or for a fluid-fluid interface or to standard central differences for bulk cells. Thus, a sharp-interface calculation that handles any kind of immersed boundaries and their interactions can be easily programmed by a few lines of code that modify a simple uniform Cartesian grid flow solver. The treatment of the convection terms as well as the pressure gradients for velocity correction proceeds in a fashion identical and consistent to that described above. Further details of implementation are described in [1, 5].

3.2 Modeling the moving contact line

The precise relationship between contact line velocity and contact angle is poorly understood. This problem is simplified by using experimentally measured [10, 12, 13] contact angles as input for the numerical model. Modeling of fluid behavior in the vicinity of a moving contact line is complicated because the no-slip boundary condition at the solid-liquid interface leads to a force singularity at the contact line [14]. This is resolved by replacing the no-slip boundary condition with a slip model [15]. In the present work, the advancing and receding contact angles ($\theta_{\text{advancing}}$, θ_{receding}) are assumed constant. The Navier slip boundary condition allowing the contact line to slip in a direction tangent to the substrate is applied in the immediate vicinity of the contact line. The local level-set field is reconstructed in this region to by fitting a parabolic curve that satisfies the contact angle condition at the solid surface while intersecting the solid surface at the contact line. During hysteresis the current model allows the surface tension to retract the fluid back decreasing the angle from $\theta_{\text{advancing}}$ to θ_{receding}.

4 Validation: Water droplet impact on a flat surface

Droplet impact on a solid surface is simulated for the conditions in [10], corresponding to Re $= 3130$, We $= 64$. Computed drop shapes and those reported by Fukai et al. [10] corresponding to the same instant of time are plotted in Figure 1. Consistent with Fukai's prediction, the spreading process ends at $t^* = 7.5$. During this spreading process, the contact angle is maintained at the advancing value of $92°$ as specified by the model. Contact angle hysteresis takes place approximately from $t^* = 5.0$ to $t^* = 7.0$. After $t^* = 7.5$, the recoil process begins and the fluid recedes from the maximum wetted radius at the specified receding contact angle ($60°$). A bulk upward motion near the axis occurs after $t^* = 28$ and oscillation of the droplet ensues after this time. The equilibrium shape of the drop is characterized by a typical

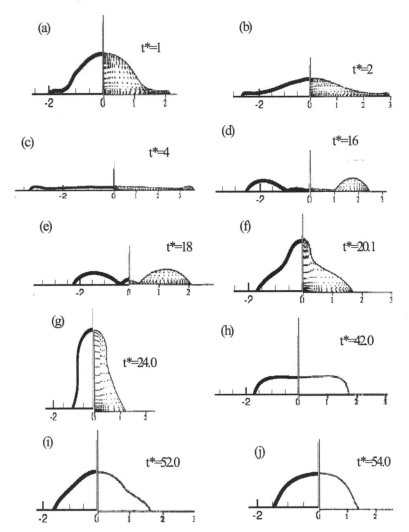

Fig. 1. Calculated droplet spreading shapes compared with numerical results from Fukai et al. [10] for droplet impact with Re = 3010 and We = 57, $\theta_{advancing}$ = 92°, $\theta_{receding}$ = 60°, θ_{static} = 75°.

sessile spherical cap drop shape with an equilibrium angle of 75°. The maximum spreading radius reported by Fukai et al. [10] for the given parameters is 3.6 while the current predicted result is 3.45. The current model obtains a drop thickness of 2.35 while Fukai et al. [10] report a value of 2.7. However, the time for the droplet to reach maximum spreading is agreement with that reported in [10]. Figure 2(a) depicts the contact angle as a function of time. During hysteresis, no fixed contact angle is imposed but the contact angle is free to be adjusted by surface tension while the contact line remains motionless.

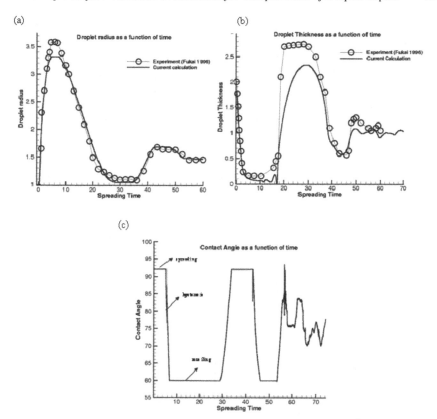

Fig. 2. (a) Calculated droplet spreading radius compared with experimental results for Re = 3010 and We = 57, $\theta_{\text{advancing}}$ = 92°, θ_{receding} = 60°, θ_{static} = 75°. (b) Calculated droplet thickness compared with experimental results for Re = 3010 and We = 57, $\theta_{\text{advancing}}$ = 92°, θ_{receding} = 60°, θ_{static} = 75°. (c) Contact angle for water droplet with Re = 3010 and We = 57, $\theta_{\text{advancing}}$ = 92°, θ_{receding} = 60°, θ_{static} = 75°.

5 Impact of droplets on arbitrarily shaped solid surfaces

Figure 3(a) shows a 2-dimensional planar simulation of droplet impact on a solid surface inclined at 45° to the horizontal corresponding to Re = 3333, We = 100, $\rho_{\text{liquid}}/\rho_{\text{gas}}$ = 1000. This case differs from the previous cases in that there are two contact lines which move with different velocities with two distinctive contact angles. When the droplet is sliding down the curved surface, the left side has an advancing (110°) and the right side has a receding (60°) contact angle. The droplet undergoes spreading, hysteresis, recoiling, as well as subsequent oscillations as it slides down the plane. It eventually reaches a static shape represented by an upper (69°) and lower (72°) contact angle. The current calculations compare qualitatively well with experimental studied reported in [16] and with VOF calculations reported in [8].

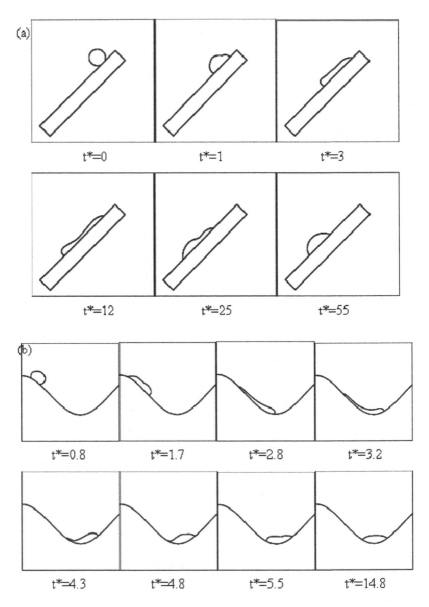

Fig. 3. (a) Calculated shapes for water droplet impacting on inclined surface with Re $= 3333$ and We $= 100$, $\theta_{\text{advancing}} = 110°$, $\theta_{\text{receding}} = 60°$. (b) Calculated shapes for water droplet impacting on curved surface with Re $= 3333$ and We $= 50$, $\theta_{\text{advancing}} = 110°$, $\theta_{\text{receding}} = 60°$.

Figure 3(b) shows droplet impact with wetting effects on an arbitrary-shaped surface corresponding to $Re = 3333$, $We = 50$, $\rho_{liquid}/\rho_{gas} = 1000$ and $\theta_{advancing} = 100°$, $\theta_{receding} = 60°$. The droplet is seen to flow down the surface to the trough, with the specified receding and advancing contact angles. It overshoots the trough due to inertia but finally settles to equilibrium in the trough with the resting contact angle values of $88.3°$ and $86.5°$ at the two contact lines.

6 Summary

A sharp-interface method is presented for the simulation of fluid-fluid interfaces interacting with solid-fluid interfaces. The framework of the method rests on a level-set representation of all interfaces allowing easy implementation of a finite-difference scheme to discretize the governing equations in the presence of interfaces such that explicit knowledge of the interface location is not necessary. The discretization scheme unifies a sharp-interface solid-fluid interface treatment with the ghost-fluid method and is used to study impact of droplets on solid surfaces.

Acknowledgements

This work was performed with support from the Computational Mechanics Branch, AFRL-MNAC, Eglin, FL (Project manager Mr. Joel Stewart) and AFOSR Computational Mathematics Division (Program manager Dr. Fariba Fahroo).

References

1. Marella, S., Krishnan, S. and Udaykumar, H.S., 2005, Sharp interface Cartesian grid method I: An easily implemented technique for 3D moving boundary computations, *Journal of Computational Physics* **210**, 1–31.
2. Kang, M. and Fedkiw, R.P., 2002, A boundary condition capturing method for multiphase incompressible flow, *Journal of Scientific Computing* **15**, 323–360.
3. Liu, X.D., Fedkiw, R.P. and Kang, M.J., 2000, A boundary condition capturing method for Poisson's equation on irregular domains, *Journal of Computational Physics* **160**(1), 151–178.
4. Fedkiw, R.L.X.D., 1998, The ghost fluid method for viscous flows, progress in numerical solutions of partial differential equations.
5. Liu, H., Krishnan, S., Marella, S. and Udaykumar, H.S., 2005, Sharp interface Cartesian grid method II: A technique for simulating droplet interactions with surfaces of arbitrary shape, *Journal of Computational Physics* **210**, 32–54.
6. Sethian, J.A. and Smereka, P., 2003, Level set methods for fluid interfaces, *Annual Review of Fluid Mechanics* **35**, 341–372.
7. Sussman, M., Smereka, P. and Osher, S.J., 1994, A level set approach for computing solutions to incompressible two-phase flow, *Journal of Computational Physics* **114**, 146.
8. Bussmann, M., Mostaghimi, J. and Chandra, S., 1999, On a three-dimensional volume tracking model of droplet impact, *Physics of Fluids* **11**(6), 1406–1417.

9. Fukai, J.Z.Z., 1993, Modeling of the deformation of a liquid droplet impinging upon a flat surface, *Physics of Fluids* **A5**(11), 2588.

10. Fukai, J., Shiiba, Y., Yamamoto, T., Miyatake, O., Poulikakos, D., Megaridis, C.M. and Zhao, Z., 1995, Wetting effects on the spreading of a liquid droplet colliding with a flat surface – Experiment and modeling, *Physics of Fluids* **7**(2), 236–247.

11. Ye, T., Mittal, R., Udaykumar, H.S. and Shyy, W., 1999, An accurate Cartesian grid method for viscous incompressible flows with complex immersed boundaries, *Journal of Computational Physics* **156**(2), 209–240.

12. Pasandideh-Fard, M., Bhola, R., Chandra, S. and Mostaghimi, J., 1998, Deposition of till droplets on a steel plate: Simulations and experiments, *International Journal of Heat and Mass Transfer* **41**(19), 2929–2945.

13. Pasandideh-Fard, M., Bussmann, M., Chandra, S. and Mostaghimi, J., 2001, Simulating droplet impact on a substrate of arbitrary shape, *Atomization and Sprays* **11**(4), 397–414.

14. Hocking, L.M., 1983, The spreading of a thin drop by gravity and capillarity, *Quarterly Journal of Mechanics and Applied Mathematics* **36**(FEB), 55–69.

15. Dussan, E.B., Rame, E. and Garoff, S., 1991, On identifying the appropriate boundary-conditions at a moving contact line – An experimental investigation, *Journal of Fluid Mechanics* **230**, 97–116.

16. Kang, B.S. and Lee, D.H., 2000, On the dynamic behavior of a liquid droplet impacting upon an inclined heated surface, *Experiments in Fluids* **29**(4), 380–387.

A Finite-Volume/Front-Tracking Method for Computations of Multiphase Flows in Complex Geometries

Metin Muradoglu, Ufuk Olgac and Arif Doruk Kayaalp

Department of Mechanical Engineering, Koc University, Rumelifeneri Yolu,
Sariyer 34Ïstanbul, Turkey; e-mail: mmuradoglu@ku.edu.tr, uolgac@ku.edu.tr,
adkayaalp@ku.edu.tr

Abstract. A finite-volume/front-tracking (FV/FT) method is developed for computations of multiphase flows in complex geometries. The front-tracking methodology is combined with a dual time-stepping based FV method. The interface between phases is represented by connected Lagrangian marker points. An efficient algorithm is developed to keep track of the marker points in curvilinear grids. The method is implemented to solve two-dimensional (plane or axisymmetric) dispersed multiphase flows and is validated for the motion of buoyancy-driven drops in a periodically constricted tube with cases where drop breakup occurs.

Key words: finite-volume/front-tracking method, dual time-stepping, dispersed multiphase flows, complex geometries.

1 Introduction

The main difficulty in simulating multiphase flows is the presence of deforming phase boundaries. Although there are a number of numerical methods developed and successfully applied to multifluid and multiphase flow problems [6, 8, 10], there is still considerable need to accurate computations of multiphase flows involving strong interactions with complex solid boundaries. Modeling these strong interactions is a challenging task faced in many engineering and scientific applications such as microfluidic systems [9], pore-scale multi-phase flow processes [4, 5] and biological systems [1, 7].

In this study, we present a FV/FT method for the computations of multiphase flows involving complex solid boundaries. The front-tracking method has many advantages such as its conceptual simplicity and small numerical diffusion. However, its main disadvantage is probably the difficulty to maintain the communication between the Lagrangian marker points and Eulerian body-fitted curvilinear or unstructured grids. To overcome this difficulty, a computationally efficient and robust tracking algorithm is developed for tracking the front marker points in body-fitted

S. Balachandar and A Prosperetti (eds), Proceedings of the IUTM Symposium on Computational Multiphase Flow, 301–310.

curvilinear grids. The tracking algorithm utilizes an auxiliary regular Cartesian grid and it can be easily extended to unstructured grids.

The finite-volume method is based on the concept of dual (or pseudo) time-stepping which provides direct coupling of the continuity and momentum equations for incompressible flows. Detailed description of the FV/FT method can be found in [11].

In the present study, the method is applied to buoyancy driven motion of drops in constricted channels studied experimentally by Hemmat and Borhan [2].

2 Mathematical formulation

The incompressible flow equations for an axisymmetric flow can be written in the cylindrical coordinates in the vector form as

$$
\frac{\partial \mathbf{q}}{\partial t} + \frac{\partial \mathbf{f}}{\partial r} + \frac{\partial \mathbf{g}}{\partial z} = \frac{\partial \mathbf{f}_v}{\partial r} + \frac{\partial \mathbf{g}_v}{\partial z} + \mathbf{h}_v + \mathbf{f}_b, \tag{1}
$$

where

$$
\mathbf{q} = \left\{ \begin{array}{c} 0 \\ r\rho v_r \\ r\rho v_z \end{array} \right\}, \quad \mathbf{f} = \left\{ \begin{array}{c} r v_r \\ r(\rho v_r^2 + p) \\ r\rho v_r v_z \end{array} \right\}, \quad \mathbf{g} = \left\{ \begin{array}{c} r v_z \\ r\rho v_r v_z \\ r(\rho v_z^2 + p) \end{array} \right\}, \tag{2}
$$

and

$$
\mathbf{f}_v = \left\{ \begin{array}{c} 0 \\ \tau_{rr} \\ \tau_{zr} \end{array} \right\}, \quad \mathbf{g}_v = \left\{ \begin{array}{c} 0 \\ \tau_{zr} \\ \tau_{zz} \end{array} \right\}, \quad \mathbf{h}_v = \left\{ \begin{array}{c} 0 \\ p - \frac{2}{r}\frac{\partial}{\partial r}(r\mu v_r) - \frac{\partial}{\partial z}(\mu v_z) \\ -\frac{\partial}{\partial z}(\mu v_z) \end{array} \right\}. \tag{3}
$$

In Equations (1)–(3), r and z are the radial and axial coordinates and t is the physical time; ρ, μ and p are the fluid density, the dynamic viscosity and pressure; v_r and v_z are the velocity components in r and z coordinate directions, respectively. The viscous stresses appearing in the viscous flux vectors are given by

$$
\tau_{rr} = 2\mu \frac{\partial r v_r}{\partial r}, \quad \tau_{zz} = 2\mu \frac{\partial r v_z}{\partial z}, \quad \tau_{zr} = \mu \left(\frac{\partial r v_r}{\partial z} + \frac{\partial r v_z}{\partial r} \right). \tag{4}
$$

The last term in Equation (1) represents the body forces resulting from the buoyancy and surface tension and is given by

$$
\mathbf{f}_b = -r(\rho_o - \rho)\mathbf{G} - \int_S r\sigma\kappa\mathbf{n}\delta(\mathbf{x} - \mathbf{x}_f)ds, \tag{5}
$$

where the first term represents the body force due to buoyancy with ρ_o and $\mathbf{G}$ being the density of ambient fluid and the gravitational acceleration, respectively. The second term in Equation (5) represents the body force due to the surface tension, and $\delta, \mathbf{x}_f, \sigma, \kappa, \mathbf{n}, S$ and ds denote the Dirac delta function, the location of the front, the surface tension coefficient, the twice of the mean curvature, the outward unit normal vector on the interface, the surface area of the interface and the surface area element of the interface, respectively.

In Equation (1), the fluids are assumed to be incompressible and the effects of heat transfer are neglected. Therefore, the density and the viscosity of a fluid particle remains constant, i.e. $D\rho/Dt = 0$ and $D\mu/Dt = 0$

3 Numerical method

As can be seen in Equation (1), the continuity equation is decoupled from the momentum equations since it does not have any time derivative term. In order to overcome this difficulty and to be able to use a time-marching solution algorithm, artificial time derivative terms are added to the flow equations in the form

$$\Gamma^{-1}\frac{\partial \mathbf{w}}{\partial \tau} + \mathbf{I}^1\frac{\partial \rho \mathbf{w}}{\partial t} + \frac{\partial \mathbf{f}}{\partial r} + \frac{\partial \mathbf{g}}{\partial z} = \frac{\partial \mathbf{f}_v}{\partial r} + \frac{\partial \mathbf{g}_v}{\partial z} + \mathbf{h}_v + \mathbf{f}_b, \tag{6}$$

where τ is the pseudo time. The solution vector $\mathbf{w}$, the incomplete identity matrix I^1 and the preconditioning matrix Γ^{-1} are given by

$$\mathbf{w} = \begin{Bmatrix} rp \\ rv_r \\ rv_z \end{Bmatrix}, \quad I^1 = \begin{bmatrix} 0 & 0 & 0 \\ 0 & 1 & 0 \\ 0 & 0 & 1 \end{bmatrix}, \quad \Gamma^{-1} = \begin{bmatrix} \frac{1}{\rho\beta^2} & 0 & 0 \\ 0 & \rho & 0 \\ 0 & 0 & \rho \end{bmatrix}, \tag{7}$$

where β is the preconditioning parameter with dimensions of velocity [11].

With the goal of treating complex geometries, Equation (6) can be transformed into a general, curvilinear coordinate system

$$\xi = \xi(r, z), \quad \eta = \eta(r, z), \tag{8}$$

and the resulting equations take the form

$$\Gamma^{-1}\frac{\partial h\mathbf{w}}{\partial \tau} + \mathbf{I}^1\frac{\partial \rho h\mathbf{w}}{\partial t} + \frac{\partial h\mathbf{F}}{\partial \xi} + \frac{\partial h\mathbf{G}}{\partial \eta} = \frac{\partial h\mathbf{F}_v}{\partial \xi} + \frac{\partial h\mathbf{G}_v}{\partial \eta} + h(\mathbf{h}_v + \mathbf{f}_b), \tag{9}$$

where $h = r_\xi z_\eta - r_\eta z_\xi$ represents the Jacobian of the transformation. The vectors $h\mathbf{F} = z_\eta \mathbf{f} - r_\eta \mathbf{g}$; $h\mathbf{G} = -z_\xi \mathbf{f} + r_\xi \mathbf{g}$ and $h\mathbf{F}_v = z_\eta \mathbf{f}_v - r_\eta \mathbf{g}_v$; $h\mathbf{G}_v = -z_\xi \mathbf{f}_v + r_\xi \mathbf{g}_v$, represent the transformed inviscid and viscous flux vectors, respectively.

A three point second order backward implicit method is used to approximate the physical time derivatives. The spatial derivatives are discretized using a finite-volume method that is equivalent to a second order finite-difference method in uniform Cartesian grid. Time integration in pseudo time is achieved by an alternating direction implicit (ADI) method. Three types of grids used in the present method are sketched in Figure 1. Conservation equations are solved on a body-fitted curvilinear grid and the interface is represented by a Lagrangian grid. To maintain efficient communication between the curvilinear and Lagrangian grids and to keep track of the marker points, an auxiliary uniform Cartesian grid is used.

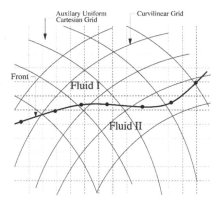

Fig. 1. Three types of grids used in the computations.

The overall solution procedure can be summarized as follows: In advancing solutions from physical time level n ($t_n = n \cdot \Delta t$) to level $n+1$, the locations of the marker points at the new time level $n + 1$ are first predicted using an explicit Euler method, i.e.,

$$\widetilde{\mathbf{X}}_p^{n+1} = \mathbf{X}_p^n + \Delta t \mathbf{V}_p^n, \tag{10}$$

where $\mathbf{X}_p$ and $\mathbf{V}_p$ denote the position of front marker points and the velocity interpolated from the neighboring curvilinear grid points onto the front point $\mathbf{X}_p$, respectively. Then the material properties and surface tension are evaluated using the predicted front position as

$$\rho^{n+1} = \rho(\widetilde{\mathbf{X}}_p^{n+1}); \quad \mu^{n+1} = \mu(\widetilde{\mathbf{X}}_p^{n+1}); \quad \mathbf{f}_b^{n+1} = \mathbf{f}_b(\widetilde{\mathbf{X}}_p^{n+1}). \tag{11}$$

The velocity and pressure fields at new physical time level $n + 1$ are then computed by solving the flow equations (Equation 9) by the FV method for a single physical time step and finally the positions of the front points are corrected as

$$\mathbf{X}_p^{n+1} = \mathbf{X}_p^n + \frac{\Delta t}{2}(\mathbf{V}_p^n + \mathbf{V}_p^{n+1}). \tag{12}$$

After this step the material properties and the body forces are re-evaluated using the corrected front position. The method is overall second order accurate both in time and space but the spatial accuracy reduces to first order near the interfaces. All terms except $\mathbf{f}_b$ in Equation (9) are treated implicitly in physical time so that the physical time is determined solely by the accuracy considerations and stability constraint mainly due to surface tension.

4 Results and discussion

The FV/FT method is applied to the buoyancy-driven motion of viscous drops through a vertical capillary with periodic corrugations studied experimentally by

Table 1. Two-phase systems used in the computations.

System	Suspending fluid	Drop fluid	μ_o (mPa·s)	μ_d (mPa·s)	ρ_o (kg/m^3)	ρ_d (kg/m^3)	σ (N/m)
GW3	glycerol-water (96.2wt%)	UCON 1145	450	530	1250	995	0.0105
GW5	glycerol-water (96.2wt%)	UCON 50HB100	450	97	1250	950	0.0057
DEGG10	diethylene-glycol (100.0wt%)	UCON 165	28	63	1110	975	0.0016
DEGG12	diethylene glycol-glycerol (63.8wt%)	UCON 285	87	115	1160	966	0.0042

Hemmat and Borhan [2]. The computational setup is sketched in Figure 2a. The capillary tube consists of a 26 cm long, periodically constricted cylindrical tube with 6 corrugations. The average internal radius of the tube is $R = 0.5$ cm, and the wavelength and amplitude of the corrugations are $h = 4$ cm and $A = 0.07$ cm, respectively. The suspending fluids are an aqueous glycerol solution (denoted by GW3 and GW5) and diethylene glycol-glycerol mixtures (denoted by DEGG10 and DEGG12). A variety of UCON oils are used as drop fluids. The properties of the drop and suspending fluids are summarized in Table 1 where the same label is used for each system as that used by Hemmat and Borhan [2]. A complete description of the experimental set up can be found in [2]. A portion of a coarse grid containing 8×416 grid cells is plotted in Figure 2b to show the overall structure of the body-fitted grid used in the simulations.

The average rise velocity of buoyant drops as well as the drop shapes are computed and the results are compared with the experimental data [2] for a range of the governing parameters, viz. the dimensionless drop size, κ, defined as the ratio of the equivalent spherical drop radius to the average capillary radius, the dimensionless corrugation amplitude, α, defined as the ratio of the amplitude of corrugation to the average capillary radius, the ratio of the drop to the suspending fluid viscosities, λ, the corresponding ratio of fluid densities γ, and the Bond number $Bo = \Delta\rho g_z R^2/\sigma$, representing the ratio of buoyancy to interfacial tension forces; $\Delta\rho$ and σ denote the density difference and interfacial tension between the drop and suspending fluid, respectively, and g_z is the gravitational acceleration.

In all the results presented in this section, the drops are initially spherical, located at $z = 1.5h$ in the ambient fluid that fully fills the cylindrical tube and is initially in the hydrostatic conditions. Symmetry boundary conditions are applied along the centerline and no-slip boundary conditions are used at top, bottom and lateral surfaces of the cylindrical tube. Drops are stationary and start rising due to buoyancy.

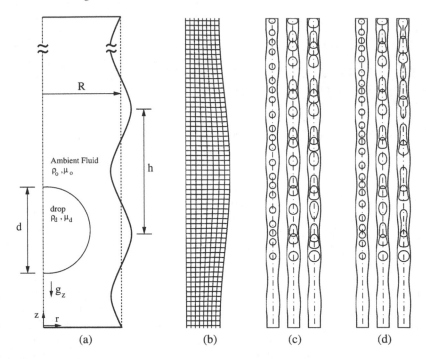

Fig. 2. (a) Schematic illustration of the computational setup for a buoyancy-driven rising drop in a constricted channel. (b) A portion of a coarse computational grid containing 8×416 cells. (c) and (d) are snapshots of buoyant drops of GW3 and DEGG12 systems, respectively, for drops sizes $\kappa = 0.54, 0.78$ and 0.92, from left to right for each system. The gap between two successive drops in each column represents the distance the drop travels at a fixed time interval and the last interface is plotted from left to right at $t^* = $ (c) 1044.4, 783.3, 783.3, (d) 2831.3, 3693.0 and 5416.4, respectively.

The results are expressed in terms of non-dimensional quantities denoted by superscript "*". The dimensionless coordinates are defined as $z^* = z/h$ and $r^* = r/R$. Time and velocity are made dimensionless with

$$T_{\text{ref}} = \frac{\mu_o}{\Delta \rho g_z R} \quad \text{and} \quad V_{\text{ref}} = \frac{\Delta \rho g_z R^2}{\mu_o},$$

respectively.

First a qualitative analysis of the shapes of the drops are shown in Figures 2c and 2d. In these figures, a sequence of images showing the evolution of the shapes of viscous drops through constricted channel are plotted for GW3 and DEGG12 systems with the non-dimensional drop sizes $\kappa = 0.54, 0.78$ and 0.92. The computations are performed on a 32×1664 grid, the physical time step is $\Delta t^* = 1.641$ and the residuals are reduced by three orders of magnitude in each sub-iteration. As can be seen in these figures, when a large drop ($\kappa > 0.7$) reaches a constriction, its leading edge follows the capillary wall contour and squeezes through the throat.

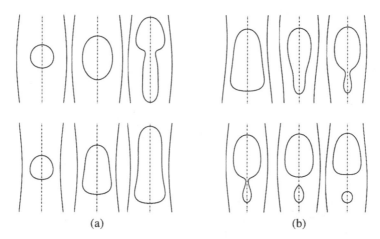

Fig. 3. (a) Snapshots of the drops at the expansion (upper plots) and at the throat (lower plots) of the constriction for the DEGG12 system for drop sizes (from left to right) $\kappa = 0.54$, 0.78 and 0.92, respectively. (b) Evolution of a GW5 drop with $\kappa = 0.90$ in the constricted channel. The drop size is greater than the critical drop size and thus the drop breakups ($\alpha = 0.14$ and $\kappa_{cr} = 0.87$).

Once the leading meniscus clears the throat, its rise velocity increases as it enters the diverging cross-section while the trailing edge of the drop remains trapped behind the throat similar to the experimental observations [2]. After a critical value of the non-dimensional drop size denoted by κ_{cr}, there occurs a neck between the leading edge and the trapped trailing edge which eventually leads to drop breakup. To better show the effects of the constrictions, the snapshots of the drops before and after the throat of the constriction are shown in Figure 3a for DEGG12 system for drop sizes $\kappa = 0.54, 0.78$ and 0.92. As can be seen in this figure, the drop shapes are smooth in all the cases indicating accuracy of the computations. A sequence of snapshots showing the breakup phenomenon is given in Figure 3b. Note that the drop shapes in Figure 3 qualitatively compare well with the experimental observations published by Hemmat and Borhan [2]. The critical values leading to drop breakup of the drop sizes is an important parameter which is given in Table 2 for various geometries together with available experimental data [3]. As can be seen in this table, the computed critical drop radius for breakup compares well with the experimental value. It is also seen that κ_{cr} reduces as the amplitude of corrugation increases as expected.

Table 2. Critical non-dimensional drop sizes for GW5 system.

κ_{cr}	Numerical	1.10	0.87	0.75	0.68	0.62	0.52
κ_{cr}	Experimental	–	0.85	–	–	–	–
α		0.07	0.14	0.21	0.28	0.35	0.50

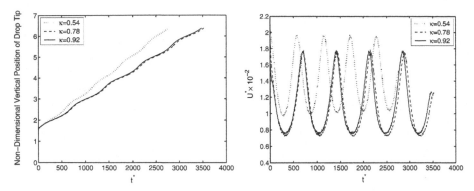

Fig. 4. The non-dimensional vertical positions (left plot) and the non-dimensional rise velocities (right plot) of the drop tip plotted against the non-dimensional time t^* for the drops of DEGG12 system with $\kappa = 0.54$, 0.78 and 0.92. Grid: 32×1664, $\Delta t^* = 1.641$.

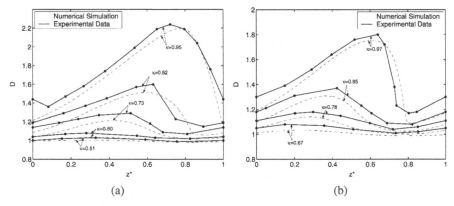

(a)   (b)

Fig. 5. The variations of the deformation parameter D with axial position of the advancing meniscus within one period of corrugation for (a) DEGG12 system and (b) GW3 system. The dashed curves are the numerical results and the solid lines are the experimental data.

Finally the vertical drop tip location scaled by the corrugation wavelength and the drop tip rise velocity scaled by the reference velocity V_{ref} are plotted against the non-dimensional time in Figure 4 for DEGG12 system for various drop sizes. The retardation effect of the constrictions is clearly seen in these figures for large drops, i.e., $\kappa > 0.7$. It is also seen that the drops quickly accelerate and reach a periodic motion in all the cases.

To demonstrate the performance of the present FV/FT method, the numerical results are compared with the experimental data. In order to qualitatively characterize the evolving shapes of drops as they pass through the corrugations, a deformation parameter denoted by D is defined as the ratio of the perimeter of the deformed drop profile to that of the equivalent spherical drop. The variations of the deformation parameter as a function of the axial position of the drop within one period of corrug-

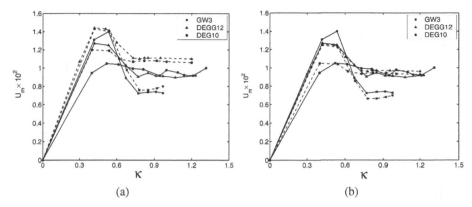

Fig. 6. Dimensionless average rise velocity as a function of drop size. The dashed curves are the numerical results and the solid lines are the experimental data. The numerical results are obtained with (a) the average tube radius of $R = 0.5$ cm and (b) the modified tube radius of $R = 0.535$ cm that yields the same Bond number given by Hemmat and Borhan [2].

ation are plotted in Figures 5a and 5b for GW3 and DEGG12 systems, respectively, and are compared with the experimental data. It can be seen in these figures that the general trend for the deformation parameter is well captured by the present computations for both GW3 and DEGG12 systems. The deformation is negligibly small for small drops, i.e., drops with $\kappa < 0.60$, and increases rapidly as the drop size gets larger. The discrepancy between the computed and the experimental results for D is partly attributed to the uncertainties in the experimental data and the inconsistency between the Bond number reported by Hemmat and Borhan [2] and the Bond number computed from the material properties and the average tube radius.

The computed average rise velocities are compared with the experimental data and the non-dimensional average rise velocity U_m is plotted against the non-dimensional drop size κ in Figure 6a. The numerical results are in a good agreement with the experimental data, i.e., the trend is well captured and the maximum error is less than 10% for all the cases. In addition, if the inconsistency between the Bond numbers mentioned above is taken into account by modifying the average tube radius to match the given Bond number in [2], the computed results match much better with the experimental data as shown in Figure 6b. In this case, the difference between the computed and experimental data reduces below a few percent.

5 Conclusions

A finite-volume/front-tracking (FV/FT) method has been developed for computations of dispersed multiphase flows in complex geometries. The method is based on the one-field formulation of the flow equations and treating the different phases as a single fluid with variable material properties. The flow equations are solved by a FV method on a body-fitted curvilinear grid and a separate Lagrangian grid is used to represent the interfaces between different phases. An efficient tracking algorithm which utilizes an auxiliary uniform Cartesian grid is developed to track the interfaces on the curvilinear grid.

The method is implemented to solve two-dimensional (plane or axisymmetric) dispersed multiphase flows and has been successfully applied to the motion and breakup of buoyancy-driven rising drops in a continuously constricted channel. Comparison with the experimental data proved the method to be successfully. It is found that the present method is a viable tool for accurate modeling of dispersed multiphase flows in complex geometries.

References

1. L. Fauci and S. Gueron (eds), 2001, *Computational Modeling in Biological Fluid Dynamics*, Springer-Verlag, New York.
2. M. Hemmat and A. Borhan, 1996, Buoyancy-driven motion of drops and bubbles in a periodically constricted capillary, *Chem. Eng. Commun.* **150**, 363.
3. M. Hemmat, 1996, The motion of drops and bubbles through sinusoidally constricted capillaries, Ph.D. Thesis, The Pennsylvania State University.
4. W.L. Olbricht and L.G. Leal, 1983, The creeping motion of immicible drops through a converging/diverging tube, *J. Fluid Mech.* **134**, 329.
5. W.L. Olbricht, 1996, Pore-scale prototypes of multiphase flow in porous media, *Anu. Rev. Fluid Mech.* **28**, 187.
6. S. Osher and R.P. Fedkiw, 2001, Level set methods: An overview, *J. Comput. Phys.* **169**(2), 463.
7. C. Pozrikidis (ed.), 2003, *Modeling and Simulation of Capsules and Biological Cells*, Chapman and Hall/CRC.
8. R. Scardovelli and S. Zaleski, 1999, Direct numerical simulation of free-surface and interfacial flow, *Anu. Rev. Fluid Mech.* **31**, 567.
9. H.A. Stone, A.D. Stroock and A. Ajdary, 2004, Engineering flows in small devices: Microfluidics toward lab-on-a-chip, *Anu. Rev. Fluid Mech.* **36**, 381.
10. G. Tryggvason, B. Bunner, A. Esmaeeli, D. Juric, N. Al-Rawahi, W. Tauber, J. Han, S. Nas and Y.-J. Jan, 2001, A front-tracking method for the computations of multiphase flow, *J. Comput. Phys.* **169**(2), 708.
11. M. Muradoglu and A.D. Kayaalp, 2005, An auxiliary grid method for computations of multiphase flows in complex geometries, *J. Comput. Phys.*, submitted.

The Effect of Surfactant on Rising Bubbles

Yoichiro Matsumoto, Touki Uda and Shu Takagi

Department of Mechanical Engineering, The University of Tokyo, Hongo, Bunkyo-ku, Tokyo 113-8656, Japan; e-mail: ymats@mech.t.u-tokyo.ac.jp, takagi@mech.t.u-tokyo.ac.jp

Abstract. The rising velocity of a spherical bubble in contaminated water can be less than a half of that in pure water. This is explained in terms of the Marangoni effect caused by the adsorption of surfactants in liquid phase on bubble surface. In this study, we conduct a numerical simulation with different surfactant species and bulk concentrations, and the dependence of their properties on the rising velocity of a bubble is analyzed through comparison with experiments. The simulation results show good agreement with the experimental ones, and the surface velocity and the concentrations are estimated. We also develop a simulation method for solving bubble deformation in the presence of a surfactant. We succeed in reproducing the conglobation effect of a bubble in surfactant solutions.

Key words: Marangoni effect, boundary-fitted grid, finite difference, adsorption desorption kinetics.

1 Introduction

It is well known that a bubble in contaminated water rises much slower than one in super-purified water (see, for example, [1]), and the rising velocity in a contaminated system can be less than half of that in a pure system. This phenomenon is explained by the Marangoni effect: when a bubble is rising, there exists a surface-concentration distribution along the bubble surface because the surfactant is swept off at the front and accumulates at the rear by advection. Due to this surfactant accumulation at the rear of the bubble surface, a variation of surface tension along the surface is developed which causes a tangential shear stress on the bubble surface. This is known as the Marangoni effect. This shear stress results in a decrease of the rising velocity of the bubble in contaminated liquid. This explanation was first given by Frumkin and Levich [2], after which many studies have been conducted on this subject.

However, a full numerical simulation using the Navier–Stokes equations with a contamination effect was not conducted until the last decade due to a lack of computer performance. There are many kinds of free-surface solvers using rectangular grid systems, such as Level Set [3], Front Tracking [4] and CIP [5]. Many of these

S. Balachandar and A. Prosperetti (eds), Proceedings of the IUTAM Symposium on Computational Multiphase Flow, 311–321.

methods are very powerful tools to simulate multiphase flows. The present problem, however, presents a very thin boundary layer of the bulk concentration of the surfactant above the bubble surface. To capture this thin boundary layer, we need to have more grid points close to the surface than are required for capturing the velocity boundary layer. This requirement is very restrictive if we use a fixed rectangular grid system.

To capture the thin surfactant boundary layer, Cuenot et al. [6] used a boundary-fitted coordinate system with sufficient grid points near the surface. A good review of this field is also given in their paper. They investigate the effects of slightly soluble surfactants on the flow around a spherical bubble. In their method, the continuity, momentum and bulk/surface concentration equations are coupled under a spherical bubble assumption. Liao and McLaughlin [7] allows the deformation of a bubble using the same governing equations as Cuenot et al. [6].

In the present study, we investigate bubbles rising through a liquid with different kinds of surfactant under different bulk concentrations. We conducted numerical simulations to investigate the steady and unsteady behaviors of the bubble. We also conducted a related experiment, using super-purified water with the addition of various amount of three kinds of surfactants (1-Pentanol, 3-Pentanol, TritonX-100). Through a comparison between the experimental results and the numerical ones, the effect of the bulk concentration and physical-chemical parameters of a surfactant on a single bubble motion is discussed.

2 Numerical methods

2.1 Governing equations

An axisymmetric grid is used with a boundary-fitted coordinate system. The grid system near the bubble is shown in Figure 1. The physical model for adsorption and desorption of surfactant molecules on the bubble surface is almost the same as Cuenot et al. [6]. The continuity equation (Equation (1)), full Navier–Stokes equation (Equation (2)), and the transport equation of surfactant concentration in the bulk liquid (Equation (3)) and that on the surface (Equation (4)) are coupled. The dimensionless forms of these equations are shown below. These differential equations are solved by the SIMPLER algorithm.

$$\nabla \cdot \mathbf{u} = 0, \tag{1}$$

$$\frac{\partial \mathbf{u}}{\partial t} + \nabla \cdot (\mathbf{uu}) = -\nabla P + \frac{1}{Re}\nabla^2\mathbf{u}, \tag{2}$$

$$\frac{\partial C}{\partial t} + \nabla \cdot (\mathbf{u}C) = \frac{1}{Pe}\nabla^2 C, \tag{3}$$

$$\frac{\partial \Gamma}{\partial t} + \nabla_s \cdot (\mathbf{u}_s\Gamma) = \frac{1}{Pe_s}\nabla_s^2\Gamma - \frac{1}{Pe}\nabla C_s \cdot \mathbf{n}, \tag{4}$$

where

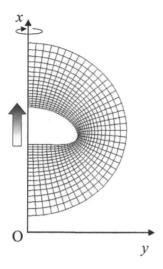

Fig. 1. Grid near the bubble.

$$\mathbf{u} = \frac{\mathbf{u}_\wp}{U_\infty}, \quad C = \frac{C_\wp}{C_\infty}, \quad \Gamma = \frac{\Gamma_\wp}{\Gamma_{\max}}, \quad Re = \frac{2\rho U_\infty R_0}{\mu}, \quad Pe = \frac{2U_\infty R_0}{D}. \quad (5)$$

The subscript $\wp$ denotes the dimensional values, and the subscript s denotes the values at the surface. R_0, ρ, μ, D, U_∞, C_∞, $\Gamma_{\max}$ denote bubble radius, liquid density, liquid viscosity, diffusion coefficient of surfactant concentration, terminal rising velocity, surfactant concentration in the far-field, and saturation value of surface concentration, respectively. Re and Pe are the Reynolds and Peclet number, respectively.

2.2 Boundary conditions

The boundary condition of the bulk concentration near the interface is:

$$-(\nabla C)_s \cdot \mathbf{n} = Ha \cdot K \cdot Pe \left\{ C_s(1 - \Gamma) - \frac{1}{La}\Gamma \right\}, \quad (6)$$

where

$$Ha = \frac{R_0 k_a C_\infty}{U_\infty}, \quad La = \frac{C_\infty}{\beta}, \quad K = \frac{\Gamma_{\max}}{2R_0 C_\infty}, \quad (7)$$

are dimensionless numbers. In this equation, diffusion flux from the bulk near the surface balances the adsorption/desorption originating from Langmuir kinetics. The adsorption of the surfactant on the bubble surface is given by the Marangoni formulation. The Marangoni stress balances the tangential stress in the liquid phase as follows:

$$\tau_{\xi\eta} = \mathbf{t} \cdot \nabla_s \sigma = -\mathbf{t} \cdot Ma \frac{\nabla_s \Gamma}{1 - \Gamma}, \quad (8)$$

where the Marangoni number (Ma) is defined by

$$Ma = \frac{R_G T \Gamma_{\max}}{\mu U_\infty}, \tag{9}$$

where R_G denotes the gas constant and T denotes the temperature. Ma expresses the ratio between the Marangoni stress and viscous stress. From these expressions, the surface tension coefficient is

$$\sigma = \sigma_0 + Ma \ln(1 - \Gamma). \tag{10}$$

In the present study, we solve the problem not only for the fixed spherical shape but also for the deformable bubble. In the case of the deformable bubble, an additional condition is required to decide the bubble shape. That is the normal stress condition, which is expressed as follows:

$$-p + \frac{2}{Re} e_{\eta\eta} = -p_g + \sigma(\kappa_{(\xi)} + \kappa_{(\phi)}), \tag{11}$$

where $\frac{2}{Re} e_{\eta\eta}$ is the dimensionless normal viscous stress, p_g is the dimensionless pressure inside a bubble, while $\kappa_{(\xi)}$ and $\kappa_{(\phi)}$ are the principal curvatures of the bubble surface.

To satisfy the above relations, an iterative procedure [8] similar to the one from Ryskin and Leal [9], which was originally developed for the steady state problem, is here extended to the unsteady problem and is used to obtain the instantaneous shape of a deformed bubble.

3 Results and discussions

3.1 Steady behavior

To verify our simulation results, we compare the computed drag coefficient and surface concentration with those of Cuenot et al. [6]. Although the Peclet number is 10^5 in their simulation, we used a lower Peclet number of 200. The effect of this on the results is discussed below.

Table 1. Simulation condition (spherical bubble).

Re	Pe	Ma	Ha	La	K
100	200	61	0.001	0.112	1

In this simulation, the bubble rising velocity is assumed to be constant, and the surfactant effect suddenly emerges at a certain instant when the bubble is at its steady state in the clean liquid system.

Figure 2(a) shows the temporal evolution of the surface concentration as a function of the angle from the front stagnation point of the bubble. In our numerical simulation, the effect of surface diffusion is larger than that of Cuenot et al., because

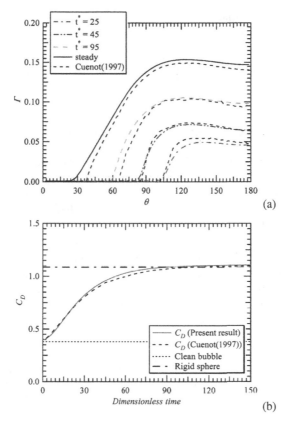

Fig. 2. Comparison with the simulation by Cuenot et al. [6] for $Pe = 200$. (a) Surface concentration as a function of angle from the front stagnant point. (b) Drag coefficient as a function of dimensionless time. The drag coefficient changes from that of a clean bubble to that of a rigid sphere.

our Peclet number is much smaller, and the discrepancy in surface concentration is recognized in the region where the surface concentration gradient is sharp. However, in most of the surface area, the distribution agrees well with Cuenot et al. including the unsteady behavior.

Figure 2(b) shows the temporal evolution of the drag coefficient as a function of dimensionless time. The drag coefficient increases to that of a rigid particle. As for the drag coefficient, good agreement with Cuenot et al. is also obtained, and it is confirmed that the force acting on a bubble is evaluated accurately even at the relatively low Peclet number of 200.

We conclude that the Peclet number does not affect the surface concentration and the rising velocity, for values of Pe larger than 200. Therefore, in the present simulations, we set the Peclet number 200, since the simulation becomes easier due to a larger grid size to resolve the boundary layer of surfactant concentration.

Table 2. Physical properties of surfactant.

Surfactant	k_a [m^3/mol.s]	β [mol/m^3]	$k_a\beta$ [1/s]	$\Gamma_{max} \times 10^{-6}$ [mol/m^2]
1-Pentanol (Fainerman and Lylik [12])	5.08	21.7	110.24	5.9
TritonX-100 (Borwankar et al. [13])	50	6.6×10^{-4}	0.033	2.9

3.2 Dependence on the adsorption/desorption properties

Figure 3(a) shows the drag coefficient for three different kinds of surfactant (1-Pentanol, 3-Pentanol, TritonX-100) in two different conditions of bulk concentration. In the case of a 3-Pentanol solution of 3.70×10^{-2} mol/m^3, there is little deviation from the drag coefficient of the clean bubble [10]. For a TritonX-100 solution, however, the bubble already shows the drag of a rigid particle [11] even at 7.10×10^{-4} mol/m^3. This is due to the difference in the adsorption/desorption kinetics between Pentanol and TritonX-100. The adsorption rate of TritonX-100 is 10 times larger than that of 1-Pentanol, and the desorption rate is 1/3000 that of 1-Pentanol. TritonX-100 being less desorbable, a surfactant molecule adsorbed in the front part of the bubble surface and transported to the rear part will remain there, producing a large surface concentration gradient. Therefore, in a TritonX-100 solution, the rising velocity of a bubble can be more easily reduced to that of a solid particle even if the bulk concentration is much lower than for Pentanol.

Figure 3(b) shows the numerical results of the surface velocity. The simulation conditions correspond to those induced by the arrows in Figure 6(a). It is found that the surface velocity of a bubble in a TritonX-100 solution is similar to that of a non-slip rigid sphere and differs from that of a bubble in a 1-Pentanol solution. For 1-Pentanol, it is shown that the surface velocity is retarded depending on the bulk concentration, that is, the surface velocity for case 2 is more retarded than that for case 3. This result explains why case 2 shows the drag coefficient closer to that of a solid particle than does case 3. It is noted that the surface velocity and the concentration distributions are very difficult to obtain through experiments but numerical simulation gives detailed information on them.

3.3 Unsteady motion of a bubble

Figure 4 shows the effect of the adsorption/desorption kinetics on the unsteady motion of a rising bubble. In these simulations, the initial bubble velocity is set to zero. Therefore, the bubble starts accelerating towards terminal velocity. There are three important processes related to the unsteady motion of a bubble:

(1) acceleration toward the terminal velocity of a clean bubble;
(2) deceleration by the effect of the slow adsorption process of surfactant;
(3) nearly steady state with a constant rising velocity.

In (1), the adsorbed surfactant is transported toward the rear stagnation point by the advection on the surface, and a large gradient of surface concentration is formed. In

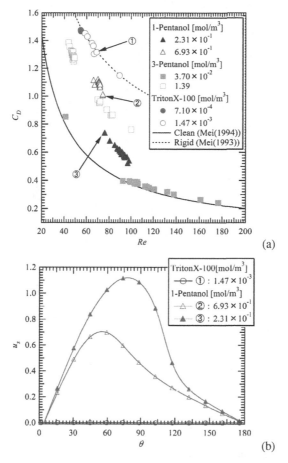

Fig. 3. Dependence on concentration and kind of surfactant. (a) Drag coefficient for various surfactants (1-Pentanol, 3-Petanol, TritonX-100). (b) Surface velocity distribution as a function of the angle from the front stagnant point.

(2), the effect of adsorption gradually begins to emerge on the rising velocity. The position of the large gradient of surface concentration moves forward and gives a larger stress on the bubble surface. This increase of stress retards the bubble surface velocity, and the bubble begins to decelerate. In (3), the drag on the bubble balances the buoyancy force and the rising velocity becomes nearly constant, although the amount of adsorption continues to increase for a while.

Figure 4(a) shows the dependence of the rising velocity on the adsorption constant and Figure 4(b) shows that on the desorption constant. The results show that if the adsorption constant is small or the desorption constant is large, then the bubble behaves like a clean bubble. Increasing the adsorption constant or decreasing the desorption coefficient reduces the terminal velocity to the value for a solid sphere. Beyond the condition which affects the terminal velocity, a qualitative difference of

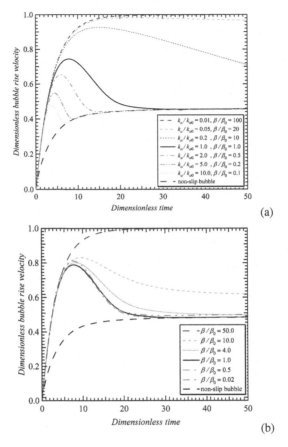

Fig. 4. Temporal evolution of the instantaneous rising velocity for (a) different adsorption constants, (b) different desorption constants.

the phenomenon between the increase of adsorption constant and the decrease of desorption constant appears. As is shown in Figure 4(a), a further increase of the adsorption constant gives a reduction of the relaxation time and the surfactant effect appears in a shorter time. On the other hand, as shown in Figure 4(b), a further decrease of the desorption constant does not reduce the relaxation time and there exists an asymptotic behavior for this case.

3.4 Deformation of a contaminated bubble

To investigate the effect of the boundary condition on the bubble surface, we conducted a numerical simulation with both a free-slip and a non-slip bubble. The non-slip bubble represents the situation in which sufficient surfactant is dissolved in the liquid. Figure 5 shows the simulation result of a deformed bubble at $Re = 10$, $We = 10$, where the Weber number We is defined as

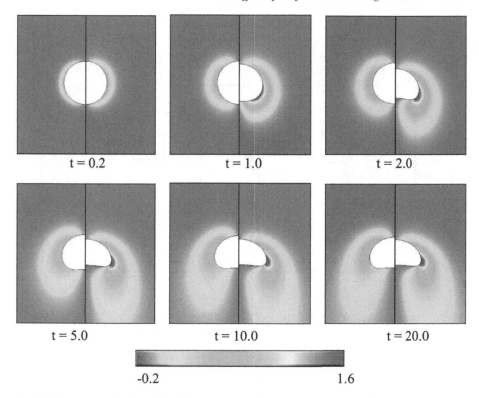

Fig. 5. Temporal evolution of bubble shape for $Re = 10$, $We = 10$. Color represents the amplitude of vorticity. (Left: non-slip bubble, Right: free-slip bubble).

$$We = \frac{2R\rho U_\infty^2}{\sigma_0}. \tag{12}$$

It is shown that a free-slip bubble deforms much more than a non-slip bubble. The tendency is observed not only in the steady state but also in the unsteady state. It must be noted that these behaviors of two bubbles correspond to the cases of different liquids, since the comparison was made for the same dimensionless number Re and We, although the non-slip bubble have a larger drag coefficient.

Next, the deformation of a bubble in a surfactant solution is discussed. The simulation conditions are shown in Table 3. This corresponds to the lower Reynolds number and the higher Weber number condition, under which the steady shape is insensitive to the Weber number in the case of a super-purified system. Here, Ma, which governs the tangential stress on a bubble surface, is changed to investigate the Marangoni effect on the bubble deformation.

The bubble shapes in the steady state are shown in Figure 6. The overall shapes of both bubbles look very similar except for the high curvature region of the sharp edge around the bubble. As is shown in Figure 6, Case 2 has a sharper shape of the bubble than Case 1 in the largest curvature part. Subject to Equation (10), different values

Table 3. Simulation conditions (deformed bubble).

Re	We	Ha	La	K	Ma
20	15	5.0×10^{-4}	0.112	1.0	Case 1: 30
					Case 2: 60

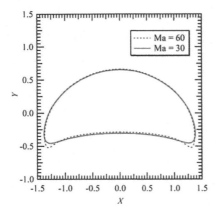 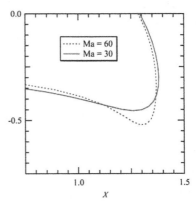

Fig. 6. Numerical result of bubble shape for two Marangoni numbers: 30 and 60. Higher Marangoni number gives the larger curvature at the edge of the bubble.

of the Ma number give different values of the surface tension coefficient as well as the Marangoni stress on a bubble. When the value of Ma is high, the reduction of the surface tension coefficient increases, and the curvature near the outer edge of the bubble becomes larger.

4 Conclusion

The effect of different kinds of surfactant on a single bubble motion in a quiescent liquid was investigated numerically taking the Marangoni effect into account. As a result, the following conclusions were obtained:

Spherical Bubbles:

(1) The terminal velocity can be easily reduced to that of a rigid sphere for surfactants like TritonX-100 which have a very low desorption constant. The difference of adsorption constant gives the difference of relaxation time to the terminal velocity. Whereas, in Pentanol solutions, the terminal velocity of the bubble is more gradually reduced than that in TritonX-100 solutions, depending on the amount of surfactant.

(2) Not only for the steady behavior of a bubble but also for the unsteady behavior, the ratio between the adsorption and desorption constants has a large effect. An increase of the adsorption rate shortens the transient overshoot process of the

rising velocity. On the contrary, a decrease of the desorption rate reduces the terminal velocity itself.

Deformed Bubbles:

(3) It is confirmed that a non-slip bubble becomes more spherical than a free-slip bubble at the same Reynolds and Weber number.

(4) In the case of a deformable bubble in surfactant solutions at the lower Re (≤ 20) and the higher We (≥ 10), the Marangoni number does not significantly affect the overall bubble shape.

References

1. Duineveld, P.C., 1995, *J. Fluid Mech.* **292**, 325–332.
2. Frumkin, A. and Levich, V., 1947, *Zhur. Fiz. Khim.* **21**, 1183 [in Russian].
3. Sussman, M., Smereka, P. and Osher, S., 1994, *J. Comput. Phys.* **114**, 146–159.
4. Unverdi, S.O. and Tryggvason G., 1992, *J. Comput. Phys.* **100**, 25–37.
5. Yabe, T., Xiao, F. and Utsumi, T., 2001, *J. Comput. Phys.* **169**, 556–593.
6. Cuenot, B., Magnaudet, J. and Spennato, B., 1997, *J. Fluid Mech.* **339**, 25–53.
7. Liao, Y. and McLaughlin, J.B., 2000, *J. Colloid Interface Sci.* **224**, 297–310.
8. Takagi, S., Prosperetti, A. and Matsumoto, Y., 1994, *Phys. Fluids* **6**, 3186–3188.
9. Ryskin, G. and Leal, L.G., 1984, Part 1, Part 2, *J. Fluid Mech.* **148**, 1–35.
10. Mei, R., Klausner, J.F. and Lawrence, C.J., 1994, *Phys. Fluids* **6**, 418–420.
11. Mei, R., 1993, *Int. J. Multiphase Flow* **19**, 509–525.
12. Fainerman, V.B. and Lylyk, S.V., 1982, *Kolloidn. Zh.* **44**, 538–544.
13. Borwankar, R.P. and Wasan, D.T., 1983, *Chem. Eng. Sci.* **38**, 1637–1649.

Numerical Simulation of Shock Propagation in Bubbly Liquids by the Front Tracking Method

Can F. Delale[1,2], Selman Nas[1] and Gretar Tryggvason[3]

[1] Faculty of Aeronautics and Astronautics, Istanbul Technical University, 34469 Maslak, Istanbul, Turkey; e-mail: delale@itu.edu.tr, nas@itu.edu.tr
[2] TÜBİTAK Feza Gürsey Institute, P.O. Box 6, 81220 Çengelköy, Istanbul, Turkey; e-mail: delale@gursey.gov.tr
[3] Department of Mechanical Engineering, Worcester Polytechnic Institute, 100 Institute Road, Worcester, MA 01609-2280, USA; e-mail: gretar@wpi.edu

1 Introduction

The propagation of shock waves in bubbly liquids has attracted great interest because of its practical importance. It has many interesting applications in petroleum and chemical engineering, biological and medical sciences, geophysics, etc. It is also of fundamental importance. Despite the fact that shock waves in a dilute bubbly liquid have been extensively investigated in the literature [1–4] using model equations, direct numerical simulations using the full Navier–Stokes equations, where the effects of viscosity and vorticity are fully accounted for, have not been treated before due to the complexity of the phenomenon. The front tracking method [5], has been rather successful in solving the full Navier–Stokes equations in the presence of a deforming phase boundary in many multi-phase flow applications. Such applications [6,7] include the collapse of a cavitation bubble near a solid wall and the formation of a toroidal bubble by a high speed micro-jet near a rigid boundary.

Here, our goal is to extend the front tracking method to be able to follow the collapse of a cluster of bubbles in a quiescent liquid inside a 2D or 3D rectangular domain excited by a pressure jump at the top. The study of a relatively simple model (say, by considering a polytropic law for the gas as a first step) by numerical simulations using the front tracking method, where bubble deformations and interactions are fully accounted for, is a useful first step in understanding the flow characteristics behind the shock. Such simulations will yield information about the magnitude of velocity and pressure fluctuations, and allow us to quantify the effect of bubble/bubble interactions and bubble deformation. Numerical simulations also make it possible to go beyond the classical Rayleigh–Plesset analysis for a dilute liquid, where the upper limit of the void fraction is only a few percent.

S. Balachandar and A. Prosperetti (eds), Proceedings of the IUTAM Symposium on Computational Multiphase Flow, 323–330.

2 Shock propagation in bubbly liquids by using the front tracking method

We consider a bubbly liquid filling a 2D or a 3D rectangular domain. The bubbles are initially assumed to be either circular (in 2D) or spherical (in 3D) in shape, uniformly distributed and in equilibrium with the surrounding liquid. A pressure jump (incident shock wave) is introduced at the top wall of the domain (with liquid inflow) and a wall boundary condition is imposed at the bottom (with no outflow). The domain is taken to be periodic in the transverse direction(s). For the numerical simulation by the front tracking method, the fluid motion is governed by the normalized unsteady Navier–Stokes equations, valid for the whole flow field. Neglecting gravity and surface tension, these equations can be written as

$$\frac{\partial(\rho\mathbf{u})}{\partial t} + \nabla \cdot (\rho\mathbf{u}\mathbf{u}) = -\nabla p + \frac{1}{(Re)}\nabla \cdot (2\mu\mathcal{D}), \tag{1}$$

where the density ρ varies in the interval $\rho_b \leq \rho \leq \rho_\ell$ and the viscosity μ varies in the interval $\mu_b \leq \mu \leq \mu_\ell$, with subscripts b and ℓ denoting the bubble and the liquid, respectively. Here, $\mathcal{D}$ is the deformation tensor, $\mathbf{u}$ is the velocity field and p is the pressure field. The Reynolds Re number is given by

$$Re = \frac{\rho'_m \sqrt{p'_m/\rho'_m} L'}{\mu'_m}, \tag{2}$$

where L' is a characteristic length of the order of the initial mean radius of the bubbles, ρ'_m and μ'_m are conveniently defined normalization values of the density and of the viscosity, both lying between those values of the liquid and of the bubble, and p'_m is a normalization pressure chosen for a characteristic speed $\sqrt{p'_m/\rho'_m}$ or for a characteristic time $L'\sqrt{\rho'_m/p'_m}$ (all primed variables are dimensional). We neglect the compressibility of the carrier liquid, taking it to be incompressible so that

$$\nabla \cdot \mathbf{u} = 0 \tag{3}$$

in the carrier liquid phase, and assuming its viscosity and density to remain constant at all times. The bubbles, on the other hand, are compressible with the pressure inside either set to a constant or varied isothermally. This imposes a moving boundary condition on the pressure field to be satisfied at the bubble/liquid interfaces and forces the imposed pressure at the top to drop to the level specified inside the bubbles. Thus, the presence of the bubbles prevents the effect of the increased pressure at the top to reach further into the bubbly mixture. The fact that the bubbles do not completely block the channel allows this effect to be felt slightly deeper, but not significantly. Equation (1) together with the incompressibility condition (3) in the liquid phase and a specified pressure in the bubbles, set equal to either a global constant or its local value resulting from the polytropic law, are solved iteratively by a conventional finite difference method on a staggered grid [5]. For the simulations with the polytropic gas law, we evaluate the area (2D) or volume (3D) of each bubble at each time step. The

Reynolds number is assumed to be of $O(1)$ in magnitude characterizing a flow field with a finite, but low Reynolds number.

Numerical simulations were carried out for both two- and three-dimensional domains. Since the pressure is specified inside the bubbles, the fluid properties there play a minor role in the evolution and, to make the computations as easy as possible, we used a density and viscosity ratio of $\rho_\ell/\rho_b = \mu_\ell/\mu_b = 10$. Tests with higher ratios using the two-dimensional domain confirmed that the results are essentially independent of these ratios. The computed results were compared with those of one-dimensional homogeneous bubbly mixture theory. In particular, a mean shock speed was found by computing the distances advanced by a constant pressure rise in the shock profile over a time interval Δt at several times during the evolution and by dividing the mean distance by Δt. The shock speeds thus obtained were compared with those calculated by the one-dimensional homogeneous bubbly liquid theory where the shock speed U_s is given by

$$U_s^2 = \frac{(1 - \beta_1)(p_1 - p_0)}{(1 - \beta_0)(\beta_0 - \beta_1)\rho_\ell},$$ (4)

using the Rankine–Hugoniot relations [1, 2] at a discontinuity connecting two regions of equilibrium states, designated by 0 and 1. In Equation (4), β_0 and β_1 are the void fractions and p_0 and p_1 are the mixture pressures of the equilibrium regions 0 and 1, respectively. To justify the comparison of the mean shock speed with Equation (4), steady-state conditions for shock propagation should be reached which requires long distances in the direction of propagation [9, 10]. Although these conditions are probably not reached over the relatively short distances of propagation in the computational domains of the present simulations, the calculated r.m.s. values of fluctuations of the instantaneous shock speeds about the mean shock speed are shown to be only within a few percent, making the comparison of the mean shock speed with Equation (4) meaningful.

3 Results and discussion

For the numerical simulations, two- and three-dimensional rectangular computational domains were considered. A grid study was conducted with a resolution ranging from 50×98 points to 122×242 points in 2D and from $18 \times 18 \times 66$ points to $34 \times 34 \times 130$ points in 3D in order to control numerical accuracy. Better resolution of the bubble/liquid interfaces were observed as the grid resolution was increased. However, the results for the shock structures and shock speeds remained almost unchanged. For the 2D case, where the gas pressure inside the bubbles is held constant, a rectangular grid containing 24 bubbles were considered. All bubbles were assumed to be initially circular in shape with the same radius $R = 0.25$. At time zero, the pressure at the top of the domain was raised by $\Delta p = 0.4$ and kept constant during the simulations.

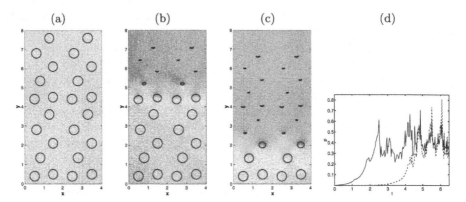

Fig. 1. Results obtained by the front tracking method showing a shock wave initially at $y = 8.0$ with strength $\Delta p = 0.4$ propagating into a quiescent bubbly liquid in a rectangular domain containing 24 bubbles, each with initial radius $R = 0.25$ where the gas pressure inside the bubble is held constant at its initial value (the density and viscosity of the liquid and of the gas are, respectively, $\rho_\ell = 2.5$; $\mu_\ell = 0.07$ and $\rho_b = 0.25$; $\mu_b = 0.007$). (a) Snapshot of the bubbly liquid at the initial time $t = 0.0$. (b) Snapshot of the bubbly liquid at time $t = 3.0$ showing the collapse of bubbles as the shock propagates (shaded areas show higher pressure zones). (c) Snapshot of the bubbly liquid at time $t = 5.5$ showing the collapse of bubbles as the shock propagates (shaded areas show higher pressure zones). (d) The pressure distribution for the bubbly shock wave at locations $y = 3.333$ (dashed line) and $y = 6.0$ (solid line) along the boundary $x = 0$.

Snapshots of the results obtained by the front tracking method showing the deformation of the collapsing bubbles and the evolution of the pressure distribution at two locations along the direction of propagation (the y-axis) in a rectangular grid with a resolution of 122×242 points are shown in Figures 1(a)–(d). The results in Figures 1(a)–(c) show that the bubbles collapse with non-circular shape (almost elliptical in the beginning) followed by a re-entrant jet before they totally disappear (the interfaces are here resolved up to a point where the top interface almost touches the bottom one). In this case, the fluctuations in the hydrodynamic variables observed in the transverse direction (x-direction) are reasonably small to justify the use of the one-dimensional homogeneous bubbly flow theory. Figure 1(d) shows the evolution of the pressure distribution of oscillating shock waves at locations $y = 3.333$ and $y = 6.0$ along the direction of propagation (y-axis) at the boundary $x = 0$, where periodic boundary conditions are imposed. The amplitude of the pressure fluctuations in this case can be as high as 0.36 at the beginning of oscillations, but they eventually decay in time. The mean shock speed obtained by the simulations using the above mentioned averaging yields a value equal to 1.103 (with r.m.s. fluctuations being less than 2%), which seems to be in good agreement with the value $U_s = 1.128$ evaluated by Equation (4) for homogeneous bubbly liquids with $p_1 - p_0 = \Delta p = 0.4$, $\rho_\ell = 2.5$, $\beta_0 = 0.1473$ and $\beta_1 = 0.0$.

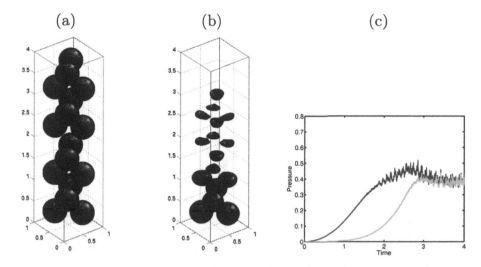

Fig. 2. Results obtained by the front tracking method showing a shock wave with strength $\Delta p = 0.4$ propagating into a quiescent bubbly liquid in a rectangular box containing 16 bubbles, each with initial radius $R = 0.25$ where the gas pressure inside the bubble is held constant at its initial value. (a) Snapshot of the bubbly liquid at the initial time $t = 0.0$. (b) Snapshot of the bubbly liquid at time $t = 3.5$ showing the non-spherical collapse of bubbles as the shock propagates. (c) The pressure distribution for the bubbly shock wave at locations $z = 2.0$ (light line) and $z = 3.0$ (dark line) along the line $x = 0$ and $y = 0$.

The 3D numerical simulations for the case where the gas pressure was held constant were carried out in a rectangular box, with a grid resolution of $34 \times 34 \times 130$ points, containing 16 bubbles. The bubbles were again taken initially in equilibrium with the quiescent liquid and spherical in shape, all with the same radius $R = 0.25$. A shock wave with strength $\Delta p = 0.4$ was incident at the top wall of the rectangular box. The results obtained by the front tracking method are shown in Figures 2(a)–(c). The non-spherical collapse of bubbles as the shock propagates can clearly be seen in Figure 2(b). An almost uniform flow field can be observed over the cross-section in the lateral direction as the shock propagates, justifying the use of Equation (4) for the one-dimensional homogeneous bubbly liquid model in this case as well. The pressure distributions at locations $z = 2.0$ and $z = 3.0$ along the line $x = 0$ and $y = 0$ in the propagation direction of the shock are plotted in Figure 2(c). The pressure at these locations oscillates with a maximum amplitude of 0.15, and the pressure fluctuations decay as the shock propagates further. The mean shock speed from the simulations using the above discussed averaging method yields a value of 0.953 (with r.m.s. fluctuations being less than 5%), which seems to agree well with the value $U_s = 0.91$ evaluated by Equation (4) for homogeneous bubbly liquids with $p_1 - p_0 = \Delta p = 0.4$, $\rho_\ell = 2.5$, $\beta_0 = 0.2618$ and $\beta_1 = 0.0$.

Finally, a 2D numerical simulation was carried out for the case where a 2D isothermal law (with the polytropic index being equal to unity) for the gas pressure was

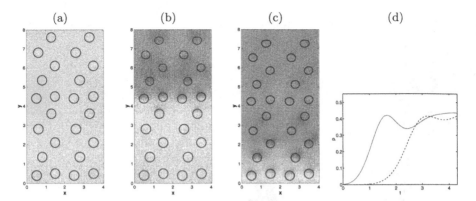

Fig. 3. Results obtained by the front tracking method showing a shock wave with strength $\Delta p = 0.4$ propagating into a quiescent bubbly liquid in a rectangular domain containing 24 bubbles, each with initial radius $R = 0.25$ where the gas pressure is varied isothermally (the density and viscosity of the liquid and of the gas are, respectively, $\rho_\ell = 2.5$; $\mu_\ell = 0.07$ and $\rho_b = 0.25$; $\mu_b = 0.007$). (a) Snapshot of the bubbly liquid at the initial time $t = 0.0$. (b) Snapshot of the bubbly liquid at time $t = 1.6$ showing collapsing bubbles behind the shock as the shock propagates (shaded areas show higher pressure zones). (c) Snapshot of the bubbly liquid at time $t = 3.6$ showing collapsing and rebounding bubbles behind the shock as the shock propagates (shaded areas show higher pressure zones). (d) The pressure distribution for the bubbly shock wave at locations $y = 3.333$ (dashed line) and $y = 6.0$ (solid line) along the boundary $x = 0$.

assumed. The same rectangular grid used for the constant pressure case containing 24 bubbles, all having the same radius $R = 0.25$, was considered for the flow simulations. The incident shock strength was also set equal to $\Delta p = 0.4$ to allow a full comparison with the case of constant gas pressure. Snapshots of the initial bubble distribution and of its evolution using a grid resolution of 122×242 points at nondimensional times $t = 1.6$ and $t = 3.6$ are shown in Figures 3(a)–(c).

When the shock front hits the boundary of the bubbles, the bubbles start to collapse isothermally and the gas pressure inside the bubbles increases resulting in a decrease in the pressure threshold for further collapse. Eventually this pressure threshold diminishes and the bubbles rebound (after a few oscillations, they would eventually reach equilibrium if the computational domain is enlarged). Therefore, in the first instance of shock propagation, only collapsing bubbles are seen (Figure 3(b)). As the shock propagates downward towards the bottom wall, the bubbles close to the top (where the shock was incident) start to rebound (Figure 3(c)). The bubbles collapse and rebound almost elliptically for this case. Again, the lateral pressure and velocity fluctuations can be neglected resulting in an almost onedimensional propagation of the shock front. The evolution of the pressure distributions at locations $y = 3.333$ and $y = 6.0$ along the boundary, where periodic boundary conditions are imposed, are shown in Figure 3(d). The profiles look much more smooth for the reasons explained above. The mean shock speed obtained from the simulations using the above mentioned averaging yields a value of 1.992 (with r.m.s. fluctuations being less than 7%). On the other hand, the shock speed U_s of one-dimensional homogeneous bubbly liquid is evaluated by Equation (4) with $\rho_\ell = 2.5$, $p_1 - p_0 = \Delta p = 0.4$, $\beta_0 = 0.1473$ and $\beta_1 = 0.1052$. The shock speed thus obtained using Equation (4) yields the value $U_s = 1.997$, which agrees well with the simulated value. Due to the increase in the gas pressure as the bubbles collapse, the bubbles collapse at a slower rate and the shock propagates faster in this case as compared to the case where the gas pressure is held constant under the same conditions.

4 Concluding remarks

The results of this investigation have shown that shock propagation in a bubbly liquid with void fractions as high as 15% to 25% can still be well described by the one-dimensional homogeneous bubbly liquid model when the gas pressure inside the bubble is kept constant or varied isothermally, irrespective of the dimensionality of the computation domain. While bubble deformation and bubble/bubble interactions are properly accounted for by the present simulations, our results do not address the effects of liquid compressibility, thermal damping and bubble fragmentation. These effects demand the solution of the compressible Navier–Stokes equations together with the energy equation both inside and outside the bubble. Although this does not seem to be possible at present times, a model equation that replaces the polytropic law for the pressure inside the bubble by an equation similar to that proposed by Prosperetti [8] to take into account the effect of thermal damping can be used as a first step. Only then, can this simple model be extended to simulate bubbly flows in more complex geometries than those examined here, such as bubbly flows through constrictions or over curved boundaries.

Acknowledgements

This work was supported in part by the US Department of Energy under grant no. DE-FG02-03ER46083 and in part by the Research Foundation of Istanbul Technical University (BAP).

References

1. Noordzij, L. and van Wijngaarden, L., 1974, *J. Fluid Mech.* **66**, 115–143.
2. Beylich, A.E. and Gülhan, A., 1990, *Phys. Fluids A* **2**, 1412–1428.
3. Watanabe, M. and Prosperetti, A., 1994, *J. Fluid Mech.* **274**, 349–381.
4. Kameda, M., Shimaura, N., Higashino, F. and Matsumoto, Y., 1998, *Phys. Fluids* **10**, 2661–2668.
5. Tryggvason, G., Bunner, B., Esmaeeli, A., Juric, D., Al-Rawahi, N., Tauber, N.W., Hans, J., Nas, S. and Jan, Y.J., 2001, *J. Comput. Phys.* **169**, 662–682.

6. Yu, P.Y., Ceccio, S.L. and Tryggvason, G., 1995, *Phys. Fluids* **7**, 2608–2616.
7. Popinet, S. and Zaleski, S., 2002, *J. Fluid Mech.* **464**, 137–163.
8. Prosperetti, A., 1991, *J. Fluid Mech.* **222**, 587–616.

Large-Eddy Simulation of Steep Water Waves

Djamel Lakehal and Petar Liovic

Institute of Energy Technology, ETH Zurich, Switzerland;
e-mail: lakehal@iet.mavt.ethz.ch

Abstract. Large-Eddy Simulation is used for the investigation of the breaking of steep water waves on a beach of constant bed slope. The method is built within a multi-fluid flow solver, in which the free surface is tracked using a Volume-of-Fluid method featuring piecewise planar interface reconstructions on a twice-as-fine mesh. The Smagorinsky sub-grid scale model is used for explicit under-resolved turbulence closure, coupled with a new scheme for turbulence decay treatment on the air-side of massively deformable free surfaces. The simulations were conducted for shear Reynolds numbers $Re_G^* \approx Re_L^* \approx 400$, based on the mean water depth. The Large-Eddy Simulation formulation in the interface tracking, single-fluid formulation is introduced for this purpose. The approach is demonstrated as a powerful tool for exploring large-scale, interfacial turbulent flows. The discussion focuses on coherent structures formation, the free surface flow effects at breaking, and form drag evolution with the surface.

1 Introduction

The traditional treatment of wave breaking in coastal engineering research using single-fluid hydro-codes has been based on free surface models imposing zero pressure boundary conditions, e.g. [1, 2]. The main flow features of wave breaking that have been explored in such research include surface rollers at the front of spilling breakers, the dynamics of water tongues while jetting forward to impact on the surface in front of the crest, flow patterns under the breaking waves, and to some extent the generation of turbulence by the wave deformations. The breaking of free-surface waves is contrasted with the breaking of interfacial waves (both owing to a convective instability) in terms of subsequent scalar mixing and air entrainment [3]. Free-surface simulations that involve sharp-interface capturing/tracking techniques are not needed for the study of scalar (density) mixing.

Water-flow simulations, on their own, provide no insight into the air flows above the free surface. The effects of winds on wave motions and breaking, and gas absorption by oceans, cannot be handled naturally without detailed coupled air-water flow solutions. Even in conventional studies of wave breaking, imposing a free surface pressure boundary condition may not be necessarily adequate. Observations of

S. Balachandar and A. Prosperetti (eds), Proceedings of the IUTAM Symposium on Computational Multiphase Flow, 331–340.

significant form drag resulting from wave motions in the stratified flow scenarios investigated by Sullivan and McWilliams [4] are applicable to wave breaking in free surface flows. In the case of steep wave propagation and breaking, form drag is not distributed uniformly across the free surface as may be implied by a zero-pressure boundary condition. In wave plunging, strong events such as the propagation of the water tongue ahead of the wave crest and its impact onto the free surface can be seriously affected by the air flow ahead of the wave. The counter-current air-water flow of Fulgosi et al. [5] shows a significant population of coherent structures on both sides of the sheared interface. The use of free surface models that ignore the details of turbulence in the air flow may lead to significant discrepancies in the the water flow solution, and, to a certain extent, in the associated interfacial scalar transfer mechanisms.

Studies of wave breaking incorporating gas-side flow solutions have been performed in the context of stratified flow studies in idealized geometries using forcing function-based wave generators. Fringer and Street [3] explored breaking dynamics in detail, and found the interface thickness to be a crucial factor in determining the size and strength of Kelvin–Helmholtz billows, and hence the prevalence of shear or convective instability. Such observations may have some analogue in the breaking of free-surface waves, but cannot be confirmed in the literature due to the lack of free-surface wave breaking studies with explicit air flow solutions.

In this paper, we investigate the dynamics of turbulence in the free surface flow solution by incorporating the flow physics pertaining to the air flow and entrainment induced by wave motion and breaking. We proceed by identifying the effects of the air flow on the free surface profile and water-flow solution. For the purpose, a novel LES approach for interfacial turbulent multi-fluid flows has been developed, based on the filtered single-fluid Navier–Stokes equations. The Smagorinsky-based eddy-viscosity model is used, coupled with a new scheme for correct damping of turbulence at deformable interfaces from the gas side. Different damping model functions are used, depending whether the flow is approaching interfaces or solid walls. The near-interface flow physics modeling is applied within a rigorously momentum conservative method, coupled with a high-order PLIC-VOF approach for large-density ratio flows.

2 Mathematical formulation

2.1 The filtered single-fluid equations

The single-field representation of two-phase flows ($k = L, G$) applies to interfacial flows with typical length scales larger than the grid size. Phase inter-penetration is not presumed, instead, interfacial jump conditions are directly incorporated by solving a single set of transport equations. The phases are identified locally using the phase indicator function C, which reduces in the continuous limit to the Heaviside function: $C = 1$ if $\mathbf{x} \in k = G$, and $C = 0$ if $\mathbf{x} \in k = L$. The local density is thus defined by $\rho = \sum_k C^k \rho^k$. Subharmonic mean is used for the treatment of viscosity.

The LES concept has been recently extended [6] for turbulent, interfacial multi-fluid flows, based on the filtered single-field transport equations. In this context the resolvable super-grid quantities $\overline{f}$ are obtained by convolution using a spatial filter G:

$$\overline{f}(x) = \int_D G(\mathbf{x} - \mathbf{x}') f(\mathbf{x}') d\mathbf{x}'. \tag{1}$$

Non-resolved sub-grid scale (SGS) components ($f' = f - \overline{f}$) are modeled. The filtered phase indicator function, $\overline{C}$, is subsequently interpreted as the resolved volume fraction, and is used to determine the filtered local density

$$\overline{\rho} = \sum_k \overline{C^k \rho^k}. \tag{2}$$

The derivation of the filtered single-field equations is based on Component-Weighted Volume Averaging procedure [7], inspired from Favre's flow averaging of compressible flows: $\widetilde{f}(\mathbf{x}) = \overline{\rho(\mathbf{x}) f(\mathbf{x})} / \overline{\rho(\mathbf{x})}$.

Performing the convolution product on the transport equations and applying the local density-based flow decomposition yields the filtered single-field transport equations for turbulent multi-fluid flows:

$$\frac{\partial \overline{\rho}}{\partial t} + \frac{\partial}{\partial x_j}(\widetilde{u}_j \overline{\rho}) = 0, \tag{3}$$

$$\frac{\partial \overline{C}}{\partial t} + \frac{\partial}{\partial x_j}(\widetilde{v}_j \overline{C}) = 0, \tag{4}$$

$$\frac{\partial \overline{\rho}\widetilde{u}_i}{\partial t} + \frac{\partial}{\partial x_j}\left(\overline{\rho}\widetilde{u}_i\widetilde{u}_j\right) = -\frac{\partial \overline{p}}{\partial x_i} + \frac{\partial}{\partial x_j}\left[\widetilde{\sigma}_{ij} - \tau_{ij}\right] + \overline{\rho}g_i + \gamma\,\overline{\kappa}\,\overline{\widehat{n}_i}\,\delta + \varepsilon, \tag{5}$$

where v_j is the interface velocity (which reduces to the fluid velocity in absence of phase change), τ_{ij} is the phasic SGS Reynolds stress tensor, defined by

$$\tau_{ij} \equiv \overline{\rho}\left(\widetilde{u_i u_j} - \widetilde{u}_i\widetilde{u}_j\right), \tag{6}$$

ε is the sum of the filtering-induced non-linearity and commutation errors, γ is the surface tension coefficient, κ is the interfacial curvature, and n_i is the unit interface normal. The surface tension is localized to the interface by the delta function δ. Derivation details can be found in [6]. Note only that ε contains the following non-resolved interfacial terms:

$$\varepsilon_d = \frac{\partial}{\partial x_j}\left[\overline{\sigma}_{ij} - \widetilde{\sigma}_{ij}\right], \tag{7}$$

$$\varepsilon_\gamma = \gamma\,\overline{\kappa\,\widehat{n}_i}\,\delta - \gamma\,\overline{\kappa}\,\overline{\widehat{n}_i}\,\delta. \tag{8}$$

These two quantities, referring to non-resolved interfacial deformations, are primarily grid dependent. Their interaction with turbulence is unclear at this stage; therefore they were neglected in this work.

2.2 Numerical schemes

The MFVOF-3D code has been developed for transient, high-density ratio multi-fluid flows. In this finite volume solver, interfaces are tracked using a 3D VOF method that is resolved on a *twice-as-fine* sub-mesh nested within the underlying solver mesh. Well-oriented interface planes in the 3D PLIC-VOF (Piecewise Linear Interface Calculation) scheme ensure free surface representations remain compact over time; more details on 3D PLIC-VOF can be found in [8]. Mesh refinement is particularly important for the tracking of high-curvature interfaces, ensuring high-order accuracy and reducing numerical surface tension effects. The code is based on a rigorous momentum-conservative formulation; accuracy and stability are ensured using VOF-augmented momentum advection, in which the convective flux densities are inferred from VOF-based material fluxes.

2.3 SGS turbulence modeling

The LES approach employed is based on explicit SGS modeling within the eddy-viscosity framework. The Smagorinsky SGS model linking the eddy-viscosity μ_t to the resolved strain rate $\overline{S}_{ij}$ takes the following form:

$$\mu_t = f_{\mu\mathsf{Int}}\,\overline{\rho}\,\left[C_s\,\overline{\Delta}\right]^2\,\sqrt{2\overline{S}_{ij}\overline{S}_{ij}} \tag{9}$$

with a value of the model coefficient $C_S = 0.1$ in the core flow. In this new multi-fluid flow context, it is understandable that the C_S value could be inferred only by reference to wall flows, or, in a certain measure, from known DNS [5] or LES [7] data. The extensive DNS study of stratified two-phase flow [5] has revealed the need for turbulence damping approaching deformable interfaces, very much in the same way as for wall flows. For low to moderate interface deformations, this DNS database suggests an exponential dependence of the model function $f_{\mu\mathsf{Int}}$ on y_{Int}^+ [9]:

$$f_{\mu\mathsf{Int}} = 1 - \exp\left[-1.3.10^{-04}y_{\mathsf{Int}}^+ - 3.6.10^{-04}y_{\mathsf{Int}}^{+\,2} - 1.08.10^{-05}y_{\mathsf{Int}}^{+\,3}\right]. \tag{10}$$

The concept of a non-dimensional "interface shear unit" y_{Int}^+, defined by analogy to "wall shear units" as

$$y_{\mathsf{Int}}^+ = U_{\mathsf{Int}}^\tau \phi^{\mathsf{RDF}}/\nu^G; \qquad U_{\mathsf{Int}}^\tau = \sqrt{\tau_{\mathsf{Int}}/\rho_G}, \tag{11}$$

requires the definition of its own interfacial gas-side shear velocity U_{Int}^τ and reconstructed distance function (RDF) ϕ^{RDF}. Given the dependence of y_{Int}^+ on local flow conditions, the width of interface support for defining the RDF may fluctuate significantly during the simulation.

2.4 Multi-physics treatment near deformable interfaces

In the RDF algorithm, the piecewise planar interface reconstructions of the 3D VOF scheme are used to generate interface markers, by extracting the coordinates of all intercepts of interface planes with the edges of its bounding mesh cell. These intercepts are stored, and the centroid of each interface plane is extracted from these intercepts and stored. For each point (i, j, k), the closest interface marker is identified, and is used to compute $\phi_{i,j,k}^{RDF}$. The use of the twice-as-fine mesh ensures that RDF computations based solely on interface-plane centroid data retain acceptable accuracy and smoothness.

The next step of the scheme is to perform a mesh sweep confined to a one-cell interface support, in which a new *list* of interface points is identified and the shear recorded. Throughout this interface support, cell-centered Points P are identified to the gas side of $\phi^{RDF} = 0$, and a normal estimate is determined at that point by $\hat{n}_P^{RDF} = \nabla\phi_P^{RDF}/|\nabla\phi_P^{RDF}|$. Tracing back along $\hat{n}_P^{RDF}$ generates the interface point coordinate estimate $(x, y, z)_{Int'}$, and an estimate of the interface velocity $\mathbf{u}_{Int'} = (u, v, w)_{Int'}$ is generated by extrapolating through the gas-sided velocity field to interface point I'; the same procedure is used to estimate the velocity at Point P. Vector resolution is used to extract the shear components of $\mathbf{u}_P$ and $\mathbf{u}_{Int'}$, and the interface shear $\tau_{Int'}$ is then computed and stored in the list.

The second mesh sweep, in which $f_{\mu Int}$ is computed, is performed over a wider cell support. Tracing back from gas-side Point P along the RDF normal $\hat{n}_P^{RDF}$ ends at a point I on the interface that doesn't necessarily coincide with any point I' in the list. An estimate of τ_{Int} at $(x, y, z)_{Int}$ is then estimated as an inverse distance-weighted interpolation estimate of $\tau_{Int'}$ values in the list. The gas-side shear velocity $U_{Int}^{\tau} = \sqrt{\tau_{Int}/\rho_G}$ is computed to finally compute the interface turbulence length scale y_{Int}^{+}. In practical applications, the grid resolution around the interface should be fine enough to resolve the interfacial viscous sublayer, in particular in the presence of interphase heat/mass exchange.

3 LES of Steep Water Waves

The fifth-order Stokes theory of Fenton [10] is the basis of the free surface initialization used here. Flow parameters were similar to those used by Christensen and Deigaard [2] in their "weak plunger" case study, albeit in our case the initial wave was of higher amplitude. Specifically, we set the mean channel depth to $d = 0.321$ m, the wave amplitude to $H = 0.12$ m, the wave period to $T = 1.4$ s, the Stokes drift velocity to $c_S = 0$ m/s, and $g = 9.81$ m/s^2. Two grid resolutions were employed $(140 \times 40 \times 20$ and $200 \times 80 \times 40)$, covering a domain size of 8 m × 0.6 m × 0.3 m. The fine grid resolution helped resolve the wall and interface viscous sublayers down to $y_{Wall}^{+} \approx y_{Int}^{+} \approx 0.1$, respectively. Only results obtained with the fine grid resolution are discussed next.

3.1 Wave breaking events

Although the problem setup corresponds (for the most part) to the "weak plunger" breaker type in [2], flow scenarios resembling the "spilling" and the "strong plunger" breaker types have also been observed. Figure 1 shows frames that correspond to a weaker plunging event, in which a tongue of water is thrown forward of the crest,

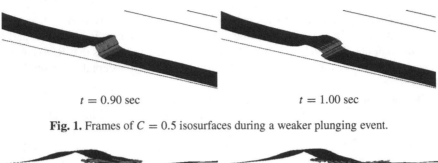

$t = 0.90$ sec $t = 1.00$ sec

Fig. 1. Frames of $C = 0.5$ isosurfaces during a weaker plunging event.

$t = 2.50$ sec $t = 2.60$ sec

$t = 2.70$ sec $t = 2.80$ sec

Fig. 2. Frames of $C = 0.5$ isosurfaces during a stronger plunging event.

before impacting with the free surface in front of the crest. In Figure 2, the tongue is thrown much further ahead of the crest. Rather than the plunger fully rebounding off the free surface, air entrainment is visible, which is consistent with penetration of the plunger. This behavior, combined with an upward rise of liquid in front of the plunger after impact, are suggestive of a stronger plunging event. A spilling event has also been captured in simulation, in which the wave rolls through the crest without any liquid being thrown ahead of the crest.

3.2 Coherent structures

Underlying coherent structures (CS) are responsible for the transport of scalars and for the production/dissipation of turbulence. The $-\lambda_2$ approach of Jeong and Hussain [11] has been used for coherent structures (CS) identification in [5] (by illustrating the structure of instantaneous vortex cores), and is used again here to provide a qualitative indication of the effect of the interfacial motion on the quasi-streamwise vortices. The analysis of various snapshots has revealed a multitude of forms, shapes and extensions of these structures resembling those in channel flow, very much dependent on the topology of the free surface; among the CS we could recognize included *sheets, hairpins* and *hockey-sticks*. In the present case CS seem to have a greater angle of inclination.

Figure 3 shows isosurfaces of $-\lambda_2 = 0.05$ applied across both phases at one instant. What is unique to this simulation is the way CS adapt to the shape of the interface, almost independently from one phase to the other. It seems that on the gas-side the CS are distributed in a relatively uniform manner, taking the form of quasi-streamwise vortices that tilt in the spanwise direction according to the orientation of the free surface. A lower CS concentration is observed over the concave part of the interface (air side), while a dense population forms on top of the convex shape,

Fig. 3. Coherent structures at $t = 0.5$ sec defined by using the CDF-based eduction of $-\lambda_2 = 0.05$: (left) air side and (right) liquid side.

where the interfacial shear is high. In fact, the genesis of CS occurs at the upslope or impact zone of the surface and project over the trough, where they lose their identity. Similar CS dynamics have already been observed by Calhoun and Street [12] in their LES of turbulent flow over a wavy surface. This is particularly remarkable in flow zones over small slope, long wave-lengths (e.g. centre of the left panel). However, in flow regions evolving over high slope, short wave-lengths, both the upsolpe and downslope of the wave are covered by CS; part of it being created by the main-flow over the upslope, the rest by the returning flow, in the opposite direction. This strong interaction between the two CS populations is responsible for their later dissipation.

The liquid-side CS analysis reveals that these are preferentially concentrated under the wave crests and near the shore, and their alignement seems to depend on the water depth, in contrast to the air-side CS. Indeed, it is important to note that the prominent structures are quasi-streamwise in shallow water flow regions (high Froude number), and more three-dimensional and isotropic in deep water zones (low Froude number). Note, too, that the particular snapshot shown may indicate that the CS are more numerous in the liquid side; other snapshots reveal the inverse scenario. The above observations lead to the conclusion that the turbulence structure is particularly sensitive to wave deformations: in particular, the production/dissipation mechanisms may in turn behave differently in the air and liquid sides. Such an analysis is beyond the scope of this paper. It seems that the evolution of the streamwise vortices differs from one phase to the other, although both are highly correlated with the wave motion.

Various authors (e.g. [12]) speculate about the role of the Taylor-Goertler inviscid instability mechanisms associated with wavy surfaces, and their exact relationship with the generation/destruction of streamwise vortices. Although the prominent structures are also quasi-streamwise in our flow, like in the flow over the wavy solid surface of [12], we also believe that the Goertler instability play an important role in the creation of vortices. That is, convex curvature regions are associated with stabilization, whereas the inception of vortices occurs in concave curvature regions.

A notable feature of the wave breaking process illustrated in Figure 2 is the formation of surface wrinkles in the plunging tongue almost normal to the spanwise direction; these are shown in more detail in figure 4(left). The wrinkles on and behind the plunging tongue are some of the larger 3D distortions of the interface in the en-

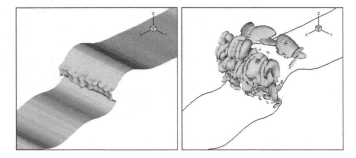

Fig. 4. (Left) wrinkles in the plunging tongue at $t = 2.32$ sec; (right) gas-side CS nestled in the wrinkles ahead of the plunging tongue.

tire flow solution. The concentration of CS in the vicinity of tongue is more detailed in Figure 4 (right) in which the CS were determined using the streamwise vorticity contours rather than $-\lambda_2 = 0.05$ isosurfaces. Figure 4 (right) shows most of the individual wrinkles on the air side to be filled with single quasi-vertical vortices, which were not observed elsewhere in the flow. The absence of significant wrinkling over most of the free surface upstream of the breaking zones points to a strong correlation between turbulence structures, including CS, and surface wrinkling. The precise mechanisms are not yet well understood.

3.3 Form drag evolution

In the case of an interface deformed by the shear imposed by the turbulent air-flow, the form drag can be defined on that side of the interface as a projection of the surface pressure onto the interface advection:

$$D_p = p\, \vec{n}\, A \cdot \vec{u}\, /|\vec{u}|, \tag{12}$$

where p is the surface pressure, $\vec{n}$ the normal vector to the interface, A is the local surface area, and the scalar product with $\vec{u}$ represents the velocity component parallel to the interface. In the present case, D_p is computed at the nearest (i, j, k) locations to the air side of the interface. In early studies dealing with fixed water wave trains [4], the form stress was integrated over the "prescribed" wave length λ. In the present study, however, waves of different slopes and wave-ages form and deform, which makes it more practical to derive the form drag using the above equation.

Figure 5 displays the instantaneous distributions of form drag over the interface (upper panel), together with the interface elevation contours on the same interface (middle panel), at two instants during the simulation. The third panel of the figure highlights the corresponding positions on the free surface where D_p is computed; the contours correspond to the gas-sided turbulence damping function $f_{\mu\mathrm{Int}}$. The figure highlights a clear variation of the form stress with interface elevation, or more precisely with the change in the surface elevation. There is a persistent trend for the form stress to be higher with wave extrema.

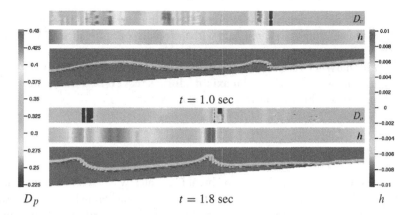

Fig. 5. Profiles of the form drag on the wave surface (top row), the corresponding elevation of the surface (middle and bottom rows), the latter is colored by $f_{\mu Int}$.

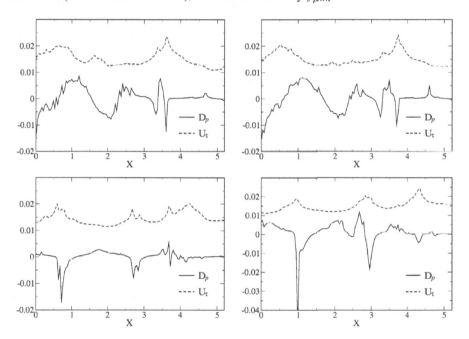

Fig. 6. Spanwise-averaged form drag and interfacial shear velocity distributions along the wave propagation direction.

The spanwise-averaged form drag and interfacial shear velocity distributions, defined by $1/z \int_{-a}^{a} D_p dz$ and $1/z \int_{-a}^{a} U_{Int}^{\tau} dz$ (where a stands for the domain width), are plotted in Figure 6 along the wave propagation direction. The various snapshots analyzed so far reveal a strong correlation between the two quantities. The form stress is shown to be significantly sensitive to the slightest variations of the interfacial shear. The profiles are almost in phase, featuring various modes with different amplitudes. This finding is in agreement with the results of Sullivan and McWilliams [4], and helps answer the question as to which velocity scale is appropriate for steep wave growth analysis. Near-zero or slightly negative form drag correspond to flow regions where the wave travels fast; maximum form stress corresponds to near standing wave scenarios, which occur in this simulation, too.

4 Conclusion

Large-eddy simulation investigation of breaking water waves has been conducted. The study has shown wave breaking events in flows initialized using Fenton's wave theory, ranging from spilling to stronger plunging events. Turbulence structure is seen to be particularly sensitive to wave deformations. The evolution of the streamwise vortices is found to be different in each phase, although both are highly correlated with the wave motion. In particular, lower CS concentration is observed over the concave part of the interface, while a dense population forms on top of the convex shape. In the liquid-side, however, CS seem to preferentially concentrate under the wave crest. The significant wrinkling at the breaking zones suggests a strong correlation between turbulence vorticity-induced structures and surface wrinkling. A clear dependency of the form stress on interface elevation is observed, with a persistent trend for rapid changes with wave extrema and interfacial shear. Ongoing research includes the investigation of breaking-waves turbulence interactions.

References

1. Zhao, Q. and Tanimoto, K., 1998, Numerical simulation of breaking waves by Large Eddy Simulation and VOF method, in *Proceedings of the 26th International Conference on Coastal Engineering*, B. Edge (ed.), ASCE.
2. Christensen, E.D. and Deigaard, R., 2001, *Coastal Eng.* **42**, 53–86.
3. Fringer, O.B. and Street, R.L., 2003, *J. Fluid Mech.* **494**, 319–353.
4. Sullivan, P.P. and McWilliams, J.C., 2002, *Phys. Fluids* **14**, 1182–1195.
5. Fulgosi, M., Lakehal, D., Banerjee, S. and De Angelis, V., 2003, *J. Fluid Mech.* **482**, 319–345.
6. Lakehal, D., 2004, DNS and LES of turbulent multifluid flows, in *Proceedings of the 3rd Symposium of Two-Phase Flow Modelling and Experimentation*, Pisa.
7. Lakehal, D., Milelli, M. and Smith, B.L., 2002, *J. Turbulence* **3**, 1–21.
8. Liovic, P., Rudman, M., Liow, J.-L., Lakehal, D. and Kothe, D., 2006, *Comp. Fluids*.
9. Lakehal, D., Reboux, S. and Liovic, P., 2004, *La Houille Blanche* **6**, 125–131.
10. Fenton, J.D., 1985, *J. Waterw. Port Coastal Ocean Eng.* **111**, 216–234.
11. Jeong, J. and Hussain, F., 1995, *J. Fluid Mech.* **285**, 69–94.
12. Calhoun, R.J. and Street, R.L., 2001, *J. Geophys. Res.* **106**, 9277–9293.

Adaptive Characteristics-Based Matching (aCBM): A Method for Interfacial Dynamics in Compressible Multiphase Flows

Robert Nourgaliev, Nam Dinh and Theo Theofanous

Center for Risk Studies and Safety, UC Santa Barbara, Santa Barbara, CA 93106-5130, USA; e-mail: robert@engr.ucsb.edu, nam@engr.ucsb.edu, theo@engr.ucsb.edu

1 Introduction

This work is motivated by the need for a high-fidelity numerical treatment of compressible flows with multimaterial interfaces. Of particular interest are immiscible flows with high Acoustic Impedance Mismatch (AIM) interfaces, such as those between gases and liquids, under both weak and strong shock wave conditions. These types of flow are encountered in numerous generically-important processes, such as propagation of shock waves in bubbly media, and interfacial instability and mixing, as well as in applications, such as explosive dispersal of liquids or solids, and atmospheric dissemination of liquid chemical agents [18]. Here, we capitalize on recent progress made by front capturing [7, 15, 16] and front tracking [5, 6] methods to deploy an adaptive mesh refinement strategy such as needed to cope with the often multiscale nature of such flows in practical settings.

The central theme, which guides the present development, addresses the need to optimize between the algorithmic complexities in advanced front capturing and front tracking methods, developed recently for high AIM interfaces, with the simplicity requirements imposed by the AMR multi-level dynamic solutions implementation [1–3, 19]. We have achieved this objective by means of relaxing the strict conservative treatment of AMR prolongation/restriction operators in the interfacial region, and by using a Natural-Neighbor-Interpolation (NNI) algorithm [17] in a characteristics-based matching (CBM) scheme at the interface [13]. The later is based on a two-fluid Riemann solver, which brings the accuracy and robustness of the front-tracking approach into the fast local level set, front-capturing [14] implementation of the CBM method. The performance of our method is demonstrated on three examples of shock wave interactions with (i) a single gas bubble, (ii) a cluster of three bubbles and (iii) a liquid drop.

S. Balachandar and A. Prosperetti (eds), Proceedings of the IUTAM Symposium on Computational Multiphase Flow, 341–352.

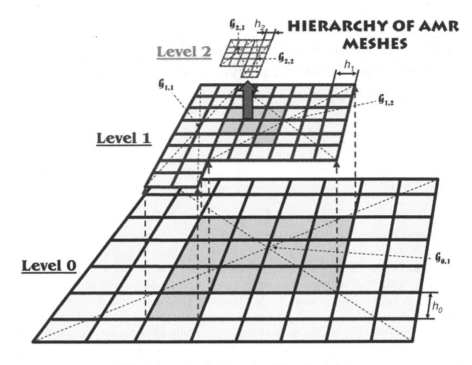

Fig. 1. Example of a hierarchy of SAMR meshes.

2 Overview of the method

Structured Adaptive Mesh Refinement (SAMR). The platform for our numerical approach is the Structured Adaptive Mesh Refinement (SAMR) methodology, originally developed by Berger and co-authors [1–3], and implemented as the SAMRAI package at Lawrence Livermore National Laboratory [19]. SAMR is based on a sequence of nested, logically rectangular meshes. Let $\underline{h}_k = \{h_0, h_1, \ldots, h_k\}|_{k=0,\ldots,L-1}$ denote a collection of mesh spacings, where h_k is a mesh spacing of a SAMR grid level k subject to $h_{k+1} \leq h_k$. A SAMR grid $\Omega^{\underline{h}_{L-1}}$ is a nested hierarchy of L grid levels $\Omega^{h_0} \supset \Omega^{h_1} \supset \cdots \supset \Omega^{h_{L-1}}$, where the coarsest grid Ω^{h_0} covers the entire computational domain. Grids are refined in both time and space, using the same mesh refinement ratio $r = \frac{h_k}{h_{k+1}}$, i.e. $\frac{\Delta t_0}{h_0} = \frac{\Delta t_1}{h_1} = \cdots = \frac{\Delta t_{L-1}}{h_{L-1}}$. Thus, the same explicit difference scheme is stable on all levels. As a consequence, more time steps are taken on the finer grids than on the coarser grids, but the smallest time step of the finest level is not imposed globally.

Each level Ω^{h_k} consists of a union of M_k logically rectangular regions, or patches, $\mathcal{G}_{k,m}|_{m=0,\ldots,M_k}$, at the same grid resolution h_k. The levels are nested, but the patches on different levels are not, and the patches on the same level may overlap, i.e. $\mathcal{G}_{k,i} \cap \mathcal{G}_{k,j} \neq 0|_{i \neq j}$. However, the requirement is that the discrete solution must be independent of how Ω^{h_k} is decomposed into patches $\mathcal{G}_{k,m}$. Thus, a point

x of the computational domain in general may exist on several grids. The solution vector $U(x)$ is taken from the finest level. If there are several equally fine grids containing the point, one can use the value from any of them, since the solution on the intersection of overlapping same-level grids is identical. Figure 1 shows an example of a hierarchy of SAMR grids with three levels of adaptation.

Time update of a solution on a hierarchy is organized in such a way, as to proceed sequentially from the coarser to finer grids. More specifically, before updating at any level, the next-coarser level solution must be already available, so as to allow the complete inter-level communications needed for populating each patch's ghost cells by interpolation in both time and space. For parallelization, all patches of a hierarchy are distributed between different processors.

Characteristics-Based Matching (CBM). The Characteristics-Based Matching method on uniform meshes was introduced in [10, 11]. There are three elements needed to incorporate the CBM into SAMR: *inter-patch communication, generation/disposal of patches*, and *time update on a patch*.

The *inter-patch communication* is necessary for synchronization of time updates on different patches. More specifically, it is required to properly populate ghost cells around each patch, before one may proceed with its update. There are three types of inter-patch communications: same-level, coarse-to-fine and fine-to-coarse. The utilities for these operations are provided by SAMRAI. In the case of gas-liquid interfaces, the conservative coarse-to-fine and fine-to-coarse inter-level communications fail and must be modified. The origin for these failures is discussed in [13]. To cure this problem, we have introduced non-conservative prolongation/restriction operators, which are applied near the interface. Algorithmic details of these operators are given in [13].

The second element of our approach is related to *adaptive generation/disposal of AMR patches* on different levels of a hierarchy. The SAMR grid may be modified at discrete times. The finest level needs to be changed most often (patches are moved, added or deleted, if required). When the level Ω^{h_k} is changed, all finer levels $\Omega^{h_j}|_{j>k}$ are changed as well, but the coarser levels $\Omega^{h_i}|_{i<k}$ may remain the same. The utilities for dynamic management of AMR patches require tagging criteria for refinement. In particular, the computational mesh needs to be refined around flow discontinuities, such as shocks and contacts, as well as around multimaterial interfaces. For interfaces, we use a distance-based criterium, and for shocks, we have a shock-detection criterium that leaves out rarefactions. Both are described in [13].

Lastly, the major components of our *patch time update strategy* include (a) a high-order-accurate Godunov-based conservative finite difference method for gas dynamics to advance the solution in the bulk fluids [10]; (b) level-set-based sharp capturing of interfaces, supplemented by localization (FLLS) and re-initialization algorithms [14] for computational efficiency and accuracy, respectively; and (c) "Characteristics-Based Matching (CBM)" for coupling solutions across the interface [10, 11]. The key features of the CBM are (i) a Riemann-solver-based coupling, which is essential for robustness in the case of high-AIM (i.e., gas-liquid) interfaces, and for accuracy in the case of very strong shock waves in both multi-gaseous and gas-liquid media; and (ii) elimination of the need for ghost fields and corresponding ghost cells, as we found this to be necessary for compatibility with AMR. The CBM is based on the generation/tracking/disposal of the subcell-interface-markers (denoted as CBM points), which exist on each patch only during one time step Δt_k. Applying the two-fluid, pseudo-multidimensional Riemann solver at CBM points, the wave structure and gas dynamics solutions at the interface are computed and applied for direct modification of numerical fluxes in the Eulerian cells near the interface. This is based on the subcell position of the interface and a flux inter-/extrapolation algorithm. The concepts of subcell markers, and two-fluid Riemann solutions, are borrowed from front-tracking methods. On the other hand, and on account of the semi-Lagrangian nature of the level set method, the natural-neighbor interpolation (NNI) procedure is used to correct the numerical solutions at Eulerian computational cells that have found themselves to change fluid occupancy (based on cell center) during the time step. These cells are denoted here as "degenerate" cells. The modification of numerical fluxes and treatment of "degenerate" cells are the substitutes for GFM's [7] ghost fields/cells and the related to them PDE- or FM-based extrapolation techniques. Algorithmic details of our patch time update approach are given in [13].

3 Propagation of shock waves in bubbly media

Two test-cases are considered in 2D: a single-bubble subjected to a planar shock and a three-bubble cluster in a cylindrically imploding shock wave.

Single-bubble collapse. Consider a 2D cylindrical air bubble, 6 mm in diameter, immersed in a water pool, initially under atmospheric conditions. The center of the bubble is located at $\mathbf{x}_b = (12, 12)$ mm in the computational domain of size 24 × 24 mm. A planar incident $M_{sh} = 1.72$ shock wave is initially located 5.4 mm to the left of the bubble center. The gas is modeled using the ideal-gas-law equation of state, with $\gamma = 1.4$. The liquid is represented with a stiffened gas equation of state, $P = (\gamma - 1)\rho i - \gamma \Pi$, where $\gamma = 4.4$ and $\Pi = 6 \cdot 10^8$. Viscous, heat transfer, and surface tension effects are neglected. Boundary conditions are periodic in the vertical direction and non-reflection at the left and right boundaries of the domain. Simulations were performed on a SAMR grid with six levels of adaptation and a refinement ratio of two, which corresponds to an effective resolution of 800 computational nodes per initial bubble diameter. The third-order Runge-Kutta TVD and the fifth-order monotonicity-preserving MP-WENO5 schemes are applied for discretization in time and space respectively. The Local Lax Friedrichs (LLF) flux splitting technique is used for numerical flux treatment. Computations are performed using CFL=0.4.

Snapshots of the bubble shape evolution are shown in Figure 2. Due to the large AIM at the water-air interface, the incident shock transmits a relatively weak shock into the air, producing a strong reflected rarefaction wave in the water. By approximately 2.3 μs, the air bubble becomes involuted, with a distinct water jet

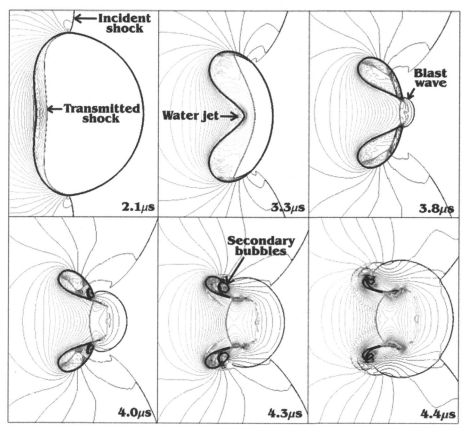

Fig. 2. Dynamics of shock-induced single-bubble collapse. The isolines represent the Mach number field (30 isolines are uniformly distributed in the range from 0 to 3.54).

formed at the centerline. At approximately 3.69 μs, the water jet hits the other side of the bubble with a velocity of 2.85 km/s, cutting the bubble in half. Upon impact, an intense blast wave, with maximum pressure of 10.1 GPa, is generated. Caused by the blast wave, secondary jets penetrate into the smaller bubbles, and finally cut the initial bubble into four pieces. The gas volume reaches its minimum at approximately 4.5 μs, starting to slowly grow after that (rebound). The maximum temperature of the gas at the moment of minimum volume is $\approx$25,000 K. Since we have not modeled heat transfer and real gas effects here, the temperature history can be regarded as qualitative. Nevertheless, the observed intense heating is consistent with the observation of luminescence in experiments by Bourne and Field [4], performed under similar conditions.

Collapse of a three-bubble cluster. Two-dimensional cylindrical gas bubbles of radius $R = 6$ mm are placed in a computational domain of size $(10 \times 10)R$, as shown in Figure 3, with one bubble in the center and two bubbles shifted from the center by

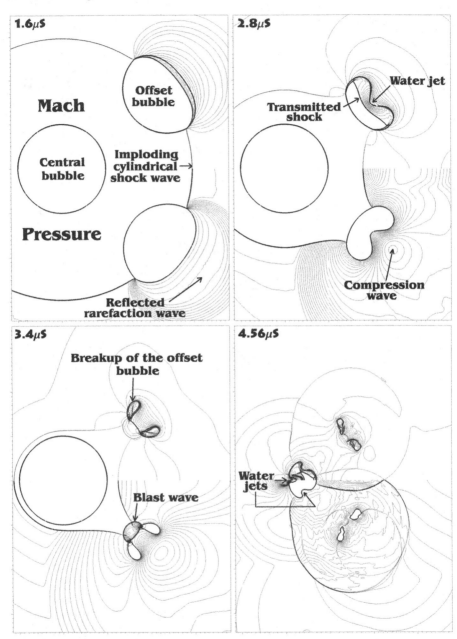

Fig. 3. Three-bubble cluster collapse under a cylindrically imploding shock wave. Mach number (top halves) and pressure (bottom halves) fields shown.

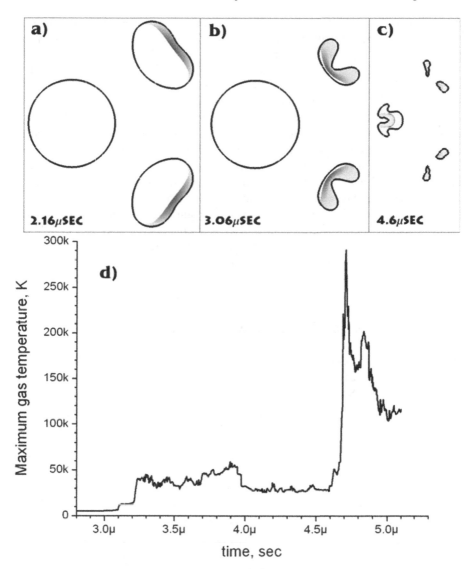

Fig. 4. (a)–(c) Samples of temperature distribution inside bubbles. (d) Transient of peak gas temperature.

$3R$. A cylindrical shock is generated setting the following pre- and post-shock conditions: $\mathbf{V}_{\text{pre-shock}} = [1 \cdot 10^5; 1000; 0; 0]^{\mathsf{T}}$ and $\mathbf{V}_{\text{post-shock}} = [4 \cdot 10^9; 1350; 0; 0]^{\mathsf{T}}$, where $\mathbf{V} = [\mathbf{P}, \rho, \mathbf{u}, \mathbf{v}]^{\mathsf{T}}$. These are chosen in such a way as to match the strength of the collapsing inwards shock at the moment of its impact upon the gas-liquid interface to the planar shock, considered in the previous example. Non-reflection boundary conditions are applied at all four boundaries of the computational domain.

A computational mesh of size 500×500 was used on the coarsest AMR level. Using 4 AMR levels with refinement ratio 2, the effective grid resolution is 800 computational nodes per initial bubble diameter.

Selected snapshots of the pressure and Mach number fields are presented in Figure 3, while the history of the gas temperature is given in Figure 4. First, the shock wave hits the "offset" (outer-layer) bubbles, causing their deformation. By $t = 3.2$ μs, two water jets are formed inside the outer bubbles, whilst the central bubble remains unaffected. Next, the outer bubbles are fragmented into two pieces each, with generation of blast waves in front of them and compression waves behind. These compression waves eventually become shock waves. By that time, the central bubble has already been hit by the primary cylindrical shock and by the shock waves due to outer-layer bubble collapses. There are three distinct jets in the central bubble, moving toward each other. The still collapsing central bubble leaves behind a growing compression wave, which is transformed into a shock wave by $t = 4.72$ μs. Finally, the central bubble collapses and generates another blast wave. At the same time, the shock waves, generated by the outer-layer bubble collapse, collide with each other, forming reflected shock waves, expanding upwards and downwards.

4 Interaction of gaseous shock wave with a liquid drop

This problem is particularly difficult, because it involves a curved gas/liquid (slow-fast) interface, and associated with this we have rather complex, and varied, shock refraction patterns. Such patterns have been investigated previously for *planar gas-gas interfaces* [9] and, recently, using methods described here, extended to *weakly shocked planar gas-liquid interfaces* [12]. It turns out that numerical schemes in general are sensitive to resolution of irregular refractions, especially in the case of gas-liquid interfaces, when the errors are easily amplified due to the stiffness of EOS for liquid [12]. The types of pattern obtained depend on the angle of incidence, and include regular refraction with reflected shock (RRR), regular refraction with reflected expansion (RRE), free precursor (FPR), free von Neumann refraction (FNR), and refraction with anomalous reflection (ARR). In the present case the angle of incidence continuously varies, as the shock propagates over the drop, and in addition we have focussing effects that interact with these patterns. Strong shocks exhibit still different and interesting behaviors, as illustrated for example in [18, figure 9] .

The problem formulation is the following. A cylindrical ($R = 3.2$ mm) liquid mass is suspended in motionless gas, under atmospheric conditions, at the center of the computational domain of size 32×32 mm. The following parameters of the stiffened equation of state are used $[\gamma; \Pi]^\mathsf{T} = [2.8; 3.036 \cdot 10^8]^\mathsf{T}$ and $[1.4; 0]$ for liquid and gas, respectively. A planar incident $M_{sh} = 1.47$ shock wave with post-shock conditions $\mathbf{V}_{\text{post-shock}} = [2.35 \cdot 10^5; 1.811; 246.24; 0]^\mathsf{T}$ is initially placed 4 mm to the left of the cylinder's center. Boundary conditions are periodic in the vertical direction and non-reflection for both left and right boundaries of the domain. A computational mesh with six AMR levels of adaptation and refinement ratio 2 was used, corresponding to the effective grid resolution of 640 computational nodes per

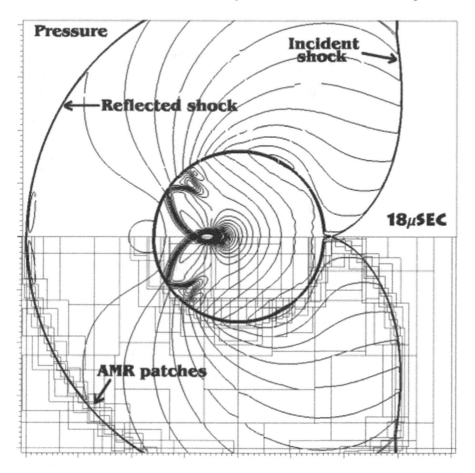

Fig. 5. Shock wave interaction with liquid drop. Sample pressure field and outline of AMR patches (bottom).

diameter. An example of layout of AMR patches together with a sample of the pressure field are shown in Figure 5. We employed the RK$_3$/LLF/MP-MUSCL$_3^{\text{MinMod}}$ scheme and CFL $= 0.2$.

The dynamics of the interaction during the first 14 μs are shown in Figure 6. When the incident shock hits the liquid, it is reflected as a shock, transmitting an acoustic pressure wave into the liquid. Early on, the incidence angle β is sufficiently small, and the refraction is regular (RRR). The intersection of the incident, reflected shocks and transmitted pressure wave is a "refraction node", which exists only during the first 2–3 μs. When the incidence angle becomes sufficiently large ($t = 4$ μs), the transmitted pressure wave "peels-off" the refraction node, transforming into a "precursor" pressure wave, and the refraction pattern corresponds to the irregular "Free Precursor Refraction (FPR)". With a further increase of the incidence angle β, the combination of the incident and reflected shocks transforms into the Mach reflec-

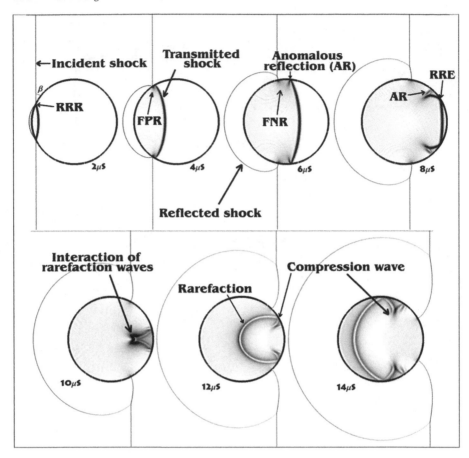

Fig. 6. Wave dynamics as depicted by numerical Schlieren, computed as $\vartheta =$ $\exp\left(-\kappa \frac{|\nabla\rho|}{|\nabla\rho|_{max}}\right)$ with $\kappa = 10^5$ and $5 \cdot 10^3$ for the liquid and gas, respectively.

tion, forming the "Free von Neumann Refraction (FNR)" irregular pattern. When the transmitted pressure wave passes the equator of the cylinder at $t = 6$ μs, it is back-refracted in a "fast-slow" configuration [12]. Since the transmitted pressure wave at this moment is nearly perpendicular to the interface, the back-refraction pattern is irregular, corresponding to the so-called anomalous reflection (AR) [8, 12]. The back-transmitted pressure wave in the gas is extremely weak, and it is indistinguishable in the plotted numerical Schlieren field. By $t = 8$ μs, the incidence angle of the back-(fast-slow)-refraction becomes smaller, which results in formation of the regular pattern with reflected rarefaction, denoted as RRE. Due to the focusing geometry of the interface, the two reflected rarefaction waves collide with each other at $t = 10$ μs, forming a strong rarefaction wave with negative pressure (cavitation possible). As the reflected rarefaction wave moves backwards, $t = 12$ μs, it interacts with the non-uniform flow field, which weakens the rarefaction, and, when the equator is passed at $t = 14$ μs, converts it into a compression wave.

5 Conclusion

The sharp capturing of high acoustic impedance interfaces under strong shock conditions in practical settings requires adaptive mesh refinement, and this in turn we found to require abandonment of (a) strict conservative treatment of the prolongation and restriction operators (needed to communicate solutions from one AMR level to another) in the immediate vicinity of the interface, and (b) extrapolation steps in populating cells found on the "other side" of the interface as needed for advanced front capturing, such as the Ghost Fluid method. In both concerns we have found alternatives that with the help of highly accurate, front-tracking-like, Riemann treatment at the interface, and the Natural Neighbor Interpolation algorithm seem to work very well, even in the highly challenging case of a shock passing over a liquid drop suspended in a gaseous medium. However, there are still problems with simulating the longer term evolution, including deformation of the interface. In this case further improvements are needed in treatment of cut cells and numerical algorithms for time advancement.

Acknowledgments

This work was supported by the Lawrence Livermore National Laboratory (ALPHA and MIX projects), the National Ground Intelligence Center (NGIC), the US Army (DURIP Project), and the JSTO/DTRA (ASOS Project). Support of Drs. Frank Handler and Glen Nakafuji at LLNL, Dr. Rick Babarsky at NGIC, Mr. Ed Conley, JBCCOM at RDEC, and Mr. John Pace and Mr. Charles Fromer of the JSTO at DTRA is gratefully acknowledged.

References

1. Berger, M.J. and Colella, P., 1989, Local adaptive mesh refinement for shock hydrodynamics, *Journal of Computational Physics* **82**, 67–84.
2. Berger, M.J. and Oliger, J., 1984, Adaptive mesh refinement for hyperbolic partial differential equations, *Journal of Computational Physics* **53**, 482–512.
3. Berger, M.J. and Rigoutsos, I., 1991, An algorithm for point clustering and grid generation, *IEEE Transactions on Systems, Man, and Cybernetics* **21**(5), 1278–1286.
4. Bourne, N.K. and Field, J.E., 1992, Shock induced collapse of single cavities in liquids, *Journal of Fluid Mechanics* **244**, 225.
5. Chern, I.L., Glimm, J., McBryan, O., Plohr, B. and Yaniv, S., 1985, Front tracking for gas dynamics, *Journal of Computational Physics* **62**, 83–110.
6. Cocchi, J.-P. and Saurel, R., 1997, A Riemann problem based method for the resolution of compressible multimaterial flows, *Journal of Computational Physics* **137**, 265–298.

7. Fedkiw, R.P., Aslam, T., Merriman, B. and Osher, S., 1999, A non-oscillatory Eulerian approach to interfaces in multimaterial flows (the Ghost Fluid Method), *Journal of Computational Physics* **152**, 457–492.
8. Grove, J.W., and Menikoff, R., 1990, Anomalous reflection of a shock wave at a fluid interface, *Journal of Fluid Mechanics* **219**, 313–336.
9. Henderson, L.-F., Colella, P. and Puckett, E.G., 1991, On the refraction of shock waves at a slow-fast gas interface, *Journal of Fluid Mechanics* **224**, 1–27.
10. Nourgaliev, R.R., Dinh, T.N. and Theofanous, T.G., 2004, The 'Characteristics-Based Matching' (CBM) method for compressible flow with moving boundaries and interfaces, *ASME Journal of Fluids Engineering* **126**, 586–604.
11. Nourgaliev, R.R., Dinh, T.N. and Theofanous, T.G., 2004, Direct numerical simulation of compressible multiphase flows: Interaction of shock waves with dispersed multimaterial media, in *Proceedings of the 5th International Conference on Multiphase Flow, ICMF'04*, Yokohama, Japan, May 30–June 4, 2004, CD-ROM, Paper No. 494.
12. Nourgaliev, R.R., Sushchikh, S.Y., Dinh, T.N. and Theofanous, T.G., 2005, Shock wave refraction patterns at interfaces, *International Journal of Multiphase Flow* **31**(9), 969–995.
13. Nourgaliev R.R., Dinh T.N., Theofanous T.G., 2006, Adaptive Characteristics-Based Matching for compressible multifluid dynamics, *Journal of Computational Physics*, in press.
14. Peng, D.P., Merriman, B., Osher, S., Zhao, H. and Kang, M., 1999, A PDE-based fast local level set method, *Journal of Computational Physics* **155**, 410–438.
15. Osher, S. and Fedkiw, R., 2003, *Level Set Methods and Dynamic Implicit Surfaces*, Applied Mathematical Sciences, Vol. 153, Springer-Verlag, New York.
16. Sethian, J.A., 1999, *Level Set Methods and Fast Marching Methods*, Cambridge University Press.
17. Sibson, R., 1981, A brief description of natural neighbor interpolation, in *Interpreting Multivariate Data*, V. Barnett (ed.), John Wiley & Sons, New York, pp. 21–36.
18. Theofanous, T., Nourgaliev, R., Li, G. and Dinh, N., 2006, Compressible multi-hydrodynamics (CMH): Breakup, mixing and dispersal of liquids/solids in high speed flows, in *Proceedings of the IUTAM Symposium on Computational Multiphase Flow*, S. Balachandar and A. Prosperetti (eds), Springer, Dordrecht, pp. 353–369.
19. Wissink, A.M., Hornung, R., Kohn, S., Smith, S. and Elliott, N., 2001, Large scale parallel structured AMR calculations using the SAMRAI framework, in *Proceedings of the SC01 Conference on High Perf. Network. & Comput.*, Denver, CO, November 10–16, 2001. Also available as LLNL Technical Report UCRL-JC-144755.

Compressible Multi-Hydrodynamics (CMH): Breakup, Mixing and Dispersal of Liquids/Solids in High Speed Flows

Theo Theofanous, Robert Nourgaliev, Guangjun Li and Nam Dinh

Center for Risk Studies and Safety, UC Santa Barbara, Santa Barbara, CA 93106-5130, USA; e-mail: theo@engr.ucsb.edu, robert@engr.ucsb.edu, li@engr.ucsb.edu, nam@engr.ucsb.edu

1 Introduction

The subject of this talk is multiphase flow under extreme conditions of pressure, pressure gradients, transients, and phase-differential speeds (kilometers per second). Single-phase as well as multi-phase shocks are important. Reflection/transmission at interfaces, including material interfaces of extremely large density ratios, and acoustic impedance mismatch (AIM) are important too. We are interested in interfacial breakup, mixing phenomena, and the eventual dispersal of the dense phase.

Processes of interest may involve pre-existing particulates, or evolving length scales via the breakup of liquid and/or solid masses. The liquids may be Newtonian or viscoelastic, the latter being rendered so, to varying degrees, by the addition of polymeric substances of varying molecular weight, cross-linking and concentration. The length scale evolution defines the degree of coupling in both momentum and energy (cooling or reaction effects, for example) between the gas and the liquid. Scales of interest range from the microscopic, where interfacial instabilities nucleate and where rupture finally occurs, to the grossly macroscopic that embody evolutions of hundreds of kilogram quantities of material over tens to hundreds of meters spatial domains. Some of the areas of application include Inertia Confinement Fusion, energetic dissemination of liquids or solids in the atmosphere, estimation of weapon effects such as fallout of nuclear explosions, and innovative designs of rocket propulsion or Internal Combustion Engines.

Our work in this area, and this talk, are motivated by a new impetus derived from national defense and homeland security needs. My purpose is to illustrate the fundamental and diverse nature of these needs, to provide highlights of an approach, and related infrastructure necessary towards meeting these needs, and to show a sampling of initial results obtained as we began to dwell into the subject. The particular problem of interest is in the area of energetic dispersal of liquids, and as one can surmise from the following, the issues involved make this problem, along with the one concerning the mixing in inertia confinement fusion (see below), make it paradigmatic

S. Balachandar and A. Prosperetti (eds), Proceedings of the IUTAM Symposium on Computational Multiphase Flow, 353–369.

of a much broader subject we wish to define as Compressible Multi-Hydrodynamics (CMH).

2 An integral problem in CMH

Consider a liquid mass, suddenly exposed, at supersonic speeds, to a gaseous atmosphere. The quantities of liquid involved may vary in the range 10^{-3}–10^5 g, the ambient (atmospheric) pressure may be anywhere between 10^{-4}–10^0 atm, the initial relative velocity may reach Mach 10, and the liquids may be viscous Newtonian, or viscoelastic, exhibiting very substantial resistance to breakup. We would like to know the liquid material disposition; that is, the spatial distribution of material quantities and length scales at the time they have reached dynamic equilibrium (stable particle sizes at terminal velocities) with the atmosphere. This, also known as the source term, is what is needed by atmospheric dispersal codes to estimate ground deposition characteristics and any consequent effects. Of special interest is the portion of the size spectrum that is above $\sim$100 μm.

Considering how little is known about the breakup of mm-size liquid droplets, a much-studied microcosm of the subject at hand, with this problem we found ourselves in essentially virgin territory. Accordingly we began by a broad/comprehensive, but cautious, scoping of the various aspects of the problem and by building the essential infrastructure for laboratory experiments and computations. At this time we have an overall approach, and we find ourselves at the initial stages of implementation. The approach is key-physics oriented, yet, our principal guide is fitness for purpose, and expect that in a flexible, continuously refocusing, and appropriately integrative effort, we will be able to meet the objective robustly, in an efficient manner, and within time constraints that are consistent with the practical needs. It should be clear at the outset that, even if affordable, a purely empirical approach, that is one based on generating and correlating data taken at field conditions (full scale, prototypic), is simply out of the question. Rather, we plan for a few field tests that will be specially-designed for final testing, hopefully validation of predictions, once the problem has been well understood.

A first-order partition of the problem can be made in terms of the material quantities involved. At the small mass extreme, the principal scaling parameters are the Weber (We), Ohnesorge (Oh), and Mach (M) numbers. For viscoelastic liquids, an additional scaling group is needed to involve the polymeric fluid relaxation time, τ_R, and the rate of strain in the induced flow, $R = \tau_R/\tau_S$. For given values of Oh, and M numbers, and a perfectly spherical (small) droplet of a Newtonian liquid *suddenly exposed* to uniform gas flow, there is a critical Weber number (We$_{CR}$) below which the drop will be accelerated (to the gas flow speed) while remaining intact. For Oh $\sim$ 0 this critical threshold for breakup is known from experiments to be $\sim$10. Also known is that under supercritical conditions (We $>$ We$_{CR}$) the resulting daughter drops are well-subcritical at the free stream conditions, however only an initial and rather narrow attempt has been made so far for quantification (see [44], and further below). The We$_{CR}$ increases with increasing Oh number, but this relationship, as well as the daughter size spectra remain to be quantified as well. Initial results (see below) suggest that the role of viscoelasticity in these matters is profoundly rich.

Larger masses would inherently entail departures from perfectly spherical shape, as well as extraneously imposed, poorly defined, and uncontrollable surface perturbations. Also, larger masses would entail, for increasing portions of the masses involved, greater and greater departures from the "sudden exposure to steady flow" scenario. In particular, we can expect shielding, other long-range interaction effects, collisions and coalescence, and more generally collective behavior, until the cloud expands to a sufficiently large dimension for it to be considered dilute, and the entities within it to be acting independently.

It is important to appreciate that breakup processes are inseparable from the collective response, and that the so-created aerodynamic history causes a spreading out, and in a sense a blunting of the severity of the interaction. On the other hand, viscoelastic liquids resist breakup, they first stretch into ligaments and sheets extensively, and this, combined with aerodynamic history, is at the essence of the profound difficulty of our problem. We expect the experiments at the small mass extreme, aided by direct numerical simulations, will yield a lower limit on the daughter drop spectra to be found in large scale events of this type. On the basis of these results, experiments involving larger liquid quantities, and with the further help of effective field modeling, we aim to address the overall mass-scaling question, over the whole range of parameters of interest, including consideration of uncertainty in such assessments.

3 Design and role of experiments

Previous experiments in aerobreakup of mm-scale droplets, were carried out in shock tubes, and focused principally on the morphology of break-up. Results were expressed in qualitative depictions of these regimes, and We number criteria for transitions. Only meager data on final size distributions exist, and only for "bag" break-up. The accepted regime sequence is "vibrational", "bag", "bag-and-stamen", "shear" or "stripping", and "catastrophic". Theoretical understanding is meager too. The "stripping" regime was postulated by Taylor in 1949 [42], who proceeded to estimate the rates of liquid removal and thus the total breakup time, and this calculation was repeated, in a slightly improved version, by Ranger and Nichols [40] and others. The catastrophic regime was proposed by Harper et al. [16] on the basis of analysis that accounted for aerodynamic deformation along with Rayleigh–Taylor (R–T) instability, on the windward-side of the interface. They defined its existence for We > 10^4 but without any basis, or even reasoning. The same regime was postulated again (without reference to any criteria for existence) very recently by Joseph et al. [20] who reconsidered the classical R–T problem (i.e. [3]) in terms of viscous potential flow theory. It is apparent now [43] that due to visualization limitations the experiments in this case mislead, and erroneously were appealed to for supporting, these partial [16], or ad hoc [20], attempts at theory. Moreover no theory, or simulation are available for the other regimes, nor for the effect of viscosity on any of these regime transitions.

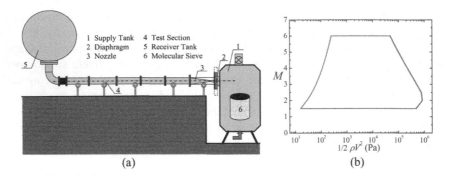

Fig. 1. (a) ALPHA facility and (b) Mach number – dynamic pressure domain of operating conditions accessible in it.

Further it is apparent that the shock tube geometry, with limited visualization access, limitations in imaging technology, and extremely limited observation times available, have prevented an approach to the equilibrium size distribution question. As we will see below the observation time requirement is particularly acute in the case of viscoelastic liquids. These then are the starting points of our consideration in the design of our experiments, as well as of our approach to instrumentation and measurement.

The ALPHA facility (Figure 1a) is a pulse, supersonic wind tunnel, capable of accessing the M-1/2 ρv^2 space shown in Figure 1b, with steady flow durations of up to 100 ms. The 4 m long, fully transparent test section, has a flow cross-sectional area of 0.2×0.23 m^2, and it has been shown able to accommodate gram-liquid quantities (a mass scale up, relative to the past work referred to above, by a factor of 10^3) without wall interference. Especially designed converging-diverging nozzles, following the rupture of a Kapton (or Mylar)-film diaphragm, accelerate the flow to the desired Mach number, while two Phantom V7 video cameras illuminated by a synchronized copper-vapor laser, record the interaction with an injected liquid mass that is hit by the gas flow while in transit. Highly resolved, in both space and time, visualization is achieved by framing rates of up to 160 kHz, each frame being exposed for only $\sim$10 ns. Special close-up arrangements allow spatial resolutions of up to 10 μm. The receiving tanks can be evacuated down to 10 Pa, and with the appropriate pressure ratio we can access operating pressures in the test section down to this level. The flow is instantaneous behind an initial shock, and the transition to somewhat lower, and steady dynamic pressure takes place within 500 μs.

Initial results from the ALPHA facility have provided the basis for establishing the following [24, 43]:

(a) There are only two principal breakup regimes for the mm-scale, low-viscosity (Oh $\sim$ 0) Newtonian liquid drops, *Piercing and Shearing* (Figure 2). In the piercing regime the gas penetrates the drop, starting with a single wave (the "bag") at We$_{CR}$, and developing to multiple piercing waves with increasing Weber number. In the shearing regime the gas flows around the drop, shearing and entraining

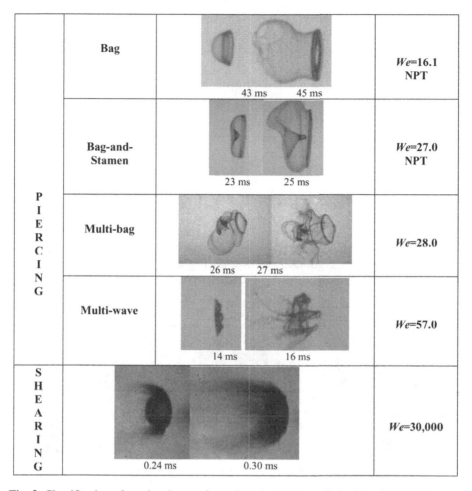

Fig. 2. Classification of aerobreakup regimes (based on [43]). All the data shown here were obtained from our ALPHA-I facility – vertical test section. The liquid is Tributyl Phosphate (TBP); water-like viscosity and surface tension. NPT means normal pressure and temperature, subsonic flow conditions. Other data were obtained at low pressure, for M = 3. Drop sizes: 2.2 to 3.8 mm.

the liquid within a rather confined two-phase layer, that exits at the drop equator. The transition between these two regimes occurs gradually, in some finite in size Weber number range, around 10^2 to 10^3. The multiple-wave piercing seems to be only accessible in the low-density, highly supersonic flow in ALPHA, where such break-up configurations were first observed.

(b) The piercing regime is dominated by Rayleigh–Taylor instabilities as confirmed by theoretical interpretation of experimental data, obtained with the low viscosity Tributyl Phosphate (TBP), and the very viscous pure Glycerin. The theory is

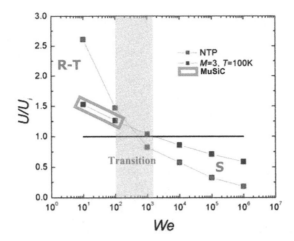

Fig. 3. Regime map for Newtonian (Oh ~ 0) drop breakup [43]. R–T: Rayleigh–Taylor *piercing*; S: *shearing*; U: Rayleigh–Taylor wave penetrating velocity; U_i: interfacial velocity due to shear. MuSiC denotes results with shear stresses obtained from DNS.

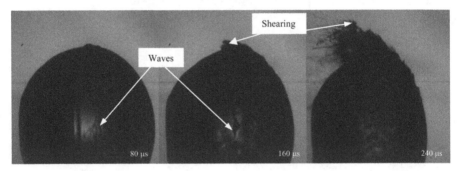

Fig. 4. Surface waves and *shearing* of a 3.5 mm TBP drop subject to M = 3 flow (We = 30,000).

based on the requirement that an odd number of half wavelengths should fit on the available forward-facing area of an appropriately flattened drop.

(c) Contrary to previous analytical results [16, 20] the asymptotic high Weber number regime is Shearing, not Piercing, and the so-called "catastrophic" regime is not physically attainable. The fallacy of these previous analytical results is to be found in ignoring *shear* as a process competing with *piercing*. Theofanous et al. [43], tipped by the ALPHA experiments, focused on this competition by comparing the penetration rate of non-linear R–T waves to the shear-induced velocities parallel to the interface. The regime map so obtained (Figure 3) was found to be in agreement with all experiments, but a still more unambiguous demonstration from still higher-resolution visualization (more recent tests in ALPHA) is given in Figure 4.

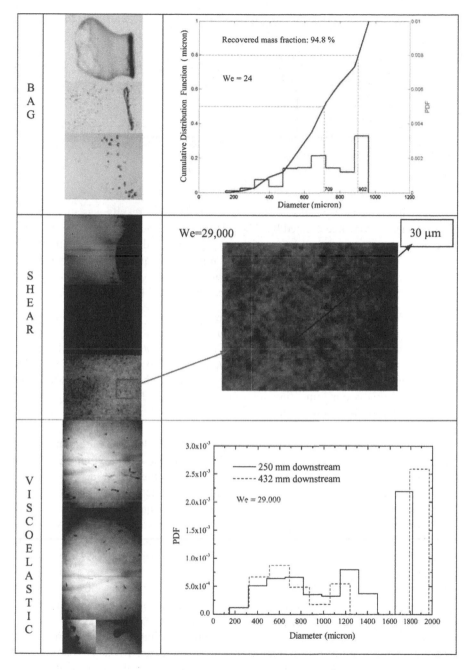

Fig. 5. Sample images and size distributions. Note the non-spherical shapes of the VE fragments.

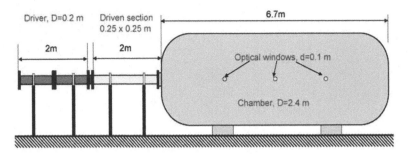

Fig. 6. Schematic of the ASOS facility.

(d) The fallacy in interpretation of previous experimental results was in failing to appreciate the mirage created by shadow imaging of the superposition of finely-entrained mist on one hand, and of the expanding gas that has been compressed behind the bow shock on the other. The mist thus acquires a significant radial velocity component, thus creating the illusion of an "explosion" (and thus the imagined "catastrophic" breakup due to penetration by R–T waves).

These ALPHA results further indicate that the daughter drop size distributions, for the first time caught in the process of them being generated, are crucially dependent on the regime of breakup. A sample to illustrate the point is given in Figure 5. The viscoelastic liquid behavior shows the importance of long flow times in ALPHA. It also shows the merit of investigating more thoroughly the small mass extreme of stability, and this is our immediate next task in ALPHA. In addition, a systematic collection of size distribution data over the range of conditions of interest is underway.

Several complementary thrusts, involving an independent measurement of size spectra, dynamic pressures that reach up to 2 million Pascal, and liquid masses that reach up to tens of gram quantities, are to be pursued in the ASOS facility, which is currently under construction. As illustrated in Figure 6, the ASOS facility is a shock tube, made at the same dimensions as the ALPHA, and equipped with a large catch tank. In distinction to normal shock tube operation, the idea here is to extend the available pulse time by operating behind the contact discontinuity, as a follow-up to the flow behind the initial shock. In this way, by using a pair of gases, Helium in the driver and air or Nitrogen in the expansion section, the dynamic pressures in these two flow regions can be matched, and thus seamlessly extend the flow duration to $\sim$3 ms, as needed for completion of fragmentation of tens of grams liquid quantities, at the upper end of the dynamic pressure range of interest ($\sim$10^6 Pa). Other features of ASOS include a double-pulse operating regime attainable in normal operation with a single gas, long term observation and measurements (radar, lidar, etc.) of particle clouds suspended in the catch chamber, and size analysis of the deposited mass on witness plates lining the inner walls.

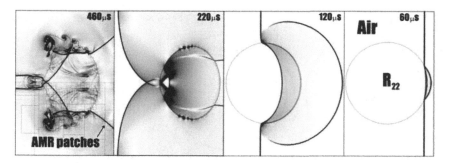

Fig. 7. Shock-induced interface instability and mixing [34]. Numerical Schlieren images and outline of AMR patches. The shock ($M_{sh} = 1.22$) propagates from right to left.

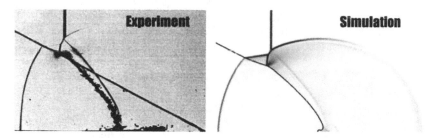

Fig. 8. Shock wave refraction pattern on a planar slow-fast, gas-gas (CO^2/CH^4) interface. Experimental [1] vs. numerical [35] Schlieren images. Twin von Neumann Reflection (TNR) pattern.

4 Direct Numerical Simulations (DNS)

Previous work with direct numerical simulations in CMH was spurned to a significant level of activity, and accomplishment, by a similar, as here, interest in interfacial breakup and mixing under the passage of a shock wave (Richtmyer–Meshkov, R–M, instability). Herein we find the first high-fidelity, front tracking approaches [6], and it is here that we find the first, and perhaps still highest fidelity super-large-scale simulations of such mixing processes [18, 44]. Yet, these are only a microcosm of the R–M problem in applications that motivate this work (fusion devises), DNS as a lone approach has come to an impasse, and it is not clear yet what modeling approach would be needed for simulating the evolution of mixing in the longer term, nor is it clear how to interface it with the DNS [9]. In addition our problem here is burdened with: (a) high acoustic impedance mismatch interfaces, and the special numerical resolution and stability issues they engender, (b) complex fields that span all flow speeds, reaching down to $M \sim 0$, (c) much broader range of time and spatial scales as our interest extends, vitally, all the way out to equilibrium, and (d) complex fluid behavior due to the presence of polymeric additives.

Other important works on front/particle tracking numerics that we must include are the contributions of Grove and Menikoff [14], Cocchi and Saurel [7] and Ball

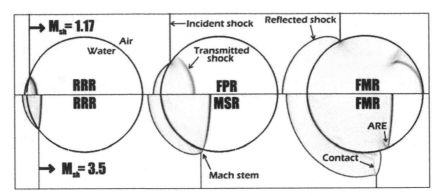

Fig. 9. Shock wave refraction patterns on a curved gas-liquid interface. Numerical Schlieren. *RRR*: Regular Refraction with Reflected shock. *FPR*: Free Precursor Refraction. *FMR*: Free precursor with Mach stem Refraction. *MSR*: Mach Stem Refraction. *ARE*: Anomalous Refraction with reflected Expansion.

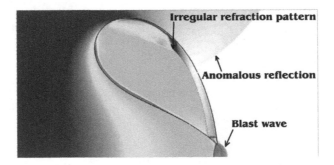

Fig. 10. Shock-induced bubble collapse [34].

et al. [2]. In parallel we also have available the important, more recently made advances in front capturing, including the Volume of Fluid [17, 28], Level Set [28, 37], γ-transport-based models [21, 38], Quasi-Conservative models [41], and the Ghost Fluid Method [13], and several further developments on them, notably on accommodating high acoustic impedance mismatch interfaces and very strong shocks [19, 26, 32], as well as the Particle Level Set [12] as a way of adding fidelity and stability to the numerical scheme.

Our approach, in this aspect of the infrastructure needs, adopts and adapts elements of these bases, and builds on them towards the creation of a comprehensive numerical package, the M̲ulti-scale S̲imulation C̲ode (MuSiC), that extends a broad but all-compatible capability from DNS to Effective Field modeling, and from incompressible to highly compressible flow conditions (within the same computation), including their seamless interfacing, in a parallel, adaptive mesh refinement environment, implemented in 3D.

Key elements of MuSiC include (a) Godunov-based compressible and pseudo-compressible high-order-accurate solvers, connected in any combination with the

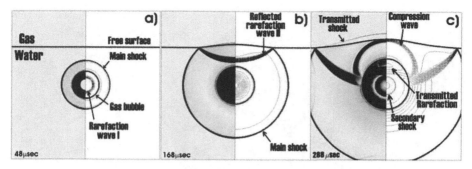

Fig. 11. Underwater explosion [34]. Dynamics of numerical Schlieren (left) and pressure (right) fields. Energy density deposited in the explosion zone is 7.163 GJ/m.

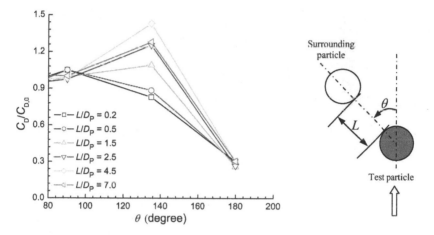

Fig. 12. Particle drag coefficients under the influence of a surrounding particle of the same diameter (D_P) located at different distances and angles as illustrated. Re = 100. C_{D0} is the drag coefficient on an isolated particle.

adaptive Characteristics-Based Matching (aCBM), for the sharp capturing of free interfaces across any media (arbitrary density and viscosity ratios, acoustic impedance, etc.), at any flow speeds [31, 34], (b) the AUSM treatment applied at DNS, as well as an effective field model of disperse multiphase flow [4, 5], again amenable to all Mach number flows, and (c) a numerical model that naturally devolves a DNS simulation into regions of dispersed flows as needed due to continuous refinement of length scales in the mixing region [9].

The status of this development in regards to DNS is illustrated by sample results as shown in Figures 7–11. Key aspects of the developments on the effective field model are discussed in the next section. On the DNS, the key considerations, challenges and requirements thereof are (a) respecting *information flow* across multimaterial interfaces, which includes taking into account high acoustic impedance mismatch and related to this a variety of complex shock refraction patterns (Figures 8

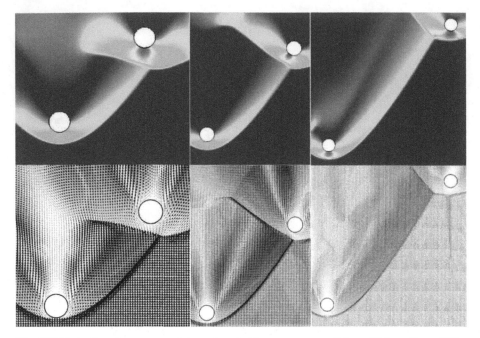

Fig. 13. Illustration of long-range interactions in disperse compressible multiphase flows [23]. Pressure (top) and velocity (bottom) fields.

and 9); (b) achieving a sufficiently sharp treatment of the interface as to accommodate truly the jump conditions in material properties and stresses [22, 36]; (c) simulating *large deformation, breakup and coalescence*, which is especially challenging in gas-liquid, slow-fast configurations, and requires high fidelity in spatial resolution and related massively parallel simulations using adaptive mesh refinement technology (Figures 7 and 10); (d) operating in a *wide range of flow speeds and compressibilities*, including simulation of supersonic and extremely slow or nearly incompressible flows within one setting, which necessitates the development of hybrid flow solvers (compressible/incompressible) (Figure 9); (e) *disparity of characteristic time scales* (acoustic and material), which necessitates the development of new, more efficient time discretization strategies (Figure 9); (f) *turbulence modeling* for compressible flows, including the development of new LES models that take into account the existence of flow and material discontinuities and are capable of working in adaptive mesh refinement environment; (g) *non-Newtonian fluid dynamics*, related constitutive descriptions, and numerical scheme implications; and (h) incorporation of micro/(nano)-scale physics, including surface tension, which raise the need to re-evaluate applicability of the existing Riemann-solver-based flux discretization schemes.

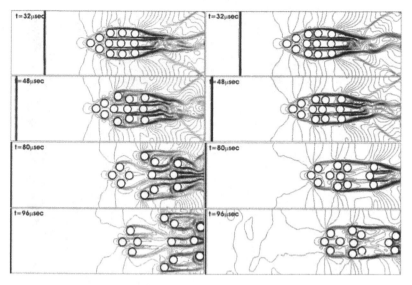

Fig. 14. Shock-induced dispersal of solid particles [33]. Dynamics of Mach field and particle displacements for elastic (left) and viscoelastic (right) particle-particle collisions.

5 Effective Field Modeling (EFM)

As noted above, the purpose of EFM in our approach is to capture and eventually predict mass scale-up; that is, collective behavior and related aerodynamic history effects. In relation to past work in formulating and solving homogenized (averaged, or multi-fluid) models (i.e. [10, 15, 30, 39]), essential new features and related challenges stem from the highly structured internal constitution of supersonic flows, as given by the simple illustrations in Figures 13 and 14. This has implications in long-range drag interaction effects, as shown in Figure 12, and on breakup behavior as depicted in Figure 15. In turn these raise the issue of what would be an appropriate homogenization approach to preserve the physics sufficiently, so that it would be amenable to a manageable constitutive description. Direct numerical simulations that examine the microcosm of collective behavior, as shown in Figure 14, experiments, and special purpose DNS for extracting drag interactions in multi-particle arrays (Figure 12 and [23]) are used to guide development of an appropriate effective field model. Of particular interest in this respect are concepts of the *Heterogeneous Multiscale Method* of E and Engquist [11].

The complementary avenue is through a direct computational framework [8] at the DNS level that can accommodate simultaneously disperse flow regions. This framework is based on the AUSM$^+$ scheme [4, 5] already extended to a simple two-fluid model, as illustrated in Figure 16.

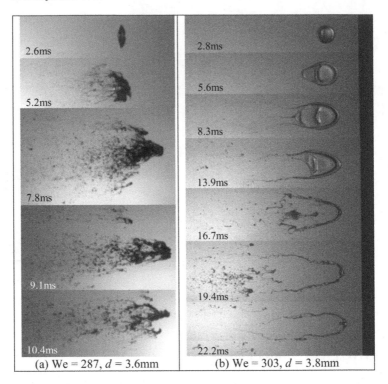

2.6ms

5.2ms

7.8ms

9.1ms

10.4ms

(a) We = 287, *d* = 3.6mm

2.8ms

5.6ms

8.3ms

13.9ms

16.7ms

19.4ms

22.2ms

(b) We = 303, *d* = 3.8mm

Fig. 15. Breakup of (a) an isolated drop and (b) of a drop behind a cylindrical rod. TBP in ALPHA.

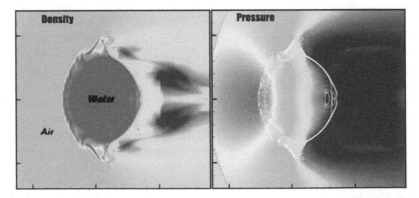

Density

Water

Air

Pressure

Fig. 16. Interaction of $M_{sh} = 6$ shock wave with a liquid drop ($D = 6.4$ mm), using the two-fluid AUSM$^+$ method [5]. Viscosity and surface tension are not taken into account.

6 Concluding remarks

The experiments, direct numerical simulations, and effective field modeling are to meet on the grounds of the interfacial area transport equation, supplemented by the source and sink terms in a way that accounts for major morphological changes, and in a way that preserves the major physics of phase interactions, until a dilute, equilibrium state has been attained.

Acknowledgments

This work was supported by the Lawrence Livermore National Laboratory (ALPHA and MIX projects), the National Ground Intelligence Center (NGIC), the US Army (DURIP Project), and the JSTO/DTRA (ASOS Project). Support of Drs. Frank Handler and Glen Nakafuji at LLNL, Dr. Rick Babarsky at NGIC, Mr. Ed Conley, JBCCOM at RDEC, and Mr. John Pace and Mr. Charles Fromer of the JSTO at DTRA is gratefully acknowledged.

References

1. Abd-El-Fattah, A.M. and Henderson, L.H., 1991, Shock waves at a slow-fast gas interface, *Journal of Fluid Mechanics* **89**(1), 79–95.
2. Ball, G.J., Howell, B.P., Leighton, T.G. and Schofield, M.J., 2000, Shock-induced collapse of a cylindrical air cavity in water: A free-Lagrange simulation, *Shock Waves* **10**, 265–276.
3. Chandrasekhar, S., 1981, *Hydrodynamic and Hydromagnetic Stability*, Dover Publication, New York, pp. 441–443.
4. Chang, C.-H. and Liou, M.-S., 2003, A new approach to the simulation of compressible multifluid flows with AUSM$^+$ scheme, in *Proceedings of the 16th AIAA Computational Fluid Dynamics Conference*, June 23–26, Orlando, FL.
5. Chang, C.-H. and Liou, M.-S., 2004, Simulation of multifluid multiphase flows with AUSM$^+$ scheme, in *Proceedings of the 3rd International Conference on Computational Fluid Dynamics, ICCFD'3*, July 12–16, Toronto, Canada.
6. Chern, I.L., Glimm, J., McBryan, O., Plohr, B. and Yaniv, S., 1985, Front tracking for gas dynamics, *Journal of Computational Physics* **62**, 83–110.
7. Cocchi, J.-P. and Saurel, R., 1997, A Riemann problem based method for the resolution of compressible multimaterial flows, *Journal of Computational Physics* **137**, 265–298.
8. Dinh, T.N., Nourgaliev, R.R. and Theofanous, T.G., 2003, On the multiscale treatment of multifluid flow, *Multiphase Science and Technology* **5**(1–4), 275–288.
9. Dinh, T.N., Nourgaliev, R.R. and Theofanous, T.G., 2005, On the numerical simulation of acceleration-driven multifluid mixing, *Multiphase Science and Technology* **17**(4), 1–32.
10. Drew, D.A. and Passman, S.L., 1998, *Theory of Multicomponent Fluids*, Springer-Verlag, New York.
11. E, W. and Engquist, B., 2003, Multiscale modeling and computation, *Notices of the AMS* **50**(9), 1062–1070.

12. Enright, D., Fedkiw, R., Ferziger, J. and Mitchell, I., 2002, A hybrid particle level set method for improved interface capturing, *Journal of Computational Physics* **183**, 83–116.

13. Fedkiw, R.P., Aslam, T., Merriman, B. and Osher, S., 1999, A non-oscillatory Eulerian approach to interfaces in multimaterial flows (the Ghost Fluid Method), *Journal of Computational Physics* **152**, 457–492.

14. Grove, J.W. and Menikoff, R., 1990, Anomalous reflection of a shock wave at a fluid interface, *Journal of Fluid Mechanics* **219**, 313–336.

15. Ishii, M., 1975, *Thermofluid Dynamic Theory of Two-Phase Flow*, Eyrolles, Paris.

16. Harper, E.Y., Grube, G.W. and Chang I.-D., 1972, On the breakup of accelerating liquid drops, *Journal of Fluid Mechanics* **52**, 565–591.

17. Henderson, L.-F., Colella, P. and Puckett, E.G., 1991, On the refraction of shock waves at a slow-fast gas interface, *Journal of Fluid Mechanics* **224**, 1–27.

18. Holmes, R.L., Dimonte, G., Fryxell, B., Gittings, M.L., Grove, J.W., Schneider, M., Sharp, D.H., Velikovich, A.L., Weaver, R.P. and Zhang, Q., 1999, Rychtmyer–Meshkov instability growth: Experiment, simulation and theory, *Journal of Fluid Mechanics* **389**, 55–79.

19. Hu, X.Y. and Khoo, B.C., 2004, An interface interaction method for compressible multifluids, *Journal of Computational Physics* **198**(1), 35–64.

20. Joseph, D.D., Belanger, J. and Beavers, G.S., 1999, Breakup of a liquid suddenly exposed to a high-speed airstream, *International Journal of Multiphase Flow* **25**, 1263–1303.

21. Karni, S., 1996, Hybrid multifluid algorithms, *SIAM Journal on Scientific Computing* **17**(5), 1019–1039.

22. Kang, M., Fedkiw, R.P. and Liu, X.-D., 2000, A boundary condition capturing method for multiphase incompressible flow, *Journal of Scientific Computing* **15**(2), 323–360.

23. Li, G.J., Nourgaliev, R.R., Dinh, T.N. and Theofanous, T.G., 2004, Particle-to-particle long-range interaction and drag in supersonic flows, AIAA 2004-1064, 42nd AIAA Aerospace Sciences Meeting and Exhibit, January 5–8, Reno, NV, USA.

24. Li, G.J., Sushchikh, S.Yu., Dinh, T.N. and Theofanous, T.G., 2004, Breakup and mixing of Newtonian liquid droplets in subsonic and supersonic gas streams, in *Proceedings of the 5th International Conference on Multiphase Flow, ICMF'04*, Yokohama, Japan, May 30–June 4, CD-Rom, Paper No. 364.

25. Liou, M.-S., 2003, A further development of the AUSM$^+$ scheme towards robust and accurate solutions for all speeds, in *Proceedings of the 16th AIAA Computational Fluid Dynamics Conference*, June 23–26, Orlando, FL.

26. Liu, T.G., Khoo, B.C. and Yeo, K.S., 2003, Ghost fluid method for strong shock impacting on material interface, *Journal of Computational Physics* **190**, 651–681.

27. Lhuillier, D., 2003, A mean-field description of two-phase flows with phase changes, *International Journal of Multiphase Flow* **29**(3), 511–525.

28. Miller, G.H. and Puckett, E.G., 1996, A high-order Godunov method for multiple condensed phases, *Journal of Computational Physics* **128**, 134–164.

29. Mulder, W., Osher, S. and Sethian, J.A., 1992, Computing interface motion in compressible gas dynamics, *Journal of Computational Physics* **100**, 209–228.

30. Nigmatulin, R.I., 1979, Spatial averaging in the mechanics of heterogeneous and dispersed systems, *International Journal of Multiphase Flow* **5**, 353–385.

31. Nourgaliev, R.R., Dinh, T.N. and Theofanous, T.G., 2004, A pseudo-compressibility method for the numerical simulation of incompressible multifluid flows, *International Journal of Multiphase Flow* **30**(7–8), 901–937.

32. Nourgaliev, R.R., Dinh, T.N. and Theofanous, T.G., 2004, Direct numerical simulation of compressible multiphase flows: Interaction of shock waves with dispersed multimaterial

media, in *Proceedings of the 5th International Conference on Multiphase Flow, ICMF'04*, Yokohama, Japan, May 30–June 4, CD-Rom, Paper No. 494.

33. Nourgaliev, R.R., Dinh, T.N. and Theofanous, T.G., 2004, On modeling of collisions in direct numerical simulation of high-speed multiphase flows, in *Proceedings of the 3rd International Conference on Computational Fluid Dynamics, ICCFD'3*, July 12–16, Toronto, Canada.

34. Nourgaliev, R.R., Dinh, T.N. and Theofanous, T.G., 2006, Adaptive characteristics-based matching for compressible multifluid dynamics, *Journal of Computational Physics*, in press.

35. Nourgaliev, R.R., Sushchikh, S.Yu., Dinh, T.N. and Theofanous, T.G., 2005, Shock wave refraction patterns at interfaces, *International Journal of Multiphase Flow* **31**(9), 969–995.

36. Nourgaliev, R.R., Dinh, T.N. and Theofanous, T.G., 2005, Sharp treatment of surface tension and viscous stresses in multifluid dynamics, AIAA 2005-5349, *17th AIAA Computational Fluid Dynamics Conference*, June 6–9, Toronto, Canada.

37. Osher, S. and Sethian, J.A., 1998, Fronts propagating with curvature-dependent speed: Algorithms based on Hamilton–Jacobi formulations, *Journal of Computational Physics* **79**, 12–49.

38. Quirk, J.J. and Karni, S., 1996, On the dynamics of a shock-bubble interaction, *Journal of Fluid Mechanics* **318**, 129–163.

39. Prosperetti, A., 2003, Two-fluid modeling and averaged equations, *Multiphase Science and Technology* **5**(1–4).

40. Ranger, A.A. and Nicholls, J.A., 1969, Aerodynamic shattering of liquid drops, *AIAA Journal* **7**, 285–290.

41. Saurel, R. and Abgrall, R., 1999, A simple method for compressible multifluid flows, *SIAM Journal on Scientific Computing* **21**(3), 1115–1145.

42. Taylor, G.I., 1963, The shape and acceleration of a drop in a high-speed air stream, in *The Scientific Papers of Sir Geoffrey Ingram Taylor, 3*, G.K. Batchelor (ed.), University Press, Cambridge.

43. Theofanous, T.G., Li, G.J. and Dinh, T.N., 2004, Aerobreakup in rarefied supersonic gas flows, *Transactions of the ASME, Journal of Fluids Engineering* **126**, 516–527.

44. Youngs, D.L., 1994, Numerical simulation of mixing by Rayleigh–Taylor and Richtmyer–Meshkov instabilities, *Laser and Particle Beams* **12**(4), 725–750.

PART V

LARGE EDDY SIMULATIONS, APPLICATIONS AND OTHER PHYSICS

On Stochastic Modeling of Heavy Particle Dispersion in Large-Eddy Simulation of Two-Phase Turbulent Flow

Babak Shotorban[1] and Farzad Mashayek[2]

[1] Center for Simulation of Advanced Rockets, University of Illinois at Urbana-Champaign, 1304 West Springfield Avenue, Urbana, IL 61801, USA; e-mail: shotorba@uiuc.edu
[2] Department of Mechanical & Industrial Engineering, University of Illinois at Chicago, 842 West Taylor Street, Chicago, IL 60607, USA; e-mail: mashayek@uic.edu

Abstract. The effect of subgrid scales on the dispersion of heavy particles could be significant especially when the subgrid energy content is not negligible and/or the particle time constant is small. In this work, a modified Langevin type equation is used to reconstruct the instantaneous velocity of the seen fluid particle which is needed in the particle momentum equation. To assess the model, a decaying isotropic turbulence is studied via *a priori* test. A good agreement between the model and DNS results is observed.

1 Introduction

Large-eddy simulation (LES) has been widely used for more than a decade to study two-phase flows in which a large number of particles are dispersed in a turbulent carrier phase [3]. The LES of single-phase turbulence by itself is a challenging task, particularly when dealing with wall boundaries at high Reynolds numbers. The presence of particles adds significantly to this challenge.

The common practice in the LES of particle-laden turbulent flows is to simulate the dispersed phase in the Lagrangian framework by individually tracking the particles and solving their Lagrangian equations. To solve these equations, the instantaneous field quantities of the carrier phase are required; however, only the resolved (filtered) quantities are available in LES. Most of the previous studies have used the resolved, instead of the instantaneous velocities to solve the particle equations, thus neglecting the effect of the subgrid scales on particles. Armenio et al. [1] and Shotorban [8] showed via *a priori* and *a posteriori* tests that this assumption could result in less accurate predictions of the dispersed phase. Furthermore, they concluded that the neglect of subgrid-scale effects on particles is more critical when the subgrid kinetic energy is significant and/or the particle time constant is small.

Only a few models are reported by which the effect of the subgrid scales on particles can be taken into account. Wang and Squires [12] proposed to model the subgrid-scale velocities using a Gaussian random variable scaled by a velocity scale

S. Balachandar and A. Prosperetti (eds), Proceedings of the IUTAM Symposium on Computational Multiphase Flow, 373–380.

obtained by the transport equation for the subgrid-scale kinetic energy. A similar model was employed by Sankaran and Menon [7] in the case of reacting droplets. The main drawback of this model is that its stochastic approach ignores the time correlation in the particle trajectory. In a model proposed in [8], it is assumed that the particle position and velocity and the velocity of the fluid particle seen by the particle (referred to as the "seen" fluid particle hereinafter) evolve based on a Langevin type equation. The statistics generated by this model were in a good agreement with those obtained by DNS for particles with small time constants. However, some discrepancy was observed between the model and DNS results for particles with large time constants. Okong'o and Bellan [4, 5] proposed an efficient deterministic method to model the subgrid-scale quantities in the droplet-laden flows. Despite good predictions for mixing layers, this model does not satisfy the Galilean invariance which is a symmetry property for the subgrid-scale components of velocities. Recently, Shotorban and Mashayek [10] proposed to model the subgrid-scale effects on particles by approximate deconvolution [11]. Using this model the represented modes in LES is reconstructed for the use in the particle momentum equations.

In this manuscript we present a modified version of our previously proposed stochastic model [8]. This modification is needed for particles with large time constants.

2 Governing equations

In the particle-laden flow considered in this study, the carrier phase is an incompressible Newtonian fluid and the dispersed phase is composed of a large number of spherical particles with equal diameters much smaller than the smallest length scale of the carrier flow. It is also assumed that the global number density of particles is small such that the effect of particles on the carrier phase can be neglected (one-way coupling assumption).

2.1 Carrier-phase equations

Considering the non-dimensional Navier–Stokes equations as the governing equations of the carrier phase and applying a spatial filter on them results in the filtered Navier–Stokes equations

$$\frac{\partial \bar{u}_i}{\partial x_i} = 0, \tag{1}$$

$$\frac{\partial \bar{u}_i}{\partial t} + \bar{u}_j \frac{\partial \bar{u}_i}{\partial x_j} = -\frac{\partial \bar{p}}{\partial x_i} + \frac{1}{\text{Re}_0} \frac{\partial^2 \bar{u}_i}{\partial x_j \partial x_j} - \frac{\partial \tau_{ij}}{\partial x_j}, \tag{2}$$

where the filter operator $\bar{\cdot}$ on any variable ϕ is defined as

$$\bar{\phi}(\mathbf{x}, t) = \int_{-\infty}^{\infty} \phi(\mathbf{x}', t) G(\mathbf{x}' - \mathbf{x}; \bar{\Delta}) d\mathbf{x}', \tag{3}$$

where G denotes the filter kernel with $\overline{\Delta}$ as the filter size. G also has the property that if $\overline{\phi} = \phi$ then ϕ is spatially uniform. In Equation (2), Re_0 is a reference Reynolds number and

$$\tau_{ij} = \overline{u_i u_j} - \overline{u}_i \overline{u}_j, \tag{4}$$

is the subgrid scale stress. The term $\overline{u_i u_j}$, appeared in Equation (4), renders the set of the carrier-phase equations unclosed. Various models are available for the closure problem in LES [6].

2.2 Dispersed-phase equations

The governing equations for particle position x_{pi}, and velocity u_{pi}, in the Lagrangian frame are, respectively,

$$\frac{dx_{pi}}{dt} = u_{pi}, \tag{5}$$

$$\frac{du_{pi}}{dt} = \frac{f_1}{\tau_p}(u_{si} - u_{pi}), \tag{6}$$

where u_{si} denotes the instantaneous velocity of the seen fluid particle, i.e. $u_{si}(t) = u_i(x_{pi}(t), t)$, $\tau_p = \text{Re}_0 \rho_p d_p^2 / 18$ is the particle time constant and $f_1 = 1 + 0.15\text{Re}_p^{0.687}$ is an empirical correlation used to modify the Stokes drag for large Re_p. The particle Reynolds number is defined as $\text{Re}_p = \text{Re}_0 \sqrt{v_{di} v_{di}} d_p$, with ρ_p, d_p and $v_{di} = u_{pi} - u_{si}$ denoting the density, diameter and the relative velocity of the particle, respectively. In Equation (6), the terms due to unsteady drag, added mass and Basset history forces are absent because their effects on particles are negligible when the density ratio of particle to fluid is high (~ 1000). Also in this work, it is assumed that the gravity force is negligible.

3 Stochastic modeling of dispersed phase

In the absence of the instantaneous velocities in LES, one needs to reconstruct them from the filtered velocities before solving Equation (6). In this work, we use a stochastic approach to model these velocities.

A Langevin type equation can be used to model the fluid particle evolution in LES

$$du_{fi} = \left[-\frac{\partial \overline{p}}{\partial x_i} + \frac{1}{\text{Re}_0} \frac{\partial^2 \overline{u}_i}{\partial x_j \partial x_j} - \frac{1}{T_L}(u_{fi} - \overline{u}_i) \right] dt + \sqrt{C_0 \varepsilon} dW_i, \tag{7}$$

where T_L is an appropriate time scale representing the time scale of the flow subgrid scales, ε is the dissipation rate of the subgrid-scale kinetic energy, and W_i is Wiener process. Equation (7) was originally proposed and implemented by Gicquel et al. [2] for the Monte-Carlo simulation of single-phase turbulence. Shotorban et al. [9] and Shotorban [8] carried out assessment studies on the application of this equation in two-phase turbulence via *a priori* and *a posteriori* tests, assuming $u_{fi} \approx u_{si}$. This assumption may not be accurate enough due to the inertial effect of the particle. The tests were conducted in decaying isotropic turbulence for particles with different time constants. Good predictions by the model were observed for particles with small time constants; however, a discrepancy between the model and DNS results was noted for large particle time constants. This discrepancy is believed to be due to the particle inertial effect which is not taken into account in Equation (7).

Another issue involved in this stochastic approach is the modeling of T_L and ε. Gicquel et al. [2] modeled T_L and ε as

$$T_L = k / \left(\frac{1}{2} + \frac{3}{4} C_0 \right) \varepsilon, \quad \varepsilon = C_\varepsilon \frac{k^{3/2}}{\overline{\Delta}}, \tag{8}$$

where k is the subgrid-scale kinetic energy. In the LES of incompressible flows, the common practice is to model only the anisotropic part of the subgrid-scale stress tensor because the deviatoric part, which represents the subgrid scale kinetic energy, can be absorbed in the pressure term. Shotorban [8] employed Yoshizawa's model [13] for subgrid-scale kinetic energy

$$k = C_I \overline{\Delta}^2 |\bar{s}|^2, \tag{9}$$

where C_I is a model constant, which can be dynamically computed. The performance of this model for calculating the subgrid-scale kinetic energy is acceptable in isotropic turbulence [8]. However, a more accurate model may be required for more complex configurations. One possibility is to solve a transport equation for the subgrid-scale kinetic energy [12].

A modification is needed in Equation (7) to make it applicable for particles with large time constants. It is shown that Equation (7) is strictly valid when $\tau_p \rightarrow 0$ which could be considered as a limit that the particle behaves similar to the fluid tracer particle. On the other hand, it can be assumed that $u_{si} \rightarrow \bar{u}_{si}$ when $\tau_p \rightarrow \infty$. Therefore, the effect of the subgrid scales on the inertial particles can be neglected at this limit. This assumption is physically sound because particles with large time constants respond mainly to the large scales of turbulence. If one assumes that T_L represents the time scale of subgrid scales in LES, then τ_p / T_L is a non-dimensional number representing the interaction between subgrid scales and particle scales. For $\tau_p / T_L \ll 1$, Equation (7) can directly be used as an equation for the evolution of the seen fluid particle. To make Equation (7) also applicable for $\tau_p / T_L \gg 1$, we modify this equation as

$$du_{si} = \left\{ -\frac{\partial \overline{p}}{\partial x_i} + \frac{1}{Re_0} \frac{\partial^2 \overline{u}_i}{\partial x_j \partial x_j} - \frac{1}{T_L} \left[1 + g \left(\frac{\tau_p}{T_L} \right) \right] (u_{si} - \overline{u}_{si}) \right\} dt$$
$$+ \sqrt{C_0 \varepsilon} dW_i, \tag{10}$$

where

$$g(\alpha) = \begin{cases} 0 & \text{if } \alpha \ll 1, \\ \infty & \text{if } \alpha \gg 1. \end{cases} \tag{11}$$

is introduced in an *ad hoc* manner. Various functions can be considered which satisfy the conditions in (11). In this work $g(\alpha) = \alpha^2$ is used.

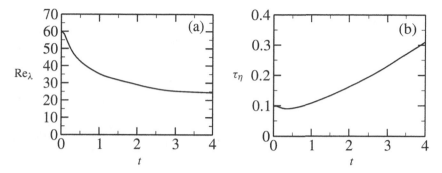

Fig. 1. (a) Variation of the Taylor-scale Reynolds number. (b) Variation of the Kolmogrov time scale.

4 Model assessment and results

To assess the performance of the model, the particle-laden decaying isotropic turbulence is considered. The carrier phase is initialized with a solenoidal random velocity using a Gaussian distribution in a box with $(2\pi)^3$ dimensions and 128^3 resolution. The initial energy distribution is $E_0(\kappa) \sim (\kappa/\kappa_m)\exp(-\kappa/\kappa_m)$ where κ is wave-number and κ_m is a wave-number for which $E_0(\kappa)$ is maximum. In this study $\kappa_m = 2.8$, the initial turbulence kinetic energy $\frac{1}{2}\langle u_i u_i \rangle = 1$ and $Re_0 = 240$. The evolution of the Taylor-scale Reynolds number and Kolmogorov time scale is presented in Figure 1. A large number of particles with $\tau_p = 0.2$ are also released with a uniform random distribution in the computational domain with initial velocities equal to their seen fluid particle velocities.

The model assessment is carried out via *a priori* test, i.e. direct numerical simulation is conducted and the instantaneous velocities are filtered at every time step to obtain the filtered velocities. The filter used is Gaussian with a filter size four times the grid spacing. Four different groups of particles are tracked simultaneously in the same simulation. Since only one-way coupling is considered and the carrier phase is not modified by particles, the statistics of all groups are independently obtained from the same realization of the carrier phase. These groups of particles are independently tracked using the instantaneous velocities obtained from DNS, the filtered velocities, the filtered velocities along with Equation (7) for the seen fluid particle and the filtered velocities along with Equation (10) for the seen fluid particles. These groups are denoted by "DNS", "filtered", "model" and "modified model", respectively.

Figure 2 shows the evolution of the turbulence kinetic energies of the particles and the seen fluid particles. The turbulence kinetic energies of both particles and the seen fluid particles, are under-predicted by "filtered" while it is over-predicted by "model". This under-prediction by "filtered" is due to the fact that in this case particles are driven by the carrier-phase filtered velocities which possess less amount of energy than the carrier-phase instantaneous velocities. It is also seen in this figure that "modified model" can precisely predict the DNS results. It is noted that the underprediction of energy at $t = 0$ in "filtered", "model" and "modified model"

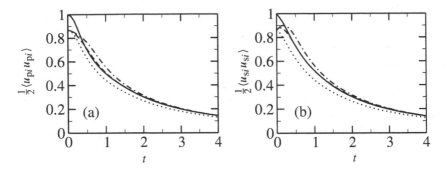

Fig. 2. (a) Turbulence kinetic energy of particles and (b) turbulence kinetic energy of the seen fluid particles for DNS (solid line), filtered (dotted line), model (dashed-dotted line) and modified model (dashed line).

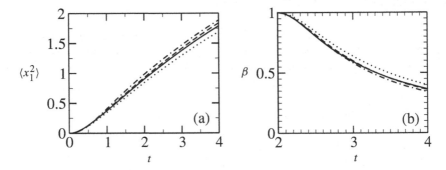

Fig. 3. (a) Mean square displacement of particles and (b) Lagrangian autocorrelation of particles for DNS (solid line), filtered (dotted line), model (dashed-dotted line) and modified model (dashed line).

cases, is due to the fact that particles are initially released with velocities equal to the filtered velocities of the carrier phase at the location of particles in these cases.

The time variation of the particle mean square displacement and Lagrangian autocorrelation are shown in Figure 3. These statistics are defined by

$$\langle x_{p1}^2 \rangle = \langle (x_{p1}(t) - x_{p1}(t_0))^2 \rangle, \tag{12}$$

and

$$\beta(t_1, t) = \frac{\langle u_{p1}(t) u_{p1}(t_1) \rangle}{\langle u_{p1}^2(t) \rangle^{1/2} \langle u_{p1}^2(t_1) \rangle^{1/2}}, \tag{13}$$

respectively. Here, $t_0 = 0$ and $t_1 = 2$ are considered. As can be seen in this figure, "modified model" predicts the DNS results better than "model" and "filtered".

5 Conclusions

The previously proposed stochastic model [8] is modified for simulation of particles with lager time constants. The main argument for this modification is that particles with large time constants interact with large scales of turbulence. Nevertheless, the model is also valid for particles with small time constants. The new model is validated in isotropic turbulence via *a priori* tests where the particle turbulence kinetic energy, the seen fluid particle turbulence kinetic energy, the mean square displacement of particles and their Lagrangian autocorrelation obtained by the new model are compared against those obtained by DNS. It is shown that the accuracy of the statistics obtained by the original model and those obtained by only the filtered velocities is less than those from the modified model.

Acknowledgements

This work was supported in part by the National Science Foundation through grant CTS-0237951 with Dr. T. J. Mountziaris as the Program Director. The Center for Simulation of Advanced Rockets is supported by the U.S. Department of Energy through the University of California under subcontract B523819.

References

1. Armenio, V., Piomelli, U. and Fiorotto, V., 1999, Effect of the subgrid scales on particles motion, *Phys. Fluids* **11**, 3030–3042.
2. Gicquel, L., Givi, P., Jaberi, F. and Pope, S., 2002, Velocity filtered density function for large eddy simulation of turbulentflows, *Phys. Fluids* **14**(3), 1196–1213.
3. Mashayek, F. and Pandya, R.V.R., 2003, Analytical description of particle/droplet-laden turbulent flows, *Prog. Energy Combust. Sci.* **29**(4), 329–378.
4. Okong'o, N. and Bellan, J., 2000, *A Priori* subgrid analysis of temporal mixing layers with evaporating droplets, *Phys. Fluids* **12**(6), 1573–1591.
5. Okong'o, N. and Bellan, J., 2004, Consistent large-eddy simulation of a temporal mixing layer laden with evaporating drops. Part 1. Direct numerical simulation, formulation and a priori analysis, *J. Fluid Mech.* **499**, 1–47.
6. Sagaut, P., 2002, *Large Eddy Simulation for Incompressible Flows: An Introduction*, 2nd edition, Springer-Verlag, Berlin.
7. Sankaran, V. and Menon, S., 2002, LES of spray combustion in swirling flows, *J. Turbulence* **3** (Art. No. 11).
8. Shotorban, B., 2005, *Modeling of Subgrid-scale Effects on Particles in Large-eddy Simulation of Turbulent Two-phase Flows*, Ph.D. Thesis, Universtiy of Illinois at Chicago, Chicago, IL.
9. Shotorban, B., Afshari, A., Jaberi, F.A. and Mashayek, F., 2004, A particle tracking algorithm for LES of two-phase flow, AIAA Paper 2004-0332.
10. Shotorban, B. and Mashayek, F., 2005, Modeling subgrid-scale effects on particles by approximate deconvolution, *Phys. Fluids* **17**(8), 081701.

11. Stolz, S., Adams, N.A. and Kleiser, L., 2001, An approximate deconvolution for large-eddy simulation with application to incompressible wall-bounded flows, *Phys. Fluids* **13**(4), 997–1015.
12. Wang, Q. and Squires, K.D., 1996, Large eddy simulation of particle-laden turbulent channel flow, *Phys. Fluids* **8**(5), 1207–1223.
13. Yoshizawa, A., 1986, Statistical theory for compressible turblent shear flows, with the application to subgrid modeling, *Phys. Fluids* **29**, 2152–2164.

Flow of Bubbly Liquids in a Vertical Pipe: Theory and Experiments

A.S. Sangani[1], S.S. Ozarkar[1], Y.H. Tsang[2] and D.L. Koch[2]

[1]*Department of Biomedical and Chemical Engineering, Syracuse University, Syracuse, NY 13244, USA*
[2]*Department of Chemical and Biomolecular Engineering, Cornell University, Ithaca, NY 14853, USA*

Abstract. Recent measurements of bubble-phase velocity, volume fraction, and velocity variance, and liquid velocity profiles for flows of bubbly liquids in a vertical pipe are shown to be in good agreement with the predictions based on averaged equations.

1 Introduction

The problem of deriving equations of motion of particles or bubbles suspended in a liquid has been the subject of many investigations over the past 40 years. Much progress has been made in the last two decades with the development of computers and efficient computational algorithms that allow numerical simulation of particle interactions. These simulations provide valuable insight into how the microstructure of the suspension depends on the nature of the flow, and how the suspension properties depend on the microstructure and microscale physics. Our analytical efforts were devoted to two special cases: (i) particles in a gas for which the particle Reynolds number is small compared with unity but the Stokes number, which is the product of particle to gas density ratio and the Reynolds number, is $O(1)$ [4, 9]; and (ii) gas bubbles in a liquid for which the bubble Reynolds number is large compared to unity but the Weber number is small such that the bubbles are approximately spherical [3, 5, 7, 8, 10]. In both cases, even though the particle scale inertial effects are significant, the particle interactions are governed by simplified forms of Navier–Stokes equations: the gas-solid suspensions can be simulated with Stokes equations of motion, and the gas-liquid suspensions by the potential flow equations. In both cases, it was possible to combine the results of numerical simulations with suitable kinetic theory to derive averaged equations of motion.

For the case of bubbly liquids a crucial assumption is the validity of the potential flow approximation. Potential flow approximation is shown to yield accurate results for a single bubble motion when the Weber number is $O(1)$ or smaller. However, its validity to bubble suspensions is not yet established. Since the conditions of large Reynolds and small Weber numbers in a liquid free of surface-active impurities are

S. Balachandar and A. Prosperetti (eds), Proceedings of the IUTAM Symposium on Computational Multiphase Flow, 381–392.

not satisfied in many of the reported experimental investigations of bubbly liquids, we began an experimental program to study flows of bubbly liquids at Cornell University. The present communication gives an overview of the theory and comparison with experiments. The experiments for vertical and inclined channel have been reported elsewhere [13, 14]; here, we shall focus on recent experiments on flows of bubbly liquids in a vertical pipe.

Section 2 reviews our past work on averaged equations of motion with a few modifications and corrections. Section 3 gives the boundary conditions for bubbly liquid flows, and Section 4 a comparison with the experiments.

2 Averaged equations

Averaged equations for non-coalescing, monodispersed, bubble suspensions under potential flow conditions were derived in [8]. The equations consist of continuity and momentum equations for the gas-liquid mixture and for the bubble phase treated as a continuum. An expression was derived for the bubble-phase stress (later corrected by Bulthuis et al. [2]) in terms of bubble velocity distribution and inter-bubble forces. The stress depends on the mean relative velocity of the bubbles and the bubble-phase temperature, defined as one-third the velocity variance of the bubbles. An equation for determining temperature was also proposed. Computations of dispersed-phase stress and other average properties of bubbly liquids have been made for two relatively simple cases: (i) flow generated by the buoyancy force acting on the bubbles [7] and (ii) bubbly liquids subjected to simple shear flow in the absence of buoyancy force [3]. Numerical simulations for the first case with periodic boundary conditions indicated that the bubbles form clusters in horizontal plane that span the unit cell width, suggesting thereby that the homogeneous state of bubbly liquids is unstable. The clustering was absent in the second case and it was shown that the kinetic theory can be used to determine constitutive relations for bubbly liquids. Averaged equations of bubbly liquids must account for the fact that the microstructure, and, hence, the average properties as well as constitutive relations differ for different imposed flows. Since the microstructure is a strong function of the magnitude of velocity fluctuations compared to the mean bubble relative velocity, Spelt and Sangani [10] carried out detailed potential flow simulations at various volume fractions and ratios of bubble-phase velocity variance and mean bubble relative velocity. These investigators also presented simplified set of equations that incorporated the results of their simulations and the kinetic theory of sheared bubbly liquids developed in [3].

The aforementioned discussion was limited to spherical bubbles. The potential flow approximation may also be used to predict virtual mass and viscous drag on a bubble at small but finite Weber numbers for which the bubble aspect ratio is less than about 2. Simulations for finite Weber numbers were carried out by Kushch et al. [5] who determined added mass and viscous drag coefficients and aspect ratio of bubbles as functions of volume fraction and Weber number. Their results were limited to a special case of randomly distributed oriented bubbles and would therefore be appropriate when the clustering is absent. Effects of clustering and bubble deformation may be combined in an ad-hoc manner as suggested in [5]. The resulting averaged equations shall be presented elsewhere. Here, we summarize the equations proposed by Spelt and Sangani [10] with a few modifications and corrections (as stated below). The averaged equations are as follows.

Equations for the gas-liquid mixture:

$$\frac{\partial U_i}{\partial x_i} = 0, \tag{1}$$

$$\frac{\partial}{\partial t}((1-\phi)U_i^L) + \frac{\partial}{\partial x_j}((1-\phi)U_i^L U_j^L) = -\frac{1}{\rho}\frac{\partial P}{\partial x_i} + (\langle\phi\rangle - \phi)g_i - \frac{1}{\rho}\frac{\partial}{\partial x_j}\hat{\Sigma}_{ij}. \tag{2}$$

Equations for the bubble-phase continuum:

$$\frac{\partial \phi}{\partial t} + \frac{\partial}{\partial x_i}(\phi V_i) = 0, \tag{3}$$

$$\frac{dI_i}{dt} = -\frac{1}{n}\frac{\partial P_{ij}}{\partial x_j} - mg_i - 12\pi\mu a C_d V_i + m\frac{DU_i}{Dt} - \gamma_{ji}I_j$$

$$+ \rho\phi\left((1+1/2C_a)V_k\frac{\partial}{\partial x_k}[\phi(1+1/2C_a)V_i]\right), \tag{4}$$

$$\frac{3}{2}(\rho/2)\phi\frac{dT}{dt} = -\frac{\partial Q_j}{\partial x_j} - P_{ij}^* e_{ij} - 36\pi\mu an(R_{\text{diss}}T - \xi V^2). \tag{5}$$

Here, U_i is the mixture velocity, U_i^L is the liquid velocity, ρ and μ are, respectively, the density and viscosity of the liquid, P is the mixture pressure, ϕ is the bubble volume fraction, n is the number density of the bubbles, $V_i = U_i^G - U_i$ is the bubble relative velocity, U_i^G is the velocity of the bubbles, I_i is the mean impulse (virtual momentum) of the bubbles due to their relative motion,

$$\frac{D}{Dt} = \frac{\partial}{\partial t} + U_j\frac{\partial}{\partial x_j}$$

is the derivative following the mixture motion,

$$\frac{d}{dt} = \frac{\partial}{\partial t} + (U_j + V_j)\frac{\partial}{\partial x_j}$$

is the derivative following the motion of the gas phase, T is the bubble-phase temperature, P_{ij} is the bubble-phase stress, $\gamma_{ij} = \partial U_i/\partial x_j$ is the mean mixture velocity gradient, e_{ij} is the rate of strain tensor given by $e_{ij} = (\gamma_{ij}^p + \gamma_{ji}^p)/2$ where $\gamma_{ij}^p = \gamma_{ij} + \partial(C_a V_i)/\partial x_j$, Q_j is the flux of fluctuation energy.

Equation (4) represents the momentum balance for the bubble phase. The second and third term on the rhs of this equation represent, respectively, the buoyancy and drag forces. The next two terms represent force due to mixture velocity variation in time and space. The force is exact when the curl of mixture velocity is zero. It also agrees with the lift force on a single spherical bubble in weak shear flow as given by Auton [1]. Finally, the last term, equivalent to the ponderomotive force in the theory of electrostatics, occurs due to interaction of dipole induced by the relative motion of the bubbles with the back flow.

Equation (5) represents the balance in fluctuation energy of bubbles' motion. It includes the sink term due to viscous energy dissipation and the source terms due to shear and the nonzero mean relative velocity of the bubbles. The last one could not be determined from numerical simulations due to excessive clustering seen in simulations of bubbles rising due to gravity [7]. We used experimental measurement of bubble-phase temperature for bubbles rising in a vertical channel in [13] to estimate ξ as given by $\xi = 0.02 + 0.45\phi$. The clustering observed in the experiments was much smaller than that observed in dynamic simulations with periodic boundary conditions, and it was conjectured that this may be due to the presence of the channel walls and the channel width-scale fluctuation motion resulting from the clusters breaking up. ξ may therefore depend on the channel width.

The bubble impulse is related to the mean bubble relative velocity by $I_i = m/2C_a V_i$ where $m = 4\pi a^3 \rho/3$ is the mass of the liquid displaced by the bubble. The added mass coefficient for spherical bubbles is estimated using

$$C_a = \frac{1 + 2\phi + \frac{9}{40}\phi A}{1 - \phi}.$$ (6)

Here, $A = V^2/T$. The spatial distribution of the bubbles becomes more uniform as A decreases. The bubble-phase stress is given by

$$P_{ij} = (\rho/2)\phi T(1 + 4\chi\phi)\delta_{ij} - [\kappa - 2/3\mu_s]e_{kk}\delta_{ij} - 2\mu_s e_{ij}.$$ (7)

In [10] an additional term referred to as the Maxwell stress was given. The force due to Maxwell stress is the ponderomotive force referred to earlier. The bubble-phase viscosity may be estimated using

$$\mu_s = \frac{8\rho}{5\pi^{1/2}}(\phi)aT^{1/2}\phi^2\chi\left[1 + \frac{\pi}{12}\left(1 + \frac{5}{8\phi\chi}\right)^2\right] + \mu\left(1 + \frac{5}{3}\phi\right).$$ (8)

Here, we have added the viscosity of the liquid to the expression given by Spelt and Sangani [10]. The kinetic theory is applicable only when the Reynolds number based on shear is large for which the transport of bubble-phase momentum by collision and fluctuations in bubble velocity is dominant. Expressions for χ, fluctuation energy flux, conductivity, viscous drag coefficient, mixture stress, and other terms may be found in [10].

3 Boundary conditions

The potential flow approximation cannot be used for large-scale dynamic simulations of bubble suspensions confined by walls as it fails in predicting bubble-wall interactions. We have used experimental observations on bubble wall interactions by Tsao and Koch [11] to determine approximate boundary conditions for the bubble-phase. The bubble motion near a wall depends on its size and the angle between the wall and the bubble velocity. In general, the component of the velocity parallel to the wall decreases as a result of bubble-wall encounter while the component perpendicular to the wall may remain same or even increase. We have carried out dynamic simulations for a model of bubble suspension in which the bubble velocity and kinetic energy associated with its motion satisfy following collision rule:

$$V_{t,+} = \alpha V_{t,-}, \quad V_+^2 = \gamma V_-^2. \tag{9}$$

Here $-$ denotes before the bubble-wall collision while $+$ denotes after the collision, t refers to the tangential component of the bubble velocity. The parameters α and γ are assumed to be independent of the angle of approach of the bubble. To isolate the effect of bubble-wall interactions, we carried out simulations in which the drag and added mass coefficients for all bubbles are identical and equal to unity. This is equivalent to neglecting hydrodynamic interactions, and if it were not for the bubble-wall interactions, all bubbles would rise with identical velocity and the bubble-phase temperature will be zero. Dynamic simulations were carried out for this model bubble suspension confined by two parallel plates. Periodic boundary conditions were used in the other two directions. Profiles of bubble velocity, temperature, stress, etc. were determined as a function of the lateral position in the channel. Results from one representative simulation are shown in Figure 1.

The profiles thus obtained using direct numerical simulations were compared with those obtained by solving the averaged bubble-phase equations presented in the previous section together with the boundary conditions

$$V_n = 0, \tag{10}$$

$$\frac{\rho \phi \chi_w (1 - \alpha) T^{1/2}}{(2\pi)^{1/2}(1 - \phi)} V_t = -\mu_s \frac{\partial V_t}{\partial x_n}, \tag{11}$$

$$k \frac{\partial T}{\partial x_n} = \frac{\phi}{2} \chi_w \left(\frac{T}{2\pi}\right)^{1/2} [(\gamma - 2\alpha + 1)V_t^2 + 4(\gamma - 1)T], \tag{12}$$

where the subscripts n and t represent the components normal and parallel to the wall, respectively, and χ_w is given by

$$\chi_w = \frac{1 + \phi + \phi^2}{(1 - \phi)^2}. \tag{13}$$

These boundary conditions were derived by assuming that the velocity distribution of bubbles near the wall is Maxwellian and determining the rate of momentum and energy flux due to bubble-wall collisions. The dashed lines in Figure 1 correspond to the predictions based on averaged equations with the above boundary conditions. These predictions agree well with the results obtained by dynamic simulation. The bubble-wall interaction is the main source of velocity fluctuations, and consequently, the bubble-phase temperature is greatest at the walls. The bubble-phase

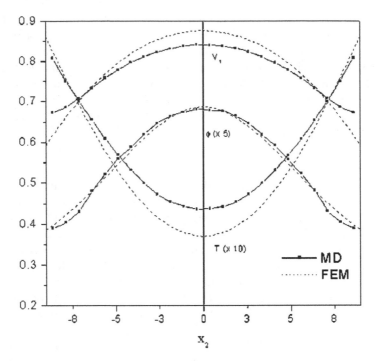

Fig. 1. Comparison between averaged equation theory (FEM) and numerical simulations (MD) for a model bubble suspension.

pressure drives the bubbles towards the center of the channel leading to a maximum in the volume fraction there.

More detailed inspection of various quantities such as shear stress and fluctuation energy flux does indicate some discrepancies. For example, the fluctuations in the velocity component parallel to the channel walls are much greater than those in the direction perpendicular to the walls. In other words, the velocity variance is anisotropic. Also the velocity distribution is non-Maxwellian, especially near the channel walls where it is significantly bimodal as the velocity distribution of the bubbles approaching the walls is quite different from those bouncing back from the wall. In [6], a kinetic theory is developed that accounts for both the anisotropy and the bimodal nature of velocity fluctuations. The resulting equations, however, are far more complicated, and therefore we shall use the simpler description based on isotropic Maxwellian distribution.

We also need to specify the boundary conditions for the mixture equations. One obvious condition is that the component of the mixture velocity normal to the rigid, nonporous walls must be zero. We allow the tangential component to have nonzero slip. Assuming that we have a layer of thickness δ of bubble-free liquid near the wall and the velocity of the liquid at the edge of the layer is U_t^L, continuity of the tangential stress at the interface of bubble suspension and the clear liquid layer requires

Fig. 2. Image of bubbles in the pipe at a typical volume fraction of 0.072. The mean equivalent diameter is 1.01 mm and the standard deviation is 0.08 mm. The mean aspect ratio is about 1.2. The terminal velocity of this size of bubble in clean water is about 20 cm/s. Re_L is 1100 (Tsang [12]).

$$U_t^L = \frac{-\delta \mu_s}{\mu} \frac{\partial U_t^L}{\partial x_n}. \tag{14}$$

We may take $\delta = \lambda a$ in the above equation to provide an effective boundary condition for the slip in the tangential component of the velocity. The predicted results appear to agree well with the experiments if we take $\lambda = 0.5$.

4 Comparison with experiments

We shall compare the theoretical predictions for a bubbly liquid flow in a vertical pipe with the experiments recently carried out by Tsang [12]. MgSO$_4$ was added to deionized water to prevent bubbles from coalescing. Figure 2 shows a photograph of a bubbly liquid. We see that bubbles are approximately spherical and uniform in size.

Averaged equations along with the boundary conditions for the flows of bubbly liquids in a vertical pipe were solved numerically using a Chebyshev pseudo-spectral collocation technique. Tsang [12] performed experiments for seven different flow conditions: four different liquid Reynolds numbers for $\langle \phi \rangle = 0.02$, two for $\langle \phi \rangle = 0.05$, and one for $\langle \phi \rangle = 0.075$. All results given below are obtained by setting $\alpha = 0.3$, $\gamma = 0.7$ in the boundary conditions given by Equations (11) and (12). The velocities are normalized by the terminal velocity of a single bubble and the radial distances by the bubble radius. The bubble radius depended on the volume fraction and the liquid Reynolds number. The bubble Reynolds number based on its terminal velocity and radius varied in the range 35–75 – a range for which one expects the potential flow based approximation to provide reasonable estimates of the bubble phase properties.

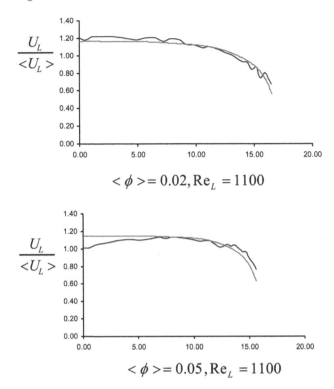

Fig. 3. Variation of ratio of liquid velocity to mean liquid velocity (smooth line represents prediction from theory).

As seen in Figure 3 the liquid velocity profiles are well predicted by the theory. The estimated liquid slip velocity and the predicted ones for all the seven flow conditions are shown in Figure 4. We see that the slip velocity increases with the Reynolds number. The shear stress in the mixture increases with the Reynolds number and this, in turn, causes the slip velocity to increase.

Tsang [12] found that the bubble-phase temperature was approximately constant across the pipe cross-section and reported average bubble-phase temperatures for each of the seven flow conditions. The results are shown in Figure 5. The agreement with the theory predictions is quite good. At higher Reynolds numbers, the shear stress is higher, and this leads to larger bubble-phase temperature. The temperature seems to be relatively insensitive to the bubble volume fraction.

Figures 6 and 7 show the results for the mean relative velocity of the bubbles as functions of Re_L and $\langle \phi \rangle$. We note that the velocities are significantly lower than what would be predicted simply by balancing the viscous drag and buoyancy, as the viscous drag coefficient, C_d, is not much greater than unity for these conditions. The lower velocities are caused by the bubble-phase shear stress gradient. The mixture momentum equation requires that the pressure gradient be balanced by the shear

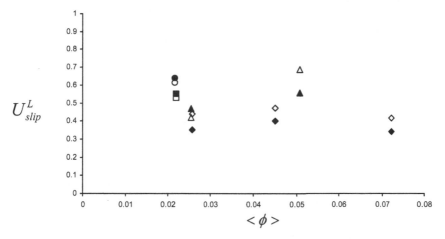

Fig. 4. Liquid slip velocity normalized by the bubble terminal velocity. Open symbols represent experiments and filled represent the theory. diamonds: $Re_L = 1100$; triangles: $Re_L = 1600$; squares: $Re_L = 2100$; circles: $Re_L = 2600$.

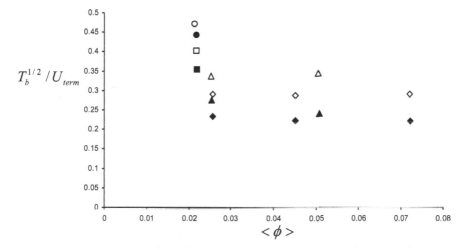

Fig. 5. Bubble-phase temperature. See caption of Figure 4 for symbols.

stress gradient. Since the same shear stress gradient also appears in the bubble-phase equation, higher the pressure gradient, higher is the shear stress force on the bubble, and consequently, lower the bubble-phase relative velocity. For the case of steady, unidirectional flow, the bubble-phase and mixture momentum equations can be combined to yield

$$\frac{\partial P}{\partial z} + \rho \langle \phi \rangle g = 12\pi \mu a n C_d V. \tag{15}$$

Thus, the effect of negative pressure gradient required for the upward liquid flow is to effectively reduce the force on the bubbles.

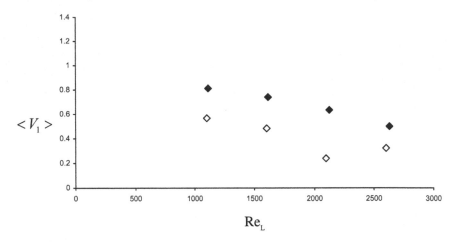

Fig. 6. Variation of mean relative velocity of bubbles normalized by the bubble terminal velocity with liquid Reynolds number, Re_L, for $\langle \phi \rangle = 0.02$. Open symbols represent experiments and filled represent the theory.

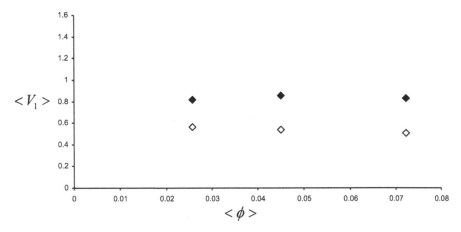

Fig. 7. Variation of mean normalized relative velocity of bubbles with volume fraction for $Re_L = 1100$.

Finally, Figure 8 shows the comparison for the bubble volume fraction profiles. The lift force causes the bubbles to accumulate near the wall. The bubble-phase pressure gradient is not large enough so that the peak in the volume fraction occurs within about one diameter of the bubbles. The experiments give peaks that are more diffused and at greater distance from the wall, especially for $\langle \phi \rangle = 0.02$. Position of the peak seems to be relatively insensitive to the bubbles-wall interaction parameters α and γ. Most probable cause for the discrepancy is the simplified boundary conditions (cf. Equations (12)–(13)) employed here. The assumption of isotropic Maxwellian velocity distribution is not valid at low volume fractions. Bimodal velocity distribution

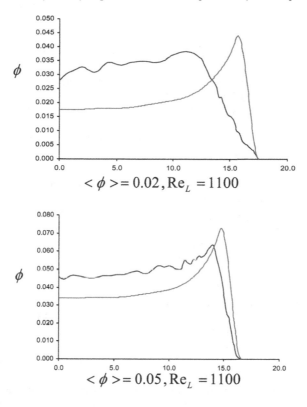

$$< \phi > = 0.02, \mathrm{Re}_L = 1100$$

$$< \phi > = 0.05, \mathrm{Re}_L = 1100$$

Fig. 8. Volume fraction profiles. Smooth lines represent theory.

is likely to cause an effective repulsive wall force that is not included in the present simplified equations.

5 Conclusions

We have presented recent experiments on flows of bubbly liquids in a vertical pipe in which the bubbles were approximately spherical and nearly monodisperse – an ideal case for comparison with the averaged equations based on the potential flow approximation. The theory predictions are shown to agree reasonably well with the experiments.

Acknowledgement

This work was supported by NASA under grant number NAG3-2427.

References

1. Auton, T.R., 1987, The lift force on a spherical body in a rotational flow, *J. Fluid Mech.* **183**, 199–218.
2. Bulthuis, H.F., Prosperetti, A. and Sangani, A.S., 1995, 'Particle stress' in disperse two-phase potential flow, *J. Fluid Mech.* **294**, 1–16.
3. Kang, S.-Y., Sangani, A.S., Tsao, H.-K. and Koch, D.L., 1997, Rheology of dense bubble suspensions, *Phys. Fluids* **9**, 1540–1570.
4. Koch, D.L. and Sangani, A.S., 1999, Particle pressure and marginal stability limits for a homogeneous monodisperse gas-fluidized bed: Kinetic theory and numerical simulations, *J. Fluid Mech.* **400**, 229–263.
5. Kushch, V.I., Sangani, A.S., Spelt, P.D.M. and Koch, D.L., 2002, Finite-Weber-number motion of bubbles through a nearly inviscid liquid, *J. Fluid Mech.* **460**, 241–280.
6. Kushch, V.I., Shah, K.D., Sangani, A.S. and Koch, D.L., 2005, Boundary conditions for bubbly liquids, in preparation.
7. Sangani, A.S. and Didwania, A.K., 1993, Dynamic simulations of flows of bubbly liquids at large Reynolds number, *J. Fluid Mech.* **250**, 307–337.
8. Sangani, A.S. and Didwania, A.K., 1993 Dispersed-phase stress tensor of bubbly liquids at large Reynolds numbers, *J. Fluid Mech.* **248**, 27–54.
9. Sangani, A.S., Mo, G., Tsao, H.-K. and Koch, D.L., 1996, Simple shear flows of dense gas-solid suspensions at finite Stokes numbers, *J. Fluid Mech.* **313**, 309–341.
10. Spelt, P.D.M. and Sangani, A.S., 1998, Properties and averaged equations for flows of bubbly liquids, *Appl. Sci. Res.* **58**, 337–386.
11. Tsao, H.-K. and Koch, D.L., 1997, Observations of high Reynolds number bubbles interacting with a rigid wall, *Phys. Fluids* **9**, 44–56.
12. Tsang, Y.H., 2005, Experimental study of millimeter-sized bubbles in electrolytic solutions – Rise velocity, coalescence, and vertical pipe flow, Ph.D. Thesis, Cornell University.
13. Zenit, R., Koch, D.L. and Sangani, A.S., 2001, Measurements of the average properties of a suspension of bubbles rising in a vertical channel, *J. Fluid Mech.* **429**, 307–342.
14. Zenit, R., Tsang, Y.H., Koch, D.L. and Sangani, A.S., 2004, Shear flow of a suspension of bubbles rising in an inclined channel, *J. Fluid Mech.* **515**, 261–292.

Effect of Particle Inertia in Particulate Density Currents

Mariano I. Cantero[1], S. Balachandar[2], Marcelo H. García[1] and James P. Ferry[3]

[1]*Department of Civil and Environmental Engineering, University of Illinois at Urbana-Champaign, 205 N. Mathews Avenue, Urbana, IL 61801, USA; e-mail: mcantero@uiuc.edu, mhgarcia@uiuc.edu*
[2]*Department of Theoretical and Applied Mechanics, University of Illinois at Urbana-Champaign, 104 S. Wright Street, Urbana, IL 61801, USA; e-mail: s-bala@uiuc.edu*
[3]*Center for the Simulation of Advanced Rockets, University of Illinois at Urbana-Champaign, 1304 W. Springfield Avenue, Urbana, IL 61801, USA; e-mail: jferry@uiuc.edu*

Abstract. In this work we address the effect of particle inertia in particulate density currents. First we introduce a novel two-fluid model based on the equilibrium Eulerian approach [6]. The resulting model captures very important physics of two-phase flows, such as preferential concentration and migration of particles down turbulence gradients (turbophoresis), which modify substantially the structure and dynamics of the flow. We solve the mathematical model with a highly accurate spectral code, capturing all the length and time scales of the flow. We present two-dimensional simulations in planar configuration for Grashof $Gr = 1.5 \times 10^6$. In the simulation results we observe the particles to migrate from the core of Kelvin–Helmholtz vortices shed from the front of the current and to accumulate in the current head, which affects the propagation speed of the front.

Key words: Density currents, gravity currents, two-phase flow, two-fluid model, equilibrium Eulerian model, spectral methods.

1 Introduction

Density (or gravity) currents are flows generated by the action of gravity over two fluids with density difference. The current may move below, above or in between of ambient fluid layers. Examples of particulate density currents are dusty thunderstorm fronts, pyroclastic flows produced in volcano eruptions, aerosol releases in the environment, flows originated by the discharge of a sediment-laden flow into the ocean or a lake and snow avalanches. Many more examples can be found in the books by Simpson [15] and Allen [2]. In most of the cases the density difference is only a few percent, however, this is enough for these currents to travel long distances and transport large amounts of sediment. Particles may settle or be re-entrained into the

S. Balachandar and A. Prosperetti (eds), Proceedings of the IUTAM Symposium on Computational Multiphase Flow, 393–402.

flow and particulate density currents are sometimes called non-conservative density currents.

Density currents have engineering, environmental and geological implications. Several studies on the accidental release of liquefied gas have been done modeling the flow as a non-conservative density current that looses mass due to evaporation [16]. In the ocean, sediment slump can trigger particulate density currents capable of traveling kilometers. These strong flows can carve submarine canyons [8] and mold the seabed producing different bed forms patterns as ripples, dunes, antidunes and gullies.

In this work we concentrate on the effect of particle inertia on the flow structure and dynamics. It is important to recognize that particles move with a velocity field which is different from the fluid. The fact that particles with finite size cannot follow exactly the fluid velocity plays a very important role and modifies substantially the structure and dynamics of the flow. In the following section we present a novel Eulerian–Eulerian mathematical model for simulating particulate density currents based in the well-accepted formalism of two-phase flow and the equilibrium Eulerian approach [6]. Then, we present two-dimensional direct numerical simulations and assess the effect of finite inertia on the current structure and front velocity.

2 Formulation of the mathematical model

We are interested in simulating buoyant flows driven by the presence of solid particles of finite inertia. In this situation particles not only modify the bulk density [13] but also move with their own velocity. Table 1 shows dimensionless parameters representing settling ($\tilde{w}$) and inertia ($\tilde{\tau}$) of sand particles of varying size (d) and initial concentration (ϕ_0) in water for different flow scales. The formal definition of these parameters is presented in the next section and for the analysis in this section it is enough to recognize that their numerical values dictates their relative importance in the mathematical model. Observe that in the cases shown in this table inertia is as important as settling.

Here we develop a new formulation that includes the role of particle inertia, in the interest to simulate gravity currents on environmental and geological scales. In this section we present an Eulerian–Eulerian model based on an asymptotic expansion of the two-phase flows equations in parameters describing the particle inertia (τ) and the particle concentration (ϕ_d). The model is formally exact to $O(\phi_d\tau + \tau^2 + \phi_d^2)$

Table 1. Sand in water, $\beta = 0.476$, $R = 1.65$.

	$d = 100 \ \mu m \ \ \phi_0 = 0.1$		$d = 100 \ \mu m \ \ \phi_0 = 0.2$		$d = 10 \ \mu m \ \ \phi_0 = 0.2$	
Gr	$\tilde{\tau}$	$\tilde{w}$	$\tilde{\tau}$	$\tilde{w}$	$\tilde{\tau}$	$\tilde{w}$
10^6	2.4×10^{-2}	7.6×10^{-2}	3.8×10^{-2}	6.1×10^{-2}	3.8×10^{-4}	6.1×10^{-4}
10^{10}	5.2×10^{-3}	1.6×10^{-2}	8.3×10^{-3}	1.3×10^{-2}	8.2×10^{-5}	1.3×10^{-4}
10^{14}	1.1×10^{-3}	3.5×10^{-3}	1.8×10^{-3}	2.8×10^{-3}	1.7×10^{-5}	2.8×10^{-5}

and consists of conservation equations for the mixture, an algebraic equation for the particle velocity and a transport equation for the particle volume fraction.

Let the indices c and d denote the continuum and disperse phases, respectively. We denote the densities, volume fractions, and velocities of each phase by ρ_c, ϕ_c, $\mathbf{u}_c$, and ρ_d, ϕ_d, $\mathbf{u}_d$, respectively. We base our formulation on the volume-averaged velocity

$$\mathbf{u}_v = \phi_c \mathbf{u}_c + \phi_d \mathbf{u}_d. \tag{1}$$

In the case of constant density phases and no mass transfer between phases the mass conservation equations are [17]

$$\frac{\partial \phi_c}{\partial t} + \nabla \cdot (\phi_c \mathbf{u}_c) = 0 \quad \text{and} \quad \frac{\partial \phi_d}{\partial t} + \nabla \cdot (\phi_d \mathbf{u}_d) = 0, \tag{2}$$

where $\phi_c + \phi_d = 1$. Observe that adding these two equations we get $\nabla \cdot \mathbf{u}_v = 0$, i.e. the volume-averaged velocity is a solenoidal field. This is a key feature that allows the use of incompressible Navier–Stokes solvers.

The process of obtaining the ensemble-averaged momentum equations [11] and their closure has been presented in detail in [17] and more recently in [12] (see also [14]). The resulting momentum equation for the flow can be expressed as

$$\frac{\partial}{\partial t}(\phi_c \rho_c \mathbf{u}_c) + \nabla \cdot (\phi_c \rho_c \, \mathbf{u}_c \otimes \mathbf{u}_c) = \phi_c \rho_c \mathbf{g} - \phi_c \nabla p + \mu_c \nabla^2 \mathbf{u}_v - \mathbf{F} \tag{3}$$
$$+ \nabla \cdot (\phi_c \mathbf{R}_c),$$

$$\frac{\partial}{\partial t}(\phi_d \rho_d \mathbf{u}_d) + \nabla \cdot (\phi_d \rho_d \, \mathbf{u}_d \otimes \mathbf{u}_d) = \phi_d \rho_d \mathbf{g} - \phi_d \nabla p + \mathbf{F} + \rho_d \nabla \cdot (\phi_d \mathbf{R}_d). \tag{4}$$

Here p is the pressure in the continuous phase, μ_c is the dynamic viscosity of the continuous phase, $\mathbf{g}$ is the gravity vector, $(\mathbf{F} - \phi_d \nabla p)$ is the net hydrodynamic interaction between phases and $\mathbf{R}_{c,d}$ are the kinematic Reynolds stresses.

In order to obtain a momentum equation based on $\mathbf{u}_v$ we operate under the assumption that $\phi_d \ll 1$, but also attempt to capture the grossest features of the dynamics in the case when ϕ_d becomes significant. A momentum equation based on the volume-averaged velocity can be obtained by adding Equations (3) and (4), which approximated to $O(\phi_d \tau + \tau^2 + \phi_d^2)$ reads

$$\rho_c \left(\frac{\partial \mathbf{u}_v}{\partial t} + \mathbf{u}_v \cdot \nabla \mathbf{u}_v \right) = \phi_d (\rho_d - \rho_c) \mathbf{g} - \nabla p + \mu_c \nabla^2 \mathbf{u}_v,$$

where the Boussinesq approximation has also been used.

The two-fluid model is completed by computing the particle velocity field from the equilibrium Eulerian approach proposed originally by Ferry and Balachandar [6]

$$\mathbf{u}_d = \mathbf{u}_c + \mathbf{w} + \frac{\tau \beta}{2} \nu_c \nabla^2 \mathbf{u}_c - \tau(1 - \beta) \left(\frac{\partial \mathbf{u}_c}{\partial t} + \mathbf{u}_c \cdot \nabla \mathbf{u}_c \right), \tag{5}$$

where $\nu_c = \mu_c / \rho_c$, τ is the particle response time and β is the density ratio factor defined by

$$\tau = \frac{d^2(\rho_d + C_m\rho_c)}{18\mu_c} \quad \text{and} \quad \beta = \frac{\rho_c + C_m\rho_c}{\rho_d + C_m\rho_c}, \tag{6}$$

respectively. Here d is the particle diameter, and the added mass coefficient is $C_m = 1/2$. The settling velocity $\mathbf{w}$ is defined as $\mathbf{w} = \tau(1 - \beta)\mathbf{g}$, and it is assumed to be $O(\tau)$ in this work, i.e. we assume that the Reynolds number based on the settling velocity and particle diameter, $\rho_c d|\mathbf{w}|/\mu_c$, is less than one.

The formula for $\mathbf{u}_d$ in terms of $\mathbf{u}_v$ then becomes

$$\mathbf{u}_d = \mathbf{u}_v + (1 - \phi_d)\mathbf{w} + \frac{\tau\beta}{2}\nu_c\nabla^2\mathbf{u}_v - \tau(1 - \beta)\left(\frac{\partial \mathbf{u}_v}{\partial t} + \mathbf{u}_v \cdot \nabla\mathbf{u}_v\right) \tag{7}$$

to $O(\tau\phi_d + \tau^2 + \phi_d^2)$.

The velocity $\mathbf{u}_d$ is used to evolve the disperse phase volume fraction ϕ_d:

$$\frac{\partial \phi_d}{\partial t} + \nabla \cdot (\phi_d\mathbf{u}_d) = \kappa\nabla^2\phi_d. \tag{8}$$

The particle diffusivity κ will be taken as a constant multiple of the continuous phase kinematic viscosity: $\kappa = \nu_c/Sc$, where Sc is the Schmidt number. Particle diffusivity is a way to account for the departure in particle motion from equilibrium prediction. Such departures arise from close interaction of particles and in general diffusivity is a function of both local particle concentration and local shear [1, 7]. However, as shown by other researchers solution of Equation (8) with little or no diffusion is numerically unstable, especially in the context of spectral simulations. Here, based on numerical considerations we simply chose $Sc = 1$ and consistently with the findings of Härtel et al. [10] we observe that the results to be presented are not sensitive to this choice.

3 Formulation of the problem

We consider the setting depicted in Figure 1. The channel is filled at one end with the mixture separated by a *gate* from the rest of the channel, which is filled with clear fluid. When the simulation begins the *gate* is released and the flow develops forming an underflow intrusion of the mixture into the clear fluid (solid line in Figure 1).

Let the half height of the channel (h_0) be the length scale, $U_0 = \sqrt{g\phi_0 R h_0}$ be the velocity scale and the initial volume fraction (ϕ_0) be the concentration scale. Here $R = (\rho_d - \rho_c)/\rho_c = 3(1 - \beta)/(2\beta)$. Then, time and pressure scales are h_0/U_0 and $\rho_c U_0^2$. Let $\tilde{}$ denote the dimensionless variables. The dimensionless equations of the model are

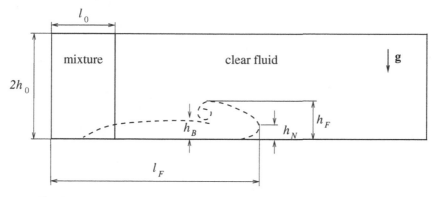

Fig. 1. Sketch of a density current with the nomenclature used in this work.

$$\frac{\partial \tilde{\mathbf{u}}_v}{\partial \tilde{t}} + \tilde{\mathbf{u}}_v \cdot \nabla \tilde{\mathbf{u}}_v = \tilde{\phi}_d \frac{\mathbf{g}}{g} - \nabla \tilde{p} + \frac{1}{\sqrt{Gr}} \nabla^2 \tilde{\mathbf{u}}_v, \tag{9}$$

$$\nabla \cdot \tilde{\mathbf{u}}_v = 0, \tag{10}$$

$$\tilde{\mathbf{u}}_d = \tilde{\mathbf{u}}_v + (1 - \phi_0 \tilde{\phi}_d)\tilde{\mathbf{w}} + \frac{\beta \tilde{\tau}}{2\sqrt{Gr}} \nabla^2 \tilde{\mathbf{u}}_v \tag{11}$$

$$- \tilde{\tau}(1 - \beta)\left(\frac{\partial \tilde{\mathbf{u}}_v}{\partial \tilde{t}} + \tilde{\mathbf{u}}_v \cdot \nabla \tilde{\mathbf{u}}_v\right), \quad \text{and}$$

$$\frac{\partial \tilde{\phi}_d}{\partial \tilde{t}} + \nabla \cdot \left(\tilde{\phi}_d \tilde{\mathbf{u}}_d\right) = \frac{1}{Sc\sqrt{Gr}} \nabla^2 \tilde{\phi}_d. \tag{12}$$

The key non-dimensional parameter that characterizes the strength of the current is the Grashof number, defined as $Gr = (U_0 h_0/\nu_c)^2$. The other two controlling parameters define the individual suspended particles in terms of particle Stokes number ($\tilde{\tau} = \tau U_0/h_0$) and non-dimensional settling velocity ($\tilde{w} = |\mathbf{w}|/U_0$). These parameters characterize the inertial and settling effects of the particle, respectively.

The dimensionless governing equations are solved using a de-aliased pseudo-spectral code [4]. Fourier expansions are employed for the flow variables in the horizontal direction (x). In the inhomogeneous vertical direction (z) Chebyshev expansion is used with Gauss–Lobatto quadrature points. The flow field is time advanced using a Crank–Nicholson scheme for diffusion terms. The advection and buoyancy terms are advanced with a third-order Runge–Kutta scheme. More details on the implementation of this numerical scheme can be found in [5]. The computational domain is a box of size $L_x = 40 \times L_z = 2$, which extends from $\tilde{x} = -20$ to $\tilde{x} = 20$ and from $\tilde{z} = 0$ to $\tilde{z} = 2$. The flow is initialized from rest with $\tilde{\phi}_d = 1$ in $\tilde{x} \in (-2, 2)$ for all $\tilde{z}$ and $\tilde{\phi}_d = 0$ otherwise. This setting of the problem generates two currents moving from the center outward. In this way we can enforce periodic boundary conditions in the horizontal direction for all variables avoiding to specify an outflow boundary condition. At the top and bottom walls no-slip conditions are enforced for velocity. For the disperse phase zero net flux is set at the top wall and zero particle resuspension is set at the bottom wall [3], i.e.

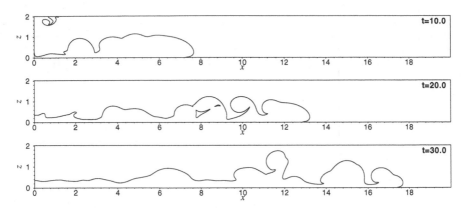

Fig. 2. Contours of particles concentration. Solid line: 0.1, dashed line:1.0. Solution for $Gr = 1.5 \times 10^6$, $\tilde{\tau} = 0.0$ and $\tilde{w} = 0.0$.

$$\tilde{w}_z \tilde{\phi}_d - \frac{1}{Sc\sqrt{Gr}} \frac{\partial \tilde{\phi}_d}{\partial \tilde{z}} = 0, \quad \text{and} \quad \frac{\partial \tilde{\phi}_d}{\partial \tilde{z}} = 0, \tag{13}$$

respectively. The solution was advanced in time until the front reached location of $\tilde{x} = 18$ to avoid the influence of finite domain size [9]. The simulations were performed using a resolution of $N_x = 1024 \times N_z = 220$. It must be mentioned that almost twice the resolution is needed for the particulate flow simulations compared to the corresponding scalar case (i.e. same dimensionless numbers with $\tilde{\tau} = 0$ and $\tilde{w} = 0$). The resolution was selected to produce a decay of 6 to 8 decades in the energy spectrum of every variable.

4 Results

Figure 2 shows the results for the limit when particles are so small that they act as a scalar field ($\tilde{w} = 0$ and $\tilde{\tau} = 0$). The flow is visualized by a contour of $\tilde{\phi}_d = 0.1$. Soon after the release an intrusion front forms with a lifted nose due to the no-slip boundary condition. As the current advances Kelvin–Helmholtz vortices are shed from the front which produce a net drag that balances the initial acceleration of the front. As a consequence, after the initial set-up of the Kelvin–Helmholtz vortices, the front moves at constant speed until the dilution in the current becomes important. Then, the current slows down and eventually dissipates.

Figures 3 and 4 show the results for currents of inertial particles with negligible settling ($\tilde{\tau} = 0.05$, $\tilde{w} = 0$) and ($\tilde{\tau} = 0.1$, $\tilde{w} = 0$) respectively. Two contours are shown in these figures: the solid line contour that corresponds to $\tilde{\phi}_d = 0.1$, and the dash line contour that corresponds to $\tilde{\phi}_d = 1$. Two main differences are observed compared to the no inertia particles case. The first one is the migration of particles away from the core of Kelvin–Helmholtz vortices, and the second one is the accumulation of particles in the front of the current, producing regions of particle

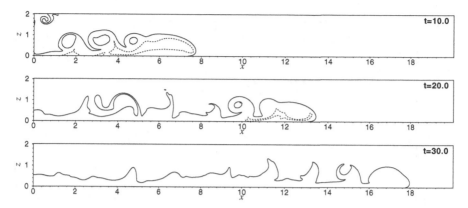

Fig. 3. Contours of particles concentration. Solid line: 0.1, dashed line:1.0. Solution for $Gr = 1.5 \times 10^6$, $\tilde{\tau} = 0.05$ and $\tilde{w} = 0.0$.

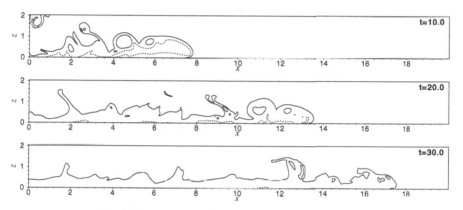

Fig. 4. Contours of particles concentration. Solid line: 0.1, dashed line:1.0. Solution for $Gr = 1.5 \times 10^6$, $\tilde{\tau} = 0.1$ and $\tilde{w} = 0.0$.

concentration greater than the initial value (i.e. $\tilde{\phi}_d > 1$). These two effects can be explained by noting that the divergence of the particles velocity field is

$$\nabla \cdot \tilde{\mathbf{u}}_d = \tilde{\tau}(1 - \beta)\left(\|\Omega\|^2 - \|S\|^2\right), \qquad (14)$$

where S and Ω are the symmetric and skew-symmetric parts of the local fluid velocity gradient tensor. Note from Equation (6) that for particles substantially heavier than the continuous phase ($\beta \rightarrow 0$), $\nabla \cdot \tilde{\mathbf{u}}_d > 0$ when $\|\Omega\| > \|S\|$, which means that particles migrate from regions of vorticity and accumulate in regions of high strain rate.

The preferential particles accumulation described above has a very important consequence in the current front velocity. Figure 5 shows the front velocity for the three different cases studied in this work. Observe that the front velocity in the phase

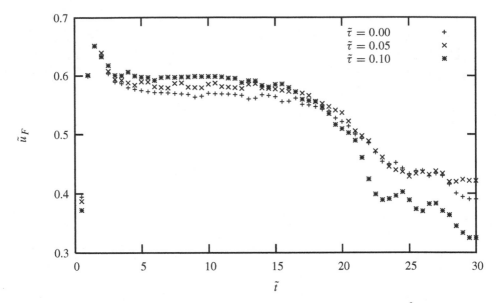

Fig. 5. Front velocity for the total time simulation. Solution for $Gr = 1.5 \times 10^6$ and $\tilde{w} = 0.0$.

of constant velocity increases for larger values of $\tilde{\tau}$. The current with $\tilde{\tau} = 0.1$ presents a front velocity 5% larger that the current with $\tilde{\tau} = 0$.

5 Summary and conclusions

We have presented a novel two-fluid model for the simulation of particulate density currents with particles of finite inertia. The model consist of conservation equations for the mixture, an algebraic equation for the particle velocity and a transport equation for the particles concentration. By the incorporation of the equilibrium Eulerian approach [6] we avoid solving a partial differential equation for the conservation of momentum of the disperse phase which constitutes a big saving in computational time.

The results presented in this work show that particle inertia has an important influence in the structure and dynamics of the flow. Particles migrate from the core of Kelvin–Helmholtz vortices and accumulate in the front of the current. As a consequence the front velocity at the initial stage of the flow is larger with increasing particle inertia. Finally, we speculate that inertia may have an important influence in the deposition patterns produced by these type of flows since the deposition flux is proportional to the particles concentration. Three-dimensional simulations are underway and will serve to address this speculation.

Acknowledgments

We gratefully acknowledge the support of the Coastal Geophysics Program of the Office of Naval Research (Award N00014-03-1-0143), the Chicago District of the US Army Corps of Engineers, and the Metropolitan Water Reclamation District of Greater Chicago. Support from the National Center for Supercomputer Applications at UIUC is also acknowledged. Mariano Cantero was supported by a Graduate Student Fellowship from the Computational Science and Engineering Program at UIUC. Computer time was provided by Dr. Fady Najjar and Dr. Nahil Sobh.

References

1. A. Acrivos, 1995, Bingham award lecture – 1994. Shear-induced particle diffusion in concentrated suspensions of noncolloidal particles, *Journal of Rheology* **39**(5), 813–826.
2. J. Allen, 1985, *Principles of Physical Sedimentology*, George Allen and Unwin Ltd.
3. M. Cantero, 2002, Theoretical and numerical modeling of turbidity currents as two-phase flows, M.S. Thesis, University of Illinois at Urbana-Champaign, Urbana, IL.
4. C. Canuto, M. Hussaini, A. Quarteroni and T. Zang, 1988, *Spectral Methods in Fluid Dynamics*, Springer-Verlag, Berlin.
5. T. Cortese and S. Balachandar, 1995, High performance spectral simulation of turbulent flows in massively parallel machines with distributed memory, *International Journal of Supercomputer Applications* **9**(3), 187–204.
6. J. Ferry and S. Balachandar, 2001, A fast eulerian method for disperse two-phase flow, *International Journal of Multiphase Flows* **27**, 1199–1226.
7. D. Foss and J. Brady, 2000, Structure, diffusion and rheology of Brownian suspensions by Stokesian Dynamics simulations, *Journal of Fluid Mechanics* **407**, 167–200.
8. Y. Fukushima, G. Parker and H. Pantin, 1985, Prediction of ignitive turbidity currents in Scripps submarine canyon, *Marine Geology* **67**, 55–81.
9. C. Härtel, L. Kleiser, M. Michaud and C. Stein, 1997, A direct numerical simulation approach to the study of intrusion fronts, *Journal of Engineering Mathematics* **32**, 103–120.
10. C. Härtel, E. Meiburg and F. Necker, 2000, Analysis and direct numerical simulation of the flow at a gravity-current head. Part 1. Flow topology and front speed for slip and no-slip boundaries, *Journal of Fluid Mechanics* **418**, 189–212.
11. D. Joseph and T. Lundgren, 1990, Ensemble averaged and mixture theory equations for incompressible fluid-particle suspensions, *International Journal of Multiphase Flows* **16**, 35–42.
12. M. Machioro, M. Tanksley and A. Prosperetti, 1999, Mixture pressure and stress in disperse two-phase flow, *International Journal of Multiphase Flows* **25**, 1395–1429.
13. F. Marble, 1970, Dynamics of dusty gases, *Annual Review of Fluid Mechanics*, 397–446.
14. A. Prosperetti, 2001, Ensemble averaging techniques for disperse flows, in *Particulate Flows Processing and Rheology*, D. Drew, D. Joseph and S. Passman (eds), Springer, New York.
15. J. Simpson, 1999, *Gravity Currents*, second edition, Cambridge University Press.

16. T. Spicer and J. Havens, 1996, Application of dispersion models to flammable cloud analyses, *Journal of Hazardous Materials* **49**, 115–124.
17. D. Zhang and A. Prosperetti, 1997, Momentum and energy equations for disperse two-phase flows and their closure for dilute suspensions, *International Journal of Multiphase Flows* **23**, 425–453.

Modeling Finite-Size Effects in LES/DNS of Two-Phase Flows

S.V. Apte[1], K. Mahesh[2] and T. Lundgren[2]

[1] *Stanford University, Stanford, CA 94305, USA; e-mail: sapte@stanford.edu*
[2] *University of Minnesota, Minneapolis, MN 55311, USA; e-mail: mahesh@aem.umn.edu,*
lundgren@aem.umn.edu

1 Introduction

Recent direct numerical simulation (DNS) of large number of solid particles interacting through a fluid medium by Joseph and collaborators [1, 2] show that a layer of heavy particles with fluid streaming above it can develop Kelvin–Helmholtz (K–H) instability waves whereas a layer of particles above a lighter fluid develops Rayleigh-Taylor instability. However, performing full DNS of millions of dispersed particles in a turbulent flow (e.g. spray combustion, liquid atomization, spray coating, fluidized bed combustion, aerosol transport) is computationally intensive. For such applications, the particle size is typically smaller than the grid-resolution used for the computation of the continuum fluid. Under these conditions, the particles are *subgrid* and some sort of subgrid modeling is necessary to simulate their motion.

The "point-particle" assumption is commonly employed where forces on the dispersed phase are computed through model coefficients. The effect of the particles on the carrier phase is represented by a force applied at the *centroid* of the particle. For dilute particle loadings with swirling, separated flows in a coaxial combustor computed using LES of point-particles, Apte et al. [3] indicated good agreement with the experimental data. However, for moderate loadings and wall-bounded flows, Segura et al. [4] have shown that the point-particle approximation fails to predict the turbulence modulation compared to experimental values. In order to capture the same level of turbulence modulation observed in experiments, it was required to artificially increase the particle loadings by an order of magnitude when using the point-particle approach [4]. In addition, if the particle size is greater than Kolmogorov scale, simple drag/lift laws used in this approach do not capture the unsteady wake effects [5, 6].

In this work we attempt to extend the point-particle approximation by accounting for the finite-size of the particles and the corresponding volume displacement (Θ_f) of the carrier phase. Accordingly, the carrier phase continuity and momentum equations are modified to include Θ_f. The formulation was originally put forth by Dukowicz [7] in the context of spray simulations. However, the particle volume fractions are often neglected owing to the increased complexity of the governing equations as

403

S. Balachandar and A. Prosperetti (eds), Proceedings of the IUTAM Symposium on Computational Multiphase Flow, 403–412.

well as numerical stiffness they impose in the dense spray regime. Several studies on dense granular flows [8–10] use this model for laminar flows. Similar formulation has been applied for bubbly flows at low bubble concentrations to investigate the effect of bubbles on drag reduction in turbulent flows [11]. However, these studies do not identify the effects of the fluid displacement by dispersed phase compared to the point-particles. In the following sections, the mathematical model and numerical scheme are described in brief. The model is applied to simulate gravitational settling and fluidization by jet to validate the numerical scheme. Next we compute plane Poisuille flow with rigid spheres at the bottom to show particle dispersal and lift.

2 Mathematical formulation

The formulation described below consists of the Eulerian fluid and Lagrangian particle equations, and accounts for the displacement of the fluid by the particles as well as the momentum exchange between them [12].

2.1 Gas-phase equations

The fluid mass for unit volume satisfies a continuity equation,

$$\frac{\partial}{\partial t}(\rho_f \Theta_f) + \nabla \cdot (\rho_f \Theta_f \mathbf{u}_f) = 0, \tag{1}$$

where ρ_f, Θ_f, and $\mathbf{u}_f$ are the fluid density, volume fraction, and velocity, respectively. This indicates that the average velocity field of the fluid phase does not satisfy the divergence-free condition even if we consider an incompressible suspending fluid. The particle volume fraction, $\Theta_p = 1 - \Theta_f$ is defined as

$$\Theta_p(\mathbf{x}_{cv}) = \sum_{k=1}^{N_p} V_{p_k} \mathcal{G}_\sigma(\mathbf{x}_{cv}, \mathbf{x}_{p_k}), \tag{2}$$

where the summation is over all particles N_p. Here $\mathbf{x}_{p_k}$ is the particle location, $\mathbf{x}_{cv}$ the centroid of a control volume, and V_{p_k} the volume of a particle. The interpolation function, $\mathcal{G}_\sigma$, effectively transfers Lagrangian quantity to give an Eulerian field (*per unit volume*, V_{cv}, of the grid cell containing the particle centroid) on the underlying grid and is defined later. The fluid momentum equation is given as

$$\frac{\partial}{\partial t}\left(\rho_f \Theta_f \mathbf{u}_f\right) + \nabla \cdot (\rho_f \Theta_f \mathbf{u}_f \mathbf{u}_f) = -\nabla(\Theta_f p) + \nabla \cdot (\mu_f \mathbf{D}_c) + \mathbf{F}, \tag{3}$$

where p is the average pressure, μ_f is the viscosity of the fluid, and $\mathbf{D}_c = \nabla \mathbf{u}_c + \nabla \mathbf{u}_c^T$ the average deformation-rate of the fluid-particle composite, $\mathbf{u}_c$ the composite velocity of the mixture [12], and $\mathbf{F}$ the force per unit volume exerted on the fluid by particles.

2.2 Dispersed-phase equations

The individual particle positions and velocities can be obtained by solving the ordinary differential equations in Lagrangian framework for each particle:

$$\frac{d}{dt}(\mathbf{x}_p) = \mathbf{u}_p; \quad m_p \frac{d}{dt}(\mathbf{u}_p) = \mathbf{F}_p, \tag{4}$$

where $\mathbf{x}_p$ is the particle position, $\mathbf{u}_p$ the particle velocity, $\mathbf{F}_p = m_p \mathbf{A}_p$ the total force acting on the particle of mass m_p, and $\mathbf{A}_p$ is the particle acceleration. This consists of the standard hydrodynamic drag force, dynamic pressure gradient, gradient of viscous stress in the fluid phase, a generalized buoyancy force, inter-particle collision and external body forces (gravity). In the present work, we assume that the particle forces consist of drag, collision and gravitational acceleration, and neglect all other terms in order to investigate the effect of the particle volume fraction. For high density ratios ($\rho_p/\rho_f \sim 1000$), these assumptions are valid [3]:

$$\mathbf{A}_p = D_p(\mathbf{u}_{f,p} - \mathbf{u}_p) - \left(1 - \frac{\rho_f}{\rho_p}\right)\mathbf{g} + \mathbf{A}_{cp}. \tag{5}$$

Here $\mathbf{A}_{cp}$ is the acceleration due to inter-particle forces and $\mathbf{u}_{f,p}$ the fluid velocity at the particle location. The standard expression for drag force, D_p, is used

$$D_p = \frac{3}{8}C_d \frac{\rho_f}{\rho_p} \frac{|\mathbf{u}_{f,p} - \mathbf{u}_p|}{R_p}, \tag{6}$$

where C_d is the drag coefficient [13],

$$C_d = \frac{24}{Re}(1 + 0.15 Re_p^{0.687})\Theta_f^{-2.65}, \quad \text{for } Re_p < 1000 \tag{7}$$

$$= 0.44\Theta_f^{-2.65}, \quad \text{for } Re_p \geq 1000; \tag{8}$$

$R_p = (3V_p/4\pi)^{1/3}$ is the particle radius. The particle Reynolds number (Re_p) is given as, $Re_p = 2\rho_f \Theta_f |\mathbf{u}_{f,p} - \mathbf{u}_p| R_p/\mu_f$. There is an indirect collective effect in this drag term: when there is a dense collection of particles passing through the fluid, the interphase momentum exchange term in Equation (3) will cause $\mathbf{u}_f$ to approach the particle velocity, $\mathbf{u}_p$, thus decreasing the drag on a particle, a drafting effect. The inter-particle collision scheme is based on the discrete element approach of Cundall and Strack as given in [9]. This is necessary to keep the particle centroids from overlapping each other. The interphase momentum transfer function per unit volume in Equation (3) is given as

$$\mathbf{F}(\mathbf{x}_{cv}) = \sum_{k=1}^{N_p} \mathcal{G}_\sigma m_{p_k} D_{p_k}(\mathbf{u}_{f,p_k} - \mathbf{u}_{p_k}). \tag{9}$$

3 Numerical method

In this work, we modify the numerical scheme for unstructured, arbitrary shaped elements developed by Mahesh et al. [14] to take into account the fluid volume fraction. On Cartesian grids in three-dimensions, bilinear interpolation functions utilizing 26 neighboring grid cells to interpolate Eulerian fields from the Lagrangian quantities have been used [8, 10]. In an effort to generalize these interpolations to unstructured, arbitrary shaped elements, we make use of a Gaussian distribution function centered at the particle centroid as an interpolation function and is given by

$$
\mathcal{G}_\sigma(\mathbf{x}_{cv}, \mathbf{x}_p) = \frac{1}{\left(\sigma\sqrt{2\pi}\right)^3} \exp\left[-\frac{\sum_{i=1}^3 [(x_{cv})_i - (x_p)_i]^2}{2\sigma^2}\right].
\tag{10}
$$

Here we assume that $V_p < V_{cv}$ and set the filter width to be equal to the longest diagonal of the control volume (CV) containing the particle. The interpolation operator is applied to all the neighbors of the CV (having at least one grid node common). Similar interpolation function has been used in the context of resolved simulations of particles [15]. In addition, $\mathcal{G}$ is normalized to satisfy $\int_{V_{cv}} \mathcal{G}_\sigma(\mathbf{x}_{cv}, \mathbf{x}_p) dV = 1$, where the integration is performed over CV and all of its immediate neighbors. The final step is necessary to enforce mass (or volume) conservation. The resulting Θ_p will be smooth and mass-conserving as the particles move from one CV to another. We use an implicit scheme for the fluid solver, however, the interphase momentum exchange terms are treated explicitly. The particle equations are integrated using third-order Runge–Kutta schemes for *ode*-solvers. At each Runge–Kutta step, the particles were re-located and the collision force was re-computed. We use the Lagrangian particle tracking algorithm developed in [3].

4 Results

4.1 Case 1: Gravitational settling

We first simulate sedimentation of solid particles under gravity in a rectangular box. Details of this case are given in Table 1. The initial parcel positions are generated randomly over the box length. These particles are then allowed to settle through the gas-medium under gravity. The dominant forces on the particles include gravity and inter-particle/particle-wall collision. As the particles hit the bottom wall of the box,

Table 1. Parameter description for gravity-dominated sedimentation.

Computational domain, $0.2 \times 0.6 \times 0.0275$ m	Grid, $10 \times 30 \times 5$
Fluid density, 1.254 kg/m^3	Particle Density, 2500 kg/m^3
Number of Parcels, 1000	Particles per parcel, 3375
Diameter of particles, 500 μm	Initial particle concentration, 0.2

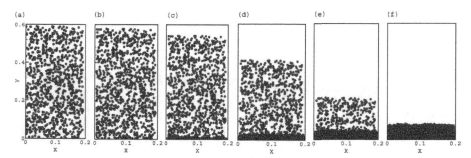

Fig. 1. Temporal evolution of particle distribution during gravity-dominated sedimentation.

they bounce back and stop the incoming layer of particles, and finally settle to a close pack limit. The upper mixture interface between the particles and the fluid is closely approximated by $h = gt^2/2$ [10]. As the particles settle the fluid in the bottom half of the box starts to move upward giving resistance to the settling particles. The evolution of the mixture interface closely follows the analytical estimate in our computation.

4.2 Case 2: Fluidization by jet

We consider the problem of fluidization of solid particles arranged in an array at the bottom of a rectangular box. Fluidization is achieved by a jet of gas from the bottom of the box. The flow parameters are given in Table 2. The particle motion is mostly dominated by the hydrodynamic drag force and collision model should not affect the overall particle motion. The collision model, however, is important in governing the particle behavior near the walls and helps prevent the volume fraction from exceeding the close-pack limit.

Figure 2 shows the position of parcels at different times during bubbling fluidization. The jet issued from the bottom wall pushes the particles away from the center region and creates a gas-bubble in the center. The particles collide with each other, against the wall and are pushed back towards the central jet along the bottom wall. They are then entrained by the jet and are levitated. This eventually divides the central bubble to form two bubbles. The particles tend to move upward and collide with the upper wall and remain levitated during future times. The computational results

Table 2. Parameter description for the simulation of fluidization by a gas jet.

Computational domain, $0.2 \times 0.6 \times 0.0275$ m	Grid, $10 \times 30 \times 5$
Gas jet velocity, 9 m/s	Jet diameter, 0.04 m
Fluid density, 1.254 kg/m^3	Particle Density, 2500 kg/m^3
Number of Parcels, 2880	Particles per parcel, 3375
Diameter of particles, 500 μm	Initial particle concentration, 0.4

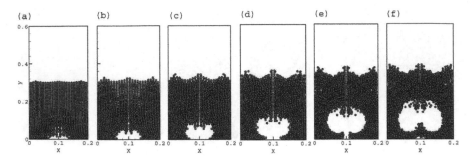

Fig. 2. Temporal evolution of particle distribution during fluidization by a gas jet. Initially all particles are uniformly arranged in layers at the bottom of the rectangular box. Air is injected through a rectangular slot at the bottom wall. Air bubbles are trapped within the particles and the growth and pattern of these bubbles are in agreement with simulations by Patankar and Joseph [9].

are in good agreement with the simulations of [9]. Similar results are reported using Eulerian–Eulerian approach in two-dimensions [16].

4.3 Case 3: Fluidization by lift

The transport of particles by fluids in coal-water slurries, hydraulically fractured rocks in oil-bearing reservoirs, bed-load transport in rivers and canals and their over-all effect on the river bed erosion etc., are important scientific and industrial issues in particulate flows. In order to understand fluidization/sedimentation in such conduits, Choi and Joseph [1] performed a DNS study of fluidization of circular cylinders (300 particles) arranged at the bottom of a channel in plane Poisuille flow. They observed that with sufficient pressure gradient across the channel, the particles initially at rest in the lower half of the channel start moving and roll over the wall. Particle rotation in a shear flow generates lift and the channel is fluidized after some time.

The flow parameters are given in Table 3. As opposed to [1], we are performing three-dimensional simulations. The particles initially at rest, accelerate and setup instability waves between the fluid and particle layers. Figure 3 shows the time-

Table 3. Parameter description for the simulation of fluidization of spherical particles in a plane Poisuille flow.

Computational domain, $63 \times 12 \times 12$ cm	Grid, $20 \times 11 \times 10$
Fluid density, 1 g/cm^3	Fluid viscosity, 1 poise
Particle Density, 10.0 g/cm^3	Diameter of particles, 0.95 cm
Number of Parcels, 3780	Particles per parcel, 1
Initial array height, 4.75 cm,	Initial centerline velocity, 360 cm/s
Pressure gradient, 20 dyne/cm^3	

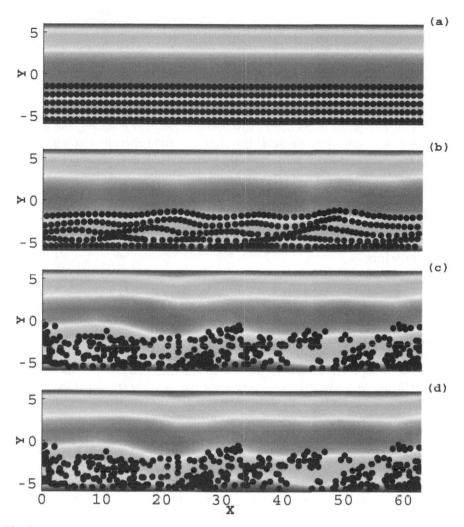

Fig. 3. Temporal evolution of axial velocity contours in a plane Poisuille flow with particles in the bottom half of the channel.

evolution of axial velocity contours in the fluid as well as particle locations in the $z = 0$ plane. As the fluid is pushed out of the control volume by motion of particles a vertical pressure gradient is created imparting vertical velocity to the particles and the channel gets fluidized. We also did several test cases, with higher grid resolution, increased density ratios to obtain similar results. With increased particle density, the inter-phase momentum exchange decelerates the fluid in the bottom half of a channel and an inflection point is created in the axial velocity profile. This eventually causes lift and particle dispersal.

It should be noted that the mechanism of lift observed in the DNS simulations is different from the one given by the model. In the DNS, the particles accelerate and rotate due to the shearing motion. This rotation of the particles in a shear flow gives lift. In the model, however, we do not consider particle rotation and the only force exerted by the fluid on the particle is through the drag law. The unsteady effects of particle motion are captured entirely through the distribution of the particle volume fraction. In the present simulations, the presence of a particle in a fluid control volume is felt through the interphase momentum transfer due to two-way coupling, and changes in the particle volume fraction field. The particle motion alters both continuity and momentum equations and in turn affects the pressure field. In the present simulation, the particles in the top layer move faster than those in the bottom layers. It is found that this gives rise to gradients in pressure in the wall normal direction. This gives vertical velocity to the fluid and causes lift of particles through the drag law.

5 Discussion

We also simulated all the above cases using the point-particle approach with collisions and compared to the present model predictions [17]. For the first case (gravitational settling), the particle evolution obtained from point-particles and the finite-size model are similar. This is mainly because, the flow is gravity and collision dominated and there is no mean fluid flow. For the second and third cases, however, the point-particle approximation gave very different results compared to the finite-size model. The patterns observed in Figure 2 are absent when simulated using point-particles. Also, for the Poisuille flow, point-particles do not predict any lift and fluidization. This indicates that two-way coupling modeled using point-particles is not sufficient to produce the effects observed in direct numerical simulations of these flows. In the present formulation where we account for the volume displacement, the particle volume fraction alters the flow evolution in three-different ways: (a) continuity equation, (b) the momentum equation, and (c) the drag force. The blocking effect of particles on the fluid phase, modeled by the continuity equation alters the fluid flow in regions of high variations in volume fraction.

These findings have several implications on LES/DNS of two-phase flows. As mentioned earlier, the point-particle approach does not reproduce the turbulence attenuation obtained by solid particles in a channel flow even at moderate loadings compared to the experimental observations [4]. For such wall-bounded flows, the particles near the wall, tend to move slowly due to their inertia thus increasing their residence time near the wall. Inter-particle and particle-wall collisions play an important role. The grid resolution in the wall-normal direction is such that the particle diameter is typically occupied by 4-5 grid cells near the wall. In addition, due to increased residence times near the wall, the local particle volume fractions become high and gradients in the volume fraction field can alter the fluid flow. Prosperetti and Zhang [18] argued that the effect of volume fraction may be more important than inter-particle collisions in the near wall regions. As shown in the above case studies, considering the fluid displaced by the particles in the continuity and momentum equations has an indirect effect of increased particle loading on the fluid phase. Segura et al. [4] had to artificially increase the particle loading to match the experimental data on turbulence modulation. This suggests that the variations in volume fraction field near the wall could account for these effects. Extension of the present formulation to LES is straight forward. The filtered equations for LES can be derived based on Favre-averaging commonly employed in variable density flows and standard subgrid closures can then be applied.

Furthermore, applications involving dense flows such as liquid-fuel atomization in automotive and aircraft engines, coal-fired combustion chambers, and fluidized beds, should account for the finite-size of the droplets/particles in order to predict the evolution of the fuel mass fractions correctly. As demonstrated by the last case above, instability waves created by dense fuel flowing in a lighter fluid can be captured by this model and will allow us to better represent the important features of primary atomization often neglected in these simulations [19].

6 Conclusions

In the present study we extend the point-particle approach typically employed in multiphase flows by accounting for the finite-size of the particles. The presence of particles affects the fluid phase continuity and momentum equations through the volume fraction field. Efficient interpolation scheme to obtain Eulerian fields from Lagrangian points on arbitrary shaped, unstructured meshes has been developed. The numerical technique has been applied to dense particulate flows such as gravitational settling and fluidization by a gaseous jet. Finally, we have shown that the present model can predict lift and fluidization of a plane channel flow with heavy particles arranged in layers at the bottom of the channel. These effects were captured entirely due to the fluid volume displaced by the particles and were not observed using the point-particle approach. Based on this study, we propose that for moderate loadings, the standard point-particle approach should be modified to account for the finite-size of the particles. Further investigations on turbulent flows at moderate to high particle loadings are necessary.

Acknowledgement

Support for this work was provided by the United States Department of Energy under the Accelerated Strategic Computing (ASC) program.

References

1. Choi, H.G. and Joseph, D.D., 2001, *J. Fluid. Mech.* **438**, 101–128.
2. Patankar, N.A., Ko, T., Choi, H.G. and Joseph, D.D., 2001, *J. Fluid Mech.* **445**, 55–76.

3. Apte, S. V., Mahesh, K., Moin, P. and Oefelein, J.C., 2003a, *Int. J. Multiphase Flow 29*, 1311–1331.
4. Segura, J.C., Eaton, J.K. and Oefelein, J.C., 2004, Predictive capabilities of particle-laden LES, Report No. TSD–156, Department of Mechanical Engineering, Stanford University.
5. Burton, T.M. and Eaton, J.K., 2003, Fully resolved simulations of particle-turbulence interaction, Report No. TSD–151, Department of Mechanical Engineering, Stanford University.
6. Bagchi, P. and Balachandar, S., 2003, *Phys. Fluids* **15**, 3496–3513.
7. Dukowicz, J.K., 1980, *J. Comput. Phys.* **35**, 229–253.
8. Patankar, N.A. and Joseph, D.D., 2001, *Int. J. Multiphase Flow* **27**, 1659–1684.
9. Patankar, N.A. and Joseph, D.D., 2001, *Int. J. Multiphase Flow* **27**, 1685–1706.
10. Snider, D.M., 2001, *J. Comput. Phys.* **170**, 523–549.
11. Ferrante, A. and Elghobashi, S., 2004, *J. Fluid Mech.* **503**, 345–355.
12. Joseph, D.D. and Lundgren, T., 1990, *Int. J. Multiphase Flow* **6**, 35–42.
13. Andrews, M.J. and O'Rourke, P., 1996, *Int. J. Multiphase Flow* **22**, 379–402.
14. Mahesh, K., Constantinescu, G. and Moin, P., 2004, *J. Comput. Phys.* **197**(1), 215–240.
15. Maxey, M.R. and Patel, B.K., 2001, *Int. J. Multiphase Flow* **27**, 1603–1626.
16. Ding, J. and Gidaspow, D., 1990, *AIChE* **36**, 523–537.
17. Apte, S.V., Mahesh, K. and Lundgren, T., 2005, *J. Fluid Mech.*, to be submitted.
18. Prosperetti, A. and Zhang, D.Z., 1995, *Theoret. Comput. Fluid Dynam.* **7**, 429–440.
19. Moin, P. and Apte, S.V., 2004, Large-eddy simulation of realistic gas-turbine combustors, AIAA Paper 2004-0330, Reno, NV.

Lagrangian Aspects to Multiphase Flows

I. Eames[1], M. Gilbertson[2], J.B. Flór[3] and V. Roig[4]

[1]Department of Mechanical Engineering, University College London, London, UK
[2]Department of Mechanical Engineering, University of Bristol, Bristol, UK
[3]LEGI, BP 53X, 38041 Grenoble Cedex 09, France
[4]Institut de Mécanique des Fluides de Toulouse, Toulouse, France

1 Introduction

A Lagrangian formulation of fluid mechanics involves following parcels of inform-ation (eddies, fluid particles, particles, droplets or bubbles) or 'identifiable pieces of matter' [1] advected by the flow. This is in contrast to an Eulerian formulation which involves keeping account of information at fixed points (such as mesh points in a numerical code). Both formulations are formally equivalent and originate from Euler, as noted by Lamb [8]. The penalty in developing a Lagrangian formulation is the large number of pieces that must be tracked in time; but it has great strengths in that it enables physical processes to be easily interpreted. Figure 1a illustrates the added complexity that may result from a Lagrangian formulation with the simplest problem: irrotational flow past a rigid cylinder. In an Eulerian framework, the flow pattern is well-known with tagged fluid elements being advected around the cylinder (Figure 1a (i)). In contrast (Figure 1a (ii)), for a Lagrangian formulation, informa-tion following fluid elements is tracked. The fluid particle trajectories are complex and must be calculated numerically. Solving a problem computationally within a Lagrangian formulation may involve following a deformable grid; while such ap-proaches are applied to non-Newtonian flows, there can be significant problems as grid elements become stretched resulting in the Jacobian of the mapping becoming zero.

Lagrangian models of a dispersed multiphase flow involves following individual elements of the dispersed phase. Figure 1b illustrates a particle interacting with an ambient flow such as, for example, a vortex. The ambient flow undisturbed by the particle is characterised by a lengthscale L_v, while the particle is characterised by a size a. Their trajectories are calculated by integrating the equations describing the action of force and torque on individual elements with time, estimated from the local velocity gradient tensor, the instantaneous velocity and acceleration of the particles (see [9]). In this paper, we describe how to set up a Lagrangian model of a dispersed phase and some of the assumptions on which they are based. Broadly, our discussion is focussed on when there is a separation of scales between L_v and the particle size.

413

S. Balachandar and A. Prosperetti (eds), Proceedings of the IUTAM Symposium on Computa-tional Multiphase Flow, 413–422.

When $a \ll L_v$, the Lagrangian description is quite mature and we show why, under certain conditions, adding together viscous and inviscid contributions is justified and apply the description to examine how particles are dispersed by coherent structures. When $a \geq L_v$, current Lagrangian models fail because the force description is no longer accurate. We provide new results showing how the force description is modified for inviscid flows and some of the important implications.

2 Small particles interacting with a large vortex ($a \ll L_v$)

When $a \ll L_v$, the dynamics of the discrete phase is determined by integrating the particle's equation of motion, using *local* estimates of the velocity and vorticity field (whether or not they are coupled to the flow of the discrete phase). We describe new results justifying the adhoc assumption that the total force on a particle may be estimated by adding together viscous and inviscid forces (for weakly straining flows). We then apply a one-way coupled Lagrangian model to study how particles are transported and dispersed in the vicinity of coherent structures such as spherical vortices.

2.1 Force and torque on individual particles

The dynamics of individual particles are determined from expressions describing the force and torque acting on them. The force on a particle moving with velocity v in a flow $u(x, t)$ is described by

$$\rho_p V \frac{dv}{dt} = -\rho_f C_m V \frac{dv}{dt} + \rho_f (1 + C_m) V \frac{Du}{Dt} + F_d + F_L + F_g, \tag{1}$$

where ρ_p, ρ_f are the density of particle, fluid, C_m is the added-mass coefficient and V is the volume of the particle. The force consists of added-mass, inertial force, buoyancy (F_g), viscous/form drag (F_d), shear-induced lift (F_L), and other forces which we do not consider here. The description of the effects of torque on particles and bubbles is quite recent, with an important contribution by Mougin and Magnaudet [11] to this area. It is important to note that the viscous force depends on the instantaneous *relative* velocity of the particles to the ambient flow while the inertial force depends on the instantaneous *local* acceleration of the flow. Although there is substantial effort computationally to justify the adhoc approach of simply adding inviscid and viscous forces together, there is little supporting theoretical work (except in the weakly inertial limit).

With a prescribed flow field $u(x, t)$, (1) can be integrated to determine how particles move through an evolving, inhomogeneous flow when low particle concentration and mass fraction are assumed. As the complexity of the flow description increases (from K.S. to D.N.S., for example), fewer terms in (1) tend to be included in the particle equation of motion. A useful approach is to drastically simplify the description of the ambient flow (e.g. by representing it in terms of coherent structures), while still capturing the salient features of the particle dynamics in turbulence.

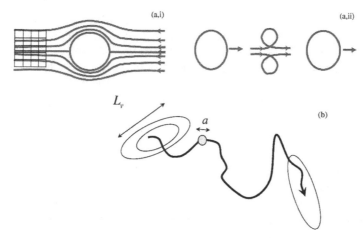

Fig. 1. (a) Schematic illustrating the difference between an Eulerian and Lagrangian description The grid on the left of the cylinder illustrates the typical Eulerian approach where information is calculated at fixed (grid) points. (b) Schematic of a particle moving near a vortex.

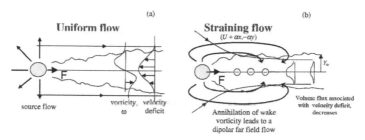

Fig. 2. Schematic (a) showing how the flow around a rigid body in a uniform flow is modified (b) by the presence of strain.

2.2 The addition of inviscid and viscous forces for straining flows

Figure 2a shows a schematic of the inertial flow past a rigid body. Positive and negative vorticity is generated on the surface of the body and advected downstream. In three-dimensions the Reynolds number of the wake flow decreases sufficiently for it to be laminar far downstream; for two-dimensional flows the Reynolds number is constant and a von Karman vortex sheet wake is generated. The wake vorticity generates a positive velocity deficit, so that fluid is transported towards the body. Sufficiently far enough downstream of the body, the volume flux Q associated with the wake tends to a constant which is related to the force on the body F_0, through $Q = F_0/\rho U$, with U the body translation velocity [2]. The presence of the volume flux means that through mass conservation the far field flow away from the wake is dominated by a source of strength Q.

    In the presence of a weak positive planar straining flow (Figure 2b), the cross-stream diffusive flux of vorticity, which normally leads to the wake spreading, is counterbalanced by the inward convective flux so that the wake spreading is ultimately arrested. Now, the positive and negative components of vorticity diffuse into one another, leading to vorticity annihilation and a gradual reduction in the strength of the velocity deficit. The volume flux in the wake decreases rapidly downstream of the body so that the flow is ultimately irrotational. In such circumstances, Betz's [2] analysis is now no longer valid. The total force on the body, evaluated using a momentum flux argument, is

$$F = -\rho_f \mu_T U \frac{du_x}{dx}, \tag{2}$$

where μ_T is the total dipole moment. The reduction of the volume flux in the wake is equivalent to a line of sinks, which combined with the source flow near the body leads to a dipolar far field, characterised by a dipole moment $\mu_W = Q/2\pi\alpha$. The local flow around the body also generates a dipolar flow with moment μ_B and speed U, so that $\mu_T = \mu_W + \mu_B$ and

$$F = -\rho_f(\mu_W + \mu_B)U\frac{du_x}{dx} = \rho_f QU - \rho_f \mu_B \frac{du_x}{dx} = F_0 - \rho_f \mu_B U \frac{du_x}{dx}. \tag{3}$$

The dipole moment associated with the flow around the body differs less than 10% from the inviscid prediction (see [7]). Thus, according to the above calculation, the addition of a viscous and inviscid force is justified, and this is supported by the numerical work of Magnaudet et al. [10].

2.3 Particle motion near coherent structures

One of the important applications of Lagrangian models is to develop closure relations, such as average settling velocity and dispersivity, for use in Eulerian models. We illustrate this technique by considering how particles move near coherent structures, in this case represented simply as a spherical vortex. Even for such a well-defined problem there are subtle ambiguities about defining even an apparently simple quantity. To illustrate this, consider inertialess particles settling in a steady flow $u(x)$ with a terminal fall velocity v_T. Their velocity will be

$$v = u - v_T \hat{x}. \tag{4}$$

If particles are introduced randomly into a bounded flow with zero mean velocity, $\langle u \rangle = 0$, we could anticipate that the average particle fall velocity

$$\langle v \rangle = \langle u \rangle - v_T \hat{x} = -v_T \hat{x}, \tag{5}$$

is unchanged, since particles are as likely to experience a positive or negative vertical velocity. But if we follow particles in time, we can see that they are always excluded from a fraction of the flow – the shadow regions – and this biased sampling by the particles means that the average fall velocity is slightly increased so that

$$\langle v \rangle = \langle u \rangle - v_T \hat{x} = -v_T(1 + \alpha)\hat{x}, \tag{6}$$

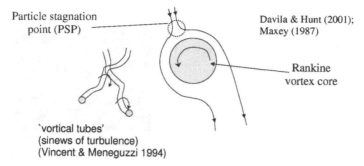

Particle stagnation
point (PSP)

Davila & Hunt (2001);
Maxey (1987)

Rankine
vortex core

'vortical tubes'
(sinews of turbulence)
(Vincent & Meneguzzi 1994)

Fig. 3. Schematic showing particles sedimenting past a Rankine vortex and illustrating the presence of a PSP.

where α is the volume fraction of the shadow regions. The particle velocity is therefore determined by the non-dimensional parameter v_T/U. Whether particles sediment faster or slower (or disperse faster or slower) in the presence of turbulence depends on the structure of the turbulence and how quickly particles respond to the flow field, which is characterised by the Stokes number $St = \tau_p/T_L$, where τ_p is the particle response time and T_L is the advective timescale associated with the flow.

To understand how the mean settling velocity of particles is influenced by vortices, Davila and Hunt [3] considered a reduced model where particles sediment past a Rankine vortex, taken to represent the elongated vortical structures observed in the DNS computations of Vincent and Meneguzzi [15] – the 'sinews of turbulence' (see Figure 3).

Davila and Hunt [3] were able to clarify the difficulty with interpreting the mean settling velocity because they considered a problem where the deviation of particle trajectories from the vertical was significant. Of particular importance is the presence of the particle stagnation points (PSPs) where the local vertical fluid velocity is equal to v_T. Shadow regions can be determined by linking together PSPs. To understand how these affect particle dispersivity, consider particles sedimenting past a vortex ring – for which some of the conditionally convergent drift integrals considered by Davila and Hunt [3] are now rendered exact. Despite the simplified nature of these problems, they can provide a clear 'mechanistic' description of particle dynamics in turbulence where the influence of PSPs dominate the bulk settling properties, and this can be supported by experimental evidence.

Figure 4a shows the computed trajectories of particles (characterised by $v_T/U = 0.1$) released above a Hill's spherical vortex translating vertically upwards. Quickly responding particles released close to the centreline are pushed around the vortex giving rise to trajectories that are also almost symmetrical up and downstream of the vortex. The radius of the shadow region is comparable to, but smaller than the size of the vortex. As the particle Stokes number increases, particles penetrate the PSP and spend an increased time within the vortex. Figure 4b shows experimental observations of particles sedimenting near a spherical vortex where $v_T/U = 0.25$.

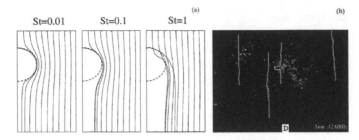

Fig. 4. (a) The influence of Stokes number on the trajectories of particles falling past a vertically rising Hill's spherical vortex [5], and (b) experimental observations.

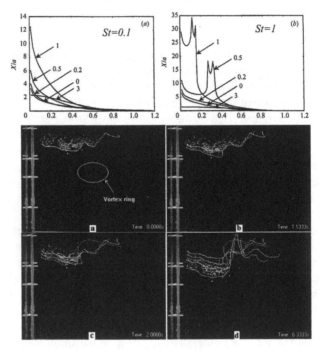

Fig. 5. (a, b) Numerical calculations of the deformation of a horizontal sheet of particles for different values of v_T/U. The right-hand figure shows the influence of increasing the particle Stokes number from $St = 0.1$ to 1 (both from [5]). The group of images below show experimental observations of a vortex passing through a descending sheet of particles for $v_T/U \approx 0.25$.

Particles close to the centreline penetrate the vortex, but still a shadow region is created.

To estimate how particles are dispersed by a vortex, it is pertinent to consider how 'conceptually' a horizontal sheet of particles is deformed by a vertically translating vortex. Figure 5a shows the permanent deformation of a horizontal sheet for a fixed Stokes number $St = 0.1$ for various v_T/U. The vertical displacement of the

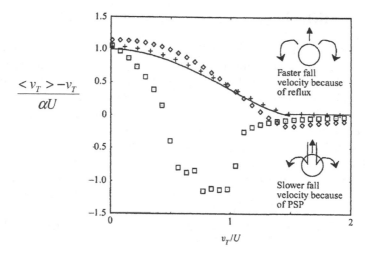

Fig. 6. The change in the average settling velocity of dense particles sedimenting through a random array of vortices is shown for varying v_T/U and $St = 0.01$ (+), 0.1 ($\diamond$), and 1 ($\square$), from [5].

sheet does not decrease monotonically with v_T, since it depends subtly on how long particles spend in the vortex. For $v_T/U \sim 1$, particles spend an increased time within the vortex leading to a faster dispersion. Figure 5b shows experimental observations of a dilute, almost horizontal, thin cloud of particles sedimenting through/around a vertical vortex. The lines drawn on the images indicate the edge of the particle cloud. The interface is strongly deformed and pushed forward by the local flow around the vortex consistent with the theoretical model.

Figure 6 shows how the mean settling velocity of particles are influenced by the vortices. Since the numerical calculations were undertaken for unbounded flows, there are some effects arising from the boundedness of a flow, most notably reflux (or return flow) contributions. These calculations highlight the subtle influence of global mass constraints on the flow, which must always be considered, and which are also dependent on whether (in computational models) the flow is doubly or singly periodic. Whether the mean settling velocity is increased or decreased by the vortices depends on the relative contribution of reflux (which pushes particles around the vortex) or particle inertia (which leads to particles spending an increased time near the PSPs).

Whether particles are dispersed quickly or slowly depends both on the fall velocity of the particles and their Stokes number. For low Stokes number, dense particles may be dispersed faster than neutrally buoyant fluid particles when $v_T/U \sim 1$, but they are dispersed much more slowly if they have a faster setting velocity. For high St, particles are dispersed much faster. The particle dispersivity tensor is always anisotropic for settling dense particles, though in this case both the particle and fluid dispersivity tensor are anisotropic. An asymptotic analysis, based on $St \ll 1$, highlights the sensitivity of horizontal dispersivity to inertia, since it varies as St^2.

3 Large particles interacting with small vortices ($a \geq L_v$)

The inertial flow past rigid particles may by turbulent or unsteady, creating flow features with a scale L_v comparable to, or even smaller than a, the characteristic size of the particles. Particles that shed vorticity will experience unsteady lift and drag forces [13], as well as those from their interaction with vortices shed from upstream particles. These forces are significant for particles whose density is comparable to, or less than, the ambient fluid. Under these circumstances, the particles are not much smaller than the local lengthscale associated with the flow, and the approximations for the force and torque generally used are not suitable. We need therefore to understand how particles move in the vicinity of coherent structures, such as vortices, which are of comparable size to or smaller than the particles. This is illustrated here using a inviscid model of a planar rigid body interacting with a distribution of point vortices. The flow generated by a collection of (free) point vortices may be interpreted in terms of image vortices required to satisfy the kinematic constraint on the surface of the body, in addition to bound vorticity representing the body itself [12].

The force on a body located at X_b and moving with velocity $U = \dot{X}_b$ is equal to the normal pressure force integrated over the surface of the body (S_b), defined by

$$\rho_b V I \cdot \dot{U} = F = \int_{S_b} p\hat{n}\,dS = -\left[\dot{I}_b + \dot{I}_i + \dot{I}_v\right], \tag{7}$$

where $\hat{n}$ is the unit vector normal to the surface of the body and directed into the body. I_b, I_v and I_i are respectively the impulse of the body, vortices and image vortices. The force on the body is therefore determined by the rate of decrease of the total impulse of the flow.

The impulse of the body, I_b, is determined by the velocity of the body and its geometry, characterised in terms of the added-mass tensor C_m, through $I_b = \rho_f C_m V \cdot U$. Equation (7) is identical to the result of Sarpkaya and Garrison [14]. Integrating (7) with respect to time gives

$$\left[(\rho_b I + \rho_f C_m)V \cdot U + I_i + I_v\right]_0^t = 0. \tag{8}$$

The above expression has a clear physical interpretation, with the sum of the momentum of the body and the total impulse of the flow being conserved. It is the simple form of (8) which enables us to study analytically the coupled dynamics of isolated bodies and singular distributions of vorticity.

To illustrate the subtlety of the coupled interaction between the exterior flow and a 'particle' we consider a dipolar vortex moving from infinity and striking a cylinder which is initially at rest. Figure 7 shows a phase diagram which distinguishes whether a cylinder acquires momentum from a vortex and moves off to infinity, or whether it is just displaced a finite distance forward. This exchange depends critically on the density of the cylinder, with dense cylinders not acquiring momentum. Such processes also occur in three dimensional flows. This illustrates the subtlety of the critical binding together of the momentum of bodies and the ambient flow, particularly when $\rho_b \leq \rho_f$, processes that are not yet included in current models.

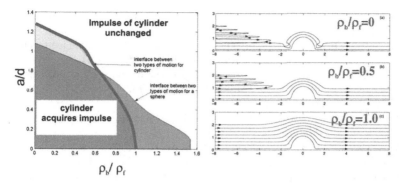

Fig. 7. The left-hand side shows a phase diagram of the interaction between a cylinder (radius a, density ρ_b) and a dipolar vortex (width $2d$), which determines whether the cylinder acquires impulse or does not. On the right-hand side, the trajectories of the dipolar vortex, for different d and ρ_b, are shown.

4 Concluding remarks

In this paper we have briefly described some of the issues related to developing and applying a Lagrangian model of a dilute two-phase flow. The essential feature of such models is following individual elements in time using semi-empirical expressions for the force and torque acting on them. The primary aim of such models is to quantify how material is moved from one place to another and to develop the necessary closure relations for Eulerian CFD models. There are still a broad number of issues which still need to be addressed in relation to Lagrangian models. When $a \ll L_v$, Lagrangian models seem to be quite mature but there is still significant progress to be made in trying to understand how inhomogeneous flow fields affect the bulk settling properties of particles. Lagrangian models for $a \geq L_v$ are in their infancy with most of the progress currently made using full resolved flow fields. Finally we note that according to Lamb [8], there should be a formal equivalence between Eulerian and Lagrangian models but this equivalence is exact only in a few instances in multiphase flow models, raising some fundamental questions about the class of models described by (1).

References

1. Batchelor, G.K., 1967, *An Introduction to Fluid Dynamics*, Cambridge University Press.
2. Betz, A., 1925, Ein Verfahren zur direkten Ermittlund des Profilwiderstandes, *Z.F.M.* **16**, 42.
3. Dávila, J. and Hunt, J.C.R., 2001, Settling of small particles near vortices and in turbulence, *J. Fluid Mech.* **440**, 117–145.
4. Drew, D.A. and Wallis, G.B., 1992, Fundamentals of two-phase flow modelling, in *Third International Workshop on Two-Phase Flow Fundamentals*, G.F. Hewitt and R.T. Lahey (eds).

5. Eames, I. and Gilbertson, M.A., 2004, The settling and dispersion of small dense particles by spherical vortices, *J. Fluid Mech.* **498**, 183–203.

6. Hill, R.J., Koch, D.L. and Ladd, A.J.C., 2001, The first effects of fluid inertia on flows in ordered and random arrays of spheres, *J. Fluid Mech.* **448**, 213–241.

7. Hunt, J.C.R. and Eames, I., 2002, The disappearance of laminar and turbulent wakes in complex and straining flows, *J. Fluid Mech.* **457**, 111–132.

8. Lamb, H., 1932, *Hydrodynamics*, Dover.

9. Magnaudet, J. and Eames, I., 2000, Dynamics of high Re bubbles in inhomogeneous flows, *Annu. Rev. Fluid Mech.* **32**, 659–708.

10. Magnaudet, J., Rivero, M. and Fabre, J., 1997, Accelerated flows past a rigid sphere or a spherical bubble. Part 1. Steady straining flow, *J. Fluid Mech.* **284**, 97–135.

11. Mougin, G. and Magnaudet, J., 2002, The generalized Kirchhoff equations and their application to the interaction of a rigid body with an arbitrary time-dependent viscous flow, *Int. J. Multiphase Flow* **28**, 1837–1851.

12. Saffman, P.G., 1992, *Vortex Dynamics*, Cambridge University Press.

13. Sarpkaya, T., 1968, An analytical study of separated flow about circular cylinders, *J. Basic Eng. ASME* **90**, 511–520.

14. Sarpkaya, T. and Garrison, C.J., 1963, Vortex formation and resistance in unsteady flow, *J. Appl. Mech.* **30**, 16–24.

15. Vincent, A. and Meneguzzi, M., 1994, The dynamics of vorticity tubes in homogeneous turbulence, *J. Fluid Mech.* **258**, 245–254.

Effect of Fluid Velocity Fluctuations on the Dynamics of a Sheared Gas-Particle Suspension

V. Kumaran

Department of Chemical Engineering, Indian Institute of Science, Bangalore 560 012, India; e-mail: kumaran@chemeng.iisc.ernet.in

Abstract. Constitutive relations are derived for a gas-particle suspension in which the particles are subject to a fluid velocity field, and experience inter-particle collisions. The flow is driven by two types of energy sources, an imposed mean shear and fluid velocity fluctuations, in the limit where the time between collisions τ_c is small compared to the viscous relaxation time τ_v, so that the dissipation of energy between collisions is small compared to the energy of a particle. Constitutive relations from the kinetic theory of dense gases are used when the flow is driven by the mean shear. The effect of fluid velocity fluctuations is incorporated using an additional diffusive term in the Boltzmann equation for the particle velocity distribution, and this leads to an additional 'diffusion' stress.

1 Introduction

In the present analysis, we focus on the effect of fluid velocity fluctuations on the velocity distribution for a sheared granular flow. The steady state velocity distributions in sheared granular flows have been typically analysed using the kinetic theory of gases [1–4]. In this analysis, there is a source of energy due to the mean shear, and dissipation due to inelastic collisions between the particles. The velocity distribution is determined by solving the Boltzmann equation, in which the 'granular temperature', which is the mean square of the velocity fluctuations, is determined by a balance between the production due to the mean shear and the dissipation due to inelastic collisions. When the coefficient of restitution of the particles is close to 1, the dissipation of energy in a collision is small compared to the energy of a particle. An asymptotic scheme can be used in which the source and dissipation of energy are neglected in the leading approximation, and the system is identical to a gas of elastic particles, for which the distribution function is a Maxwell-Boltzmann distribution function. The steady state velocity distribution has been determined using kinetic theory, and there have been systematic derivations of kinetic equations up to Burnett order starting from the Boltzmann equation [5–7].

The dynamics of the particles in a suspension is governed by the Reynolds number (Re), which is the ratio of fluid inertia and viscosity, and the Stokes number (St),

423

S. Balachandar and A. Prosperetti (eds), Proceedings of the IUTAM Symposium on Computational Multiphase Flow, 423–431.

which is the ratio of particle inertia and fluid viscosity. The drag force exerted by the fluid on a particle depends on the Reynolds number, and is given by the linear Stokes drag law when the Reynolds number is small. Since the particle density is usually three orders of magnitude larger than the fluid density, particle inertia could be significant ($St \gg 1$) even when the Reynolds number is small. In this case, particles interact due to solid body collisions or due to hydrodynamic interactions mediated by the suspending gas. In the limit of low Reynolds number, hydrodynamic interactions can be analysed using the linear Stokes equations. Koch [8] showed that particle collisions are dominant for $St \gg \phi^{(-3/2)}$, where ϕ is the volume fraction, and the velocity distribution is close to a Maxwell distribution if the coefficient of restitution is close to 1. The velocity distribution function for a bidisperse particle-gas suspension settling under gravity was determined by Kumaran and Koch [9, 10], and it was found that the distribution function is close to a Maxwell–Boltzmann distribution when the time between collisions is small compared to the viscous relaxation time. Tsao and Koch [11] analysed the distribution function for the shear flow of a gas-solid suspension, and reported that dynamical states with different fluctuating velocities could coexist at the same particle volume fraction and mean strain rate. The effect of hydrodynamic interactions on the shear flow of a particle suspension was considered by Sangani et al. [12], using numerical simulations and asymptotic analysis in the low Reynolds number and $O(1)$ Stokes number limit. The Stokes flow interactions between particles were modified to incorporate the breakdown of the lubrication theory when the gap thickness is of the same magnitude as the mean free path. Asymptotic studies were carried out in the limit of nearly elastic collisions $(1 - e) \ll 1$, where e is the coefficient of restitution for particle collisions, as well as for high Stokes numbers. The results of a moment expansion for finite Stokes number was found to be in good agreement with numerical simulations.

In the present analysis, constitutive relations are derived for a gas-particle suspension subjected to turbulent fluctuations in the absence of hydrodynamic interactions. Two sources of fluctuating energy, an imposed shear flow and fluid velocity fluctuations, are examined in the present analysis, while the dissipation is due to viscous drag which is described by the Stokes drag law. The collisions are considered to be elastic for simplicity, and the drag force used here is assumed to be a linear function of the difference between the particle and fluid velocities. When the source of energy is due to the mean shear, the constitutive relations used here are similar to those used in the dynamics of granular materials, with an additional force on the particles due to the viscous drag exerted by the fluid. The effect of fluid velocity fluctuations generated due to fluid turbulence is incorporated using a very simple model in the present analysis in the specific limit where the time scale for the fluid velocity fluctuations is small compared to the viscous relaxation time of the particles, so that the acceleration of the particles due to the fluid velocity fluctuations can be modeled as a Gaussian white noise. For simplicity, it is assumed that the statistics of the fluid velocity fluctuations are known, and the particle motion does not affect the fluid velocity fluctuations. Though this is not expected to provide quantitatively accurate results, it does provide some indication about the qualitative effect of fluid velocity fluctuations on the growth rates of the perturbations. In the model, the particle acceleration $\mathbf{a}$ is separated into fluid drag and the effect of turbulent velocity fluctuations, $\mathbf{a} = -(\mathbf{u}/\tau_v) + (\mathbf{v}'/\tau_v)$, where $\mathbf{u}$ is the difference between the particle velocity and the mean fluid velocity, $\mathbf{v}'$ is the turbulent velocity fluctuation in the fluid, and τ_v is the viscous relaxation time.

In addition to the viscous relaxation and collision times, there is an additional time scale for the fluid velocity fluctuations, τ_f, which is is integral time (u_f/λ), where u_f is the magnitude of the fluid velocity fluctuations and λ is the Taylor microscale. For $\tau_v \ll \tau_f$ and $\tau_c \ll \tau_f$, the particle trajectories follow the fluid streamlines, and the transport of particles is similar to the turbulent diffusion of a passive scalar. The present analysis is restricted to the opposite limit $\tau_v \gg \tau_f$, and $\tau_c \gg \tau_f$, where the the change of particle velocity is small for time scales comparable to the fluid integral time, and the particles experience a fluctuating force due to the fluid turbulence, in addition to the fluid drag. The parameter regime considered here is applicable to practical situations. The viscous relaxation time for particles in a gas scales as $10^8 R^2$ seconds for particles with a density of 10^3 kg/m^3, where R is the radius in meters. An upper bound on the fluid integral time scale can be taken as (L/U), where L, the length of the largest eddies, is the same as the macroscopic scale, and U is the magnitude of the velocity (it should be noted that the strain rate increases and the turnover time decreases with a decrease in the eddy size). For macroscopic systems with $(L/U) \sim 1$ s^{-1}, which corresponds to a Reynolds number of about 10^5, the integral time is small compared to the viscous relaxation time for particles with radius larger than 100μm.

In order to obtain analytical expressions for the stress tensor, it is necessary to assume that the fluctuating force due to the fluid velocity is distributed as a Gaussian white noise distribution, so that it is possible to write a Fokker–Planck equation for the particle velocity distribution which is equivalent to the microscopic equation for the evolution of the particle velocity. This equation for the distribution function contains a term that provides the diffusion of particles in velocity space due to the fluid velocity fluctuations, in addition to the rate of change of distribution function due to particle collisions from the Boltzmann equation. The diffusion coefficient D_{ij} scales as $(v_f^2 \tau_f/\tau_v^2)$, and has units of (length^2/time^3). It can be easily inferred that the rate of increase of energy due to the diffusion in the velocity coordinates is proportional to ρD_{ii}, the isotropic part of the diffusion tensor. Therefore, the production of particle fluctuating energy due to fluid velocity fluctuations is large compared to the shear production for $D_{ii} \gg (T^{1/2}\bar{G}^2/(\rho d^2))$, and the temperature is given by $T = \tau_v D_{ii}$ in this case. In this case, the ratio of the collision time and the viscous relaxation time is $(\rho d^2 D_{ii}^{1/2} \tau_v^{3/2})^{-1}$, and the collision time is small compared to the viscous relaxation time for $(\rho d^2 D_{ii}^{1/2} \tau_v^{3/2}) \gg 1$. The analysis in Section 2 shows that the deviatoric part of the stress tensor is proportional to $\tau_c D_{ii}$, and is $O(\tau_c/\tau_v)$ smaller than the isotropic pressure.

2 Constitutive relations

The system consists of a suspension of elastic particles of diameter d in a fluid with viscosity η subjected to a uniform shear flow with strain rate $\bar{G}$. A Cartesian coordinate system is used, where the mean velocity is in the x direction, the velocity gradient is in the y direction and the vorticity is in the z direction. The mean velocity of the fluid and particles are equal, and the effect of the fluid on the particles is modeled by a linear drag law. In this section, the mass and length dimensions are scaled by the particle mass and diameter in all the quantities.

The fluid velocity fluctuations are assumed to be uncorrelated over time scales comparable to the collision time (time between collisions) if the correlation time of the velocity fluctuations is small compared to the time between collisions, or if the mean free path is large compared to the correlation length of the turbulent eddies that cause the velocity fluctuations. The drag force is considered to be a linear function of the difference between the particle and fluid velocities. For the present purposes, the fluid velocity fluctuation is separated into a mean velocity and a fluctuating component due to the turbulent fluctuations. If the difference between the particle velocity and the mean velocity of the fluid is $\mathbf{u}$, and the fluctuating velocity is $\mathbf{v}$, the particle acceleration is defined as

$$\mathbf{a} = -\frac{\mathbf{u}}{\tau_v} + \frac{\mathbf{v'}}{\tau_v}. \tag{1}$$

The evolution of the velocity with time is then given by

$$\mathbf{u}(t) = \exp\left(-t/\tau_v\right)\mathbf{u}(0) + \frac{1}{\tau_v}\exp\left(-t/\tau_v\right)\int_0^t dt' \exp\left(t'/\tau_v\right)\mathbf{v'}(t'). \tag{2}$$

The rate of change of the second moment of the velocity distribution of the particles can be easily determined by taking the tensor product of (2) and the velocity, and averaging over the turbulent velocity fluctuations,

$$\frac{d\mathbf{uu}}{dt} = -\frac{2\mathbf{uu}}{\tau_v} + \exp\left(-t/\tau_v\right)\frac{\langle\mathbf{u}(0)\mathbf{v'}(t) + \mathbf{v'}(t)\mathbf{u}(0)\rangle}{\tau_v}$$
$$+ \frac{1}{\tau_v^2}\exp\left(-t/\tau_v\right)\int_0^t dt' \exp\left(t'/\tau_v\right)\langle\mathbf{v'}(t)\mathbf{v'}(t') + \mathbf{v'}(t')\mathbf{v'}(t)\rangle, \tag{3}$$

where $\langle \ \rangle$ is an average over all realisations of the fluid velocity fluctuations, and the average $\langle\mathbf{v'}\rangle$ is zero. The rate of change of the particle energy is

$$\frac{d(u^2/2)}{dt} = -\frac{u^2}{\tau_v} + \frac{1}{\tau_v^2}\exp\left(-t/\tau_v\right)\int_0^t dt' \exp\left(t'/\tau_v\right)\langle\mathbf{v'}(t).\mathbf{v'}(t')\rangle. \tag{4}$$

If a single exponential form is used for the correlation function $\langle\mathbf{v'}(t).\mathbf{v'}(t')\rangle = v_f^2 \exp\left(-|t - t'|/\tau_f\right)$, where τ_f is the correlation time for the fluid velocity field, the rate of change of energy is

$$\frac{d(u^2/2)}{dt} = -\frac{u^2}{\tau_v} + \frac{v_f^2}{\tau_v^2(\tau_v^{-1} - \tau_f^{-1})}(\exp(-(t/\tau_f)) - \exp(-(t/\tau_v)))$$

$$\approx -\frac{u^2}{\tau_v} + \frac{\tau_f v_f^2}{\tau_v^2}, \tag{5}$$

for $t \sim \tau_c$, $\tau_f \ll \tau_c$ and $\tau_v \gg \tau_c$. If the granular temperature T is the mean square of the particle velocity fluctuations, the rate of production of energy scales as $\tau_f v_f^2/\tau_v^2$, while the rate of dissipation of energy due to drag is $O(T/\tau_v)$, and so the temperature scales as $T \sim (v_f^2 \tau_f/\tau_v) \ll v_f^2$. Even though the results in (5) were calculated for a specific model for the decay of fluid velocity correlations, it can be inferred that the same scaling for T is valid for other models for the decay of fluid velocity correlations for $\tau_f \ll \tau_v$. If the time between collisions is small compared to the viscous relaxation time, the change in energy over the collision time is $O(T\tau_c/\tau_v)$, which is small compared to the energy of a particle. This provides the opportunity to use a kinetic theory approach where the leading order distribution function is given by the Maxwell-Boltzmann distribution.

To proceed further analytically, it is necessary to assume that the The distribution for the fluid velocity fluctuations is assumed to be a Gaussian, so that the Boltzmann equation for the velocity distribution function contains a diffusive term, similar to that in the Fokker–Planck equation, with a tensor diffusivity $\mathbf{D}$ given by

$$\mathbf{D} = \frac{1}{\tau_v^2} \int_0^\infty dt' \langle \mathbf{v}'(t')\mathbf{v}'(0)\rangle. \tag{6}$$

Note that $\mathbf{D}$ is a symmetric matrix, and has dimensions of (length^2/time^3), since this is a diffusion coefficient for the velocity distribution. When the fluid velocity fluctuations are driven by a shear flow in the x–y plane, the components D_{xz} and D_{yz} are zero because the probability distribution for the velocity fluctuations in the z direction is an even function of v_z'. In addition, if the fluid velocity fluctuations are driven by an imposed shear flow with positive S_{ry}, where S_{ry} is the xy component of the symmetric part of the rate of deformation tensor, then $\langle v_x' v_y'\rangle$ is negative, and so the component D_{xy} of the diffusivity tensor is negative.

The Boltzmann equation for the distribution function, with the additional diffusion term and the drag force exerted on the particles, is

$$\frac{\partial f(\mathbf{u})}{\partial t} - \mathbf{G} :(\mathbf{u}\nabla_{\mathbf{u}} f(\mathbf{u})) - \frac{1}{\tau_v}\mathbf{u}.\nabla_{\mathbf{u}} f(\mathbf{u}) - \mathbf{D} : \nabla_{\mathbf{u}}\nabla_{\mathbf{u}} f(\mathbf{u}) = \frac{\partial_c f(\mathbf{u})}{\partial t}, \tag{7}$$

where $f(\mathbf{u})$ is the distribution function, which is defined so that $f(\mathbf{u})d\mathbf{u}$ is the probability of finding a particle in the volume $d\mathbf{u}$ about $\mathbf{u}$ in velocity space, and $\nabla_{\mathbf{u}}$ is the gradient operator in velocity space. The first term on the left side of Equation (7) is the rate of change of distribution function, the second is the change in the distribution function due to the mean shear flow exerted on the particles, and $\mathbf{G} = \nabla\mathbf{U}$ is the strain rate. The third represents the effect of the drag force on the particles and the fourth is due to the fluctuating gas velocity. The term on the right side is the 'collision integral' which is the rate of change of the distribution function due to particle collisions. In the present analysis, this term is represented by the Boltzmann collision integral [13]

$$\frac{\partial_c f(\mathbf{u})}{\partial t} = \rho \chi(\phi) \int d\mathbf{k} \int d\mathbf{u}^\dagger (f(\mathbf{u}_b) f(\mathbf{u}_b^\dagger) - f(\mathbf{u}) f(\mathbf{u}^\dagger)) \mathbf{w}.\mathbf{k}. \qquad (8)$$

In Equation (8), $\mathbf{u}_b$ and $\mathbf{u}_b^\dagger$ are the velocities of a pair of particles before collision so that the post collisional velocities are $\mathbf{u}$ and $\mathbf{u}^\dagger$, $\mathbf{k}$ is the unit vector in the direction of the line joining the centers of particles at collision, $\mathbf{w} = \mathbf{u} - \mathbf{u}^\dagger$ is the velocity difference between the particles, $\chi(\phi)$ is the pair distribution function, and the above integral is carried out for $\mathbf{w}.\mathbf{k} \geq 0$ so that the particles approach each other prior to collisions.

The simplest procedure for obtaining a constitutive relation from the Boltzmann equation involves the use of a second moment closure approximation for the distribution function,

$$f = \frac{1}{(2\pi)^{3/2} \mathrm{Det}(\mathbf{T})^{1/2}} \exp\left(-\frac{\mathbf{u}.\mathbf{T}^{-1}.\mathbf{u}}{2}\right). \qquad (9)$$

This distribution function is inserted into the Boltzmann equation, multiplied by the $\mathbf{uu}$, and integrated over velocity space in order to determine the second order tensor $\mathbf{T}$. The stress is then determined from the distribution function. It is difficult to obtain analytical solutions for the Boltzmann equation due to the non-local nature of the collision integral in velocity space, and an asymptotic expansion is used in the ratio of the collision and viscous relaxation time. The leading order solution for the distribution is calculated assuming that the energy is conserved in collisions, since the dissipation of energy between successive collisions is small compared to the energy of a particle for $\tau_c \ll \tau_v$. The leading order solution is a Maxwell–Boltzmann distribution, and the leading approximation $\mathbf{T}^{(0)} = T\mathbf{I}$ is isotropic, where $\mathbf{I}$ is the identity tensor. However, the value of T is not determined in the leading approximation, and is determined from the first correction to the energy balance equation. The first correction $\mathbf{T}^{(1)}$ is determined from the first correction to the deviatoric part of the second moment balance equation, and the viscous stress is determined from $\mathbf{T}^{(1)}$. Though this procedure does not give an exact result, due to an assumption regarding the specific form of the distribution function, it is known that the result for the viscosity obtained from this procedure is in error by about 1.2% when compared to that obtained by a more exact procedure [13]. When a similar procedure is applied to the modified Boltzmann Equation (7) which contains the term due to diffusion in velocity space, it can easily be seen that the diffusion coefficient $\mathbf{D}$ is an inhomogeneous term in the resulting equation, which is independent of $\mathbf{T}$ and $\mathbf{G}$. This provides an additional 'diffusion' stress due to the diffusion of particles in velocity space.

The equations for the density (ρ) and velocity (U_i) fields for the particle phase are of the form,

$$\partial_t \rho + \nabla(\rho \mathbf{U}) = 0, \qquad (10)$$

$$\rho(\partial_t U_i + (\mathbf{U}.\nabla)\mathbf{U}) = \nabla.\sigma - R(\phi)(\mathbf{U} - \mathbf{V}), \tag{11}$$

where $\mathbf{U}$ is the mean velocity of the particle phase, $\mathbf{V}$ is the mean velocity of the fluid, $\partial_t \equiv (\partial/\partial t)$, σ is the stress tensor, ϕ is the volume fraction, and $R(\phi)$ is the drag coefficient, which is (ρ/τ_v) in the dilute limit. The last term on the right side of (11) is the drag force exerted by the fluid on the particles. At moderate and high density, the factor $R(\phi)$ incorporates the variation in the drag force with particle volume fraction due to interaction between particles.

Expressions for the pressure, viscosity and viscometric coefficients from the kinetic theory of dense gases are used in the present analysis, suitably augmented by terms that arise from the diffusion of particles in velocity space. The stress tensor σ for the particle phase is given by the constitutive relation for a 'Newtonian' fluid,

$$\sigma = -p\mathbf{I} - \mathbf{E} + \mu(\nabla\mathbf{U} + (\nabla\mathbf{U})^T - (2/3)\mathbf{I}\nabla.\mathbf{U}) + \mu_b\mathbf{I}\nabla.\mathbf{U}, \tag{12}$$

where p is the particle pressure, and μ and μ_b are the particle phase shear and bulk viscosities respectively. The pressure and viscosities for the particle phase depend on the particle density and the temperature T [13],

$$p = \rho T(1 + 4\phi\chi(\phi)), \tag{13}$$

$$\mu(\phi) = \mu_\phi(\phi)T^{1/2}, \tag{14}$$

$$\mu_b(\phi) = \mu_{b\phi}(\phi)T^{1/2}, \tag{15}$$

where μ_ϕ and $\mu_{b\phi}$ are functions of the volume fraction,

$$\mu_\phi = \frac{5}{16\sqrt{\pi}\chi(\phi)}\left(1 + \frac{8\phi\chi(\phi)}{5}\right)^2 + \frac{48\phi^2\chi(\phi)}{5\pi^{3/2}}, \tag{16}$$

$$\mu_{b\phi} = \frac{16\phi^2\chi(\phi)}{\pi^{3/2}}, \tag{17}$$

and ϕ is the volume fraction of the particles. The contribution to the stress tensor due to the fluid velocity fluctuations, $\mathbf{E}$, is given by

$$\mathbf{E} = \frac{5(\mathbf{D} - (\mathbf{I}/3)\mathrm{Tr}(\mathbf{D}))}{8\sqrt{\pi}\chi T^{1/2}}\left(1 + \frac{8\phi\chi}{5}\right). \tag{18}$$

In order to evaluate the above contribution to the stress tensor, it is necessary to determine the correction to the Boltzmann Equation (7) due to the diffusion in velocity space, D_{ij}, using an asymptotic expansion in the ratio of the collision and viscous relaxation times. This calculation is algebraically complicated, and so the details are not provided here. However, the form of the stress E_{ij}, correct to within multiplicative constants, can be determined as follows. The first correction to the second moment T_{ij} in (9) due to the diffusion in velocity space is $O(\tau_c(\mathbf{D} - (\mathbf{I}/3)\mathrm{Tr}(\mathbf{D})))$. It should be noted that the trace of the first correction $\mathbf{T}^{(1)}$ is zero, without loss of generality, because the leading approximation is isotropic when the collision time is small compared to the viscous relaxation time. The first correction to the stress is proportional to $(-\rho \mathbf{T}^{(1)})$, where $\mathbf{T}^{(1)}$ is the first correction to $\mathbf{T}$, in the dilute (kinetic) regime [13] where stress is transmitted due to the physical motion of particles. The additional factor $(8\phi\chi/5)$ in Equation (18) accounts for the 'collisional' contribution to the stress at high densities [13].

The equation for the particle temperature has the form

$$\rho C_v \frac{DT}{Dt} + p\nabla\mathbf{U} - \rho\mathrm{Tr}(\mathbf{D}) + \mathbf{E} : \nabla\mathbf{U} - 2\mu(\phi, T)(\nabla\mathbf{U}){:}(\nabla\mathbf{U})$$
$$- \mu_b(\phi, T)(\nabla.\mathbf{U})^2 + 2C_v R(\phi)T + \nabla.\mathbf{q} = 0, \tag{19}$$

where $C_v = (3/2)$ is the specific heat at constant volume, the heat flux is given by

$$\mathbf{q} = -K\nabla T, \tag{20}$$

where K, the thermal conductivity, is

$$K(\phi) = T^{1/2}\left[\frac{75}{64\sqrt{\pi}\chi(\phi)}\left(1 + \frac{12\pi\chi(\phi)}{5}\right)^2 + \frac{15\phi^2\chi(\phi)}{2\sqrt{\pi}}\right]. \tag{21}$$

The second term on the left side of (19) is the rate of change of energy due to compression or expansion, while the third term on the left is the source of energy due to the fluid velocity fluctuations. The fourth term on the left is the work done due to the diffusion stress E_{ij}, while the fifth and sixth terms contribute to the rate of increase of particle energy due to viscous dissipation. The seventh term on the left is the rate of dissipation of energy of the particles due to the drag force. As noted after Equation (6), if S_{xy} is positive for a shear flow in the x–y plane, then D_{xy} and E_{xy} are negative, and so the production of energy due to the diffusion stress has the same sign as the production due to the viscous stress for the particle phase.

3 Conclusions

The important conclusion from this analysis is the effect of fluid velocity fluctuations on the constitutive relation for a gas-particle suspension. Though a very simple model was used for the fluid velocity fluctuations in the present analysis, the results indicate that there is an additional 'diffusion' stress in the expression for the stress tensor due to spatial variations in the correlation function for the fluid velocity fluctuations. In addition, there is an additional source of energy in the suspension due to the fluid velocity fluctuations, which is balanced by the dissipation due to inelastic collisions or due to viscous drag. This could have a significant effect on the dynamics of the suspension; in particular, this contribution is known to stabilise the uniform state of a sheared suspension which is unstable in the absence of fluid velocity fluctuations [14].

References

1. S.B. Savage and D.J. Jeffrey, 1981, The stress tensor in a granular flow at high shear rates, *J. Fluid Mech.*, **110**, 255–272.
2. J.T. Jenkins and S.B. Savage, 1983, A theory for the rapid flow of identical, smooth, nearly elastic particles, *J. Fluid Mech.*, **130**, 186–202.
3. C.K.K. Lun, S.B. Savage, D.J. Jeffrey and N. Chepurnity, 1984, Kinetic theories for granular flow: inelastic particles in Couette flow and slightly inelastic particles in a general flow field, *J. Fluid Mech.*, **140**, 223–256.
4. J.T. Jenkins and M.W. Richman, 1985, Grad's 13 – Moment system for a dense gas of inelastic spheres, *Arch. Rat. Mech. Anal.*, **87**, 355–377.
5. N. Sela, I. Goldhirsch and S. H. Noskowicz, 1996, Kinetic theoretical study of a simply sheared two dimensional granular gas to Burnett order, *Phys. Fluids*, **8**, 2337.
6. N. Sela and I. Goldhirsch, 1998, Hydrodynamic equations for rapid flows of smooth inelastic spheres, to Burnett order, *J. Fluid Mech.*, **361**, 41–74.
7. V. Kumaran, 2004, Constitutive relations and linear stability of a sheared granular flow, *J. Fluid Mech.*, **506**, 1.
8. D.L. Koch, 1990, Kinetic theory for a monodisperse gas-solid suspension, *Phys. Fluids A*, **6**, 2894–2899.
9. V. Kumaran and D.L. Koch, 1993, Properties of a bidisperse particle - gas suspension. Part 1. Collision time small compared to viscous relaxation time, *J. Fluid Mech.*, **247**, 623–642.
10. V. Kumaran and D.L. Koch, 1993, Properties of a bidisperse particle - gas suspension. Part 2. Viscous relaxation time small compared to collision relaxation time, *J. Fluid Mech.*, **247**, 643–660.
11. H.-K. Tsao and D.L. Koch, 1995, Shear flows of a dilute gas-solid suspension, *J. Fluid Mech.*, **296**, 211–245.
12. A.S. Sangani, G. Mo, H.-K. Tsao and D.L. Koch, 1996, Simple shear flows of dense gas-solid suspensions at finite Stokes numbers, *J. Fluid Mech.*, **313**, 309–341.
13. S. Chapman and T.G. Cowling, 1970, *The Mathematical Theory of Non-Uniform Gases*, Cambridge University Press, London.
14. V. Kumaran, 2003, Stability of a sheared particle suspension, *Phys. Fluids*, **15**, 3625.

Prediction of Particle Laden Turbulent Channel Flow Using One-Dimensional Turbulence

John R. Schmidt[1], Jost O.L. Wendt[2] and Alan R. Kerstein[3]

[1] *Army Research Lab, Aberdeen Proving Ground, MD 21005, USA*
[2] *The University of Arizona, Tucson, AZ 85721, USA*
[3] *Sandia National Labs, Livermore, CA 94551, USA*

Abstract. This paper presents a method for integrating two-phase flow into the vector formulation of the One-Dimensional Turbulence model (ODT) without the introduction of any additional free parameters into the model. ODT is an unsteady turbulent flow simulation model implemented on a one-dimensional domain, representing flow evolution as observed along a line of sight through a 3D turbulent flow. Overturning motions representing individual eddies are implemented as instantaneous rearrangement events. Particles are simulated in a turbulent channel using one-way coupling.

Numerical simulations were run with turbulent friction Reynolds numbers, Re_τ, 180 and 640. Validation was achieved by comparing wall-normal profiles of particle statistics with DNS, LES, and experiments.

1 Introduction

Particle transport in turbulent flows is of immense importance in engineering and scientific disciplines. Because it is so widespread in nature, environmental scientists need to study and understand it for weather and pollution control. Examples range from volcanic dust dispersion in the atmosphere, to the formation of clouds, to the entrainment of pharmaceuticals into air. The development and validation of a two-phase flow submodel for vector ODT would prove advantageous for the advancement of one-dimensional (inexpensive) turbulent two-phase modeling.

2 Overview of the One-Dimensional Turbulence (ODT) model

Kerstein [2] developed a one-dimensional Monte Carlo modeling technique for turbulent mixing of velocity and scalar fields. A subsequent extension [3] keeps three velocity components on the ODT domain. This allows for the introduction of an ODT analogy of pressure scrambling. The fields defined on the one-dimensional domain evolve by two mechanisms: molecular diffusion and a stochastic process representing advection. The ODT approach represents turbulent advection by a random sequence

S. Balachandar and A. Prosperetti (eds), Proceedings of the IUTAM Symposium on Computational Multiphase Flow, 433441.

of "eddy" maps applied to a one-dimensional computational domain. Profiles of the velocity components (u_i) and the advected scalars evolve on this domain. Equations for the turbulent flow field are not solved explicitly, rather the viscous and diffusive equations are solved,

$$\frac{\partial u_i(y,t)}{\partial t} = \nu \frac{\partial^2 u_i(y,t)}{\partial y^2} - \frac{1}{\rho}\frac{dp}{dx}, \quad \frac{\partial \theta(y,t)}{\partial t} = \kappa \frac{\partial^2 \theta(y,t)}{\partial y^2}, \tag{1}$$

where t is time, ν is the kinematic viscosity, ρ is the density, θ can be any advected scalar (e.g. temperature or species concentration) and κ is the corresponding diffusion coefficient. Note that the dp/dx term in Equation (1) is an imposed mean pressure gradient in the streamwise direction.

In order for ODT to be used there must be a minimum of one homogeneous direction. There should be a predominantly streamwise direction. The one dimension in the ODT model is transverse to the mean flow. The ODT model implements triplet maps or eddies as instantaneous rearrangements of the velocity $u_i(y,t)$ field.

The events representing advection may be interpreted as the model analogue of individual turbulent eddies. However, this interpretation is not essential to the analysis; it merely provides an intuitive basis for presenting the model. Essentially each "eddy event" has three properties: a length scale, a time scale, τ, and a measure of kinetic energy.

The vector (3-component) form of ODT has eddy events consisting of two mathematical operations. The first is a measure-preserving map representing the fluid motions of a turbulent eddy. The other is a modification of the velocity profiles in order to account for energy transfers between velocity components.

$$u_i(y) \rightarrow u_i(f(y)) + c_i K(y), \quad \theta(y) \rightarrow \theta(f(y)). \tag{2}$$

The fluid at location $f(y)$ is moved to location y by the mapping operation. This mapping is the vector ODT analog of the advection operator $v \cdot \mathrm{grad}$ of the Navier–Stokes equations. This mapping is applied to all fluid properties. The additional term $c_i K(y)$ which is only applied to the velocity components is the ODT analogue of pressure-induced energy redistribution among the velocity components. This also takes care of velocity changes due to pressure gradients or body forces.

The triplet map has a starting point y_0 and a length l which are sampled randomly from an eddy distribution rate. The mapping rule $y \rightarrow \bar{y}$ for a triplet map is given by

$$\bar{y} = \begin{cases} y_0 + 1/3\,(y - y_0) & y_0 < \bar{y} < y_0 + 1/3\,l \\ y_0 + 2/3\,l - 1/3\,(y - y_0) & y_0 + 1/3\,l < \bar{l} < y_0 + 2/3\,l \\ y_0 + 2/3\,l + 1/3\,(y - y_0) & y_0 + 2/3\,l < \bar{y} < y_0 + l \\ y & \text{otherwise} \end{cases} \tag{3}$$

where $\bar{y}$ is the y profile after the instantaneous rearrangement.

The desired attribute of the triplet map is to provide a means of mimicking the increase in strain intensity, the decrease in strain length scale and the increase in mixing due to eddies in physical turbulent flow. This mapping rule assures that closest neighbors after the spatially discretized mapping event were no more than three cells (or fluid elements) apart before the mapping event. Hence the increased strain rate and shortening length scale is attained without undue introduction of discontinuities. Using the continuous analog to describe the triplet map: the original scalar profile is reduced by a factor of three, and a copy is placed in both the first third and the last third of the eddy domain. For the middle third, the reduced image is inverted.

Though there are other mappings which could be used, this implementation of the triplet map is the simplest which obeys three key physical conditions: (1) measure preservation (the non-local analog of vanishing velocity divergence; this property is manifestly satisfied in the discrete numerical implementation, in which the map is a permutation of equal-volume fluid cells on the 1D domain); (2) continuity (no introduction of discontinuities by the mapping operation); and (3) scale locality (at most order-unity changes in property gradients). The first two conditions are fundamental properties of incompressible fluid motion. The third is based on the principle that length-scale reduction in a turbulent cascade occurs by a sequence of small steps (corresponding to turbulent eddies), causing down-scale energy transfer to be effectively local in wavenumber.

Fluid parcels or elements are moved instantaneously during the triplet map from one y location to another. The momentum and passive scalar properties of each fluid particle or cell are preserved and remain unaffected (at first) by the instantaneous rearrangement. Subsequently energy redistribution among the three velocity components is implemented. This is represented by the c_i term in Equation (2). In Equation (2) the K term is a kernel function that is defined as $K(y) = y - f(y)$. Hence its value is equal to the distance the local fluid element is displaced. Therefore it is by definition non-zero only within the eddy interval l. The kernel integrates to zero so that the "eddy event" does not change the total (y-integrated) momentum of individual velocity components.

As mentioned above, each eddy event has a time, a length scale and a measure of kinetic energy associated with it. The kinetic energy of an individual velocity component i is

$$E_i \equiv \frac{1}{2}\rho \int u_i^2(y)\,\mathrm{d}y. \tag{4}$$

(The density ρ, assumed constant, is defined here as mass per unit length.) The amplitudes c_i in Equation (2) are determined for each individual eddy subject to two constraints: (1) the total kinetic energy remains constant, and (2) the energy removed from any individual velocity component by the kernel mechanism cannot exceed the energy available for extraction [3].

In ODT eddy events are instantaneous in time and occur with frequencies comparable to the turnover frequencies of corresponding turbulent eddies. Events are therefore determined by sampling from an event-rate distribution that reflects the physics governing eddy turnovers.

Flow properties (e.g. velocity variations) affect the eddy rate distribution and the successful eddy events (based on sampling using a rejection method) affect the velocity distribution (and passive scalars). This creates a feedback by increasing the strain rate which allows more triplet maps to occur. The event rate, λ, is shown to be

$$\lambda(y_0, l; t) \equiv \frac{C}{l^2 \tau(y_0, l; t)} = \frac{C\nu}{l^4} \sqrt{\left(\frac{u_{2,K}l}{\nu}\right)^2 + \alpha \sum_j T_{2,j} \left(\frac{u_{j,K}l}{\nu}\right)^2 - Z}, \quad (5)$$

where the matrix T assures invariance under axis rotation. All other new variables are defined shortly. If the quantity in the radical of Equation (5) is negative, the eddy is deemed to be suppressed by viscous damping and λ is taken to be zero for that eddy. In the square root term the quantities preceding Z involve groups that have the form of a Reynolds number. As such Z can be viewed as a parameter controlling the critical Reynolds number for eddy turnover.

There are three free parameters in the ODT model proper: C, α, and Z. The free parameter C determines the strength of the turbulence, hence C allows fine adjustments to the eddy rate distribution. The transfer coefficient α determines the degree of kinetic energy exchange among components. [The matrix T in Equation (5) depends on α.] The viscous cutoff parameter Z determines the smallest eddy size for given local strain conditions.

These three parameters, along with the initial and boundary conditions of the flow and the physical conditions of the fluid (density, viscosity, etc.), constitute the complete inputs for the vector ODT model proper. The three parameter values used here were set for single-phase channel flow [9]. Somewhat different values are preferred for free-shear flows [3]. No additional free parameters are needed for this two-phase flow application.

3 Two-phase flow addition to the ODT model

One way coupling is achieved by following motions of the particle as dictated by the particle drag law. ODT has all three velocity components (on the one-dimensional domain). As such particle trajectories are integrated in all three directions, but the particles are required to stay on the ODT domain.

The authors implement this drag coupling directly, using the vector wall-normal fluid velocity profile evolved by ODT, but lateral motion of fluid parcels (displacement by eddy events) and velocity of fluid parcels are distinct in ODT, so this procedure violates physical requirements such as correct representation of the marker-particle or tracer-particle limit. That is to say, a tracer particle does not necessarily follow the fluid cell it is in. A fluid cell only moves in the y direction during triplet maps, however, a tracer particle would likely move out of a particular fluid cell between triplet maps because the fluid velocity in the wall-normal direction v is generally non-zero.

The Bassett–Boussinesq–Oseen equation (see [11]) describes the equation of motion for a spherical particle suspended in a fluid. If the density of the particle is much greater than the density of the air, and the diameter of the particle is smaller than the smallest turbulent eddy scale, the only significant forces on the particle are gravity, F_G, and the drag force,

$$\frac{d\tilde{V}}{dt} = \frac{\tilde{F}_G}{m_p} + (\tilde{u} - \tilde{V})\frac{f}{\tau_p}, \quad \frac{d\tilde{X}}{dt} = \tilde{V}, \tag{6}$$

where tilde denotes a vector quantity, V is particle velocity, X is particle position, and m_p is particle mass. The aerodynamic response time, τ_p, the non-linear correction factor, f, [7], and the particle Reynolds number are given by

$$\tau_p = \frac{2\rho_p r_p^2}{9\mu_g}, \tag{7}$$

$$f = \begin{cases} 1 & \text{Stokes law,} \\ 1 + 0.15\text{Re}^{0.687} & \text{non-linear,} \end{cases} \quad \text{Re} = \frac{\rho r_p |\tilde{V} - \tilde{u}|}{\mu_g}, \tag{8}$$

where the subscript p denotes properties of the particle, μ_g is the viscosity of the gas, and r_p is the radius of the particle. Non-linearity of the drag force is significant for a particle Reynolds number near or greater than one.

4 Results, conclusions, and future work

4.1 Mean and rms velocities at Re_τ of 180

A series of runs were made to compare to a DNS [6]. As such the simulations used the same gravitational constant as the DNS. Runs were performed simulating 70 μm copper with a Stokes number (with fluid time scale based on wall units), $\tau_p^+ = 790$; 50 μm glass with a $\tau_p^+ = 120$; and 25 μm lycopodium spores with a $\tau_p^+ = 10$. Within each of these particle categories the velocity statistics for the mean streamwise velocity, U, the rms streamwise velocity, U', the wall normal rms velocity, V', and the spanwise rms velocity, W', were obtained as a function of wall-normal location y. Wang and Squires [12] published a LES of the same case (U, U', V', and W'). Their results are included for comparison.

The results for the velocity statistics of copper, glass, and lycopodium compared to the DNS and LES are shown in Figure 1. All velocities are scaled by the friction velocity U_τ to make them non-dimensional. The ODT is about as good a match to the DNS as the LES is. Figure 1a shows U^+ for all three particles. The rms statistics for copper, glass, and lycopodium are shown in Figures 1b, 1c, and 1d, respectively.

4.2 Mean and rms velocities at Re_τ of 640

A series of runs were made to compare to the experiments of Kulick et al. [4]. They measured U, U', and V' across a turbulent channel. Runs were performed simulating 70 μm copper with a $\tau_p^+ = 2400$; 50 μm glass, $\tau_p^+ = 350$; and 25 μm lycopodium spores, $\tau_p^+ = 31$. Within each of these particle categories the velocity statistics for U, U', V', and W' were measured across the channel width. Wang and Squires [12] published a LES of the same experiment. Their results are included for comparison.

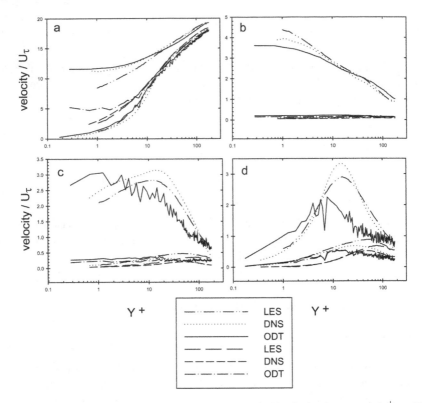

Fig. 1. ODT model wall-normal profiles of wall-normalized velocity for copper ($\tau_p^+ = 790$), glass ($\tau_p^+ = 120$), and lycopodium spores ($\tau_p^+ = 10$) for $\mathrm{Re}_\tau = 180$, compared to the DNS of Rouson and Eaton [6] and the LES of Wang and Squires [12]. (a) Mean streamwise velocity U for all three. Note, the copper and lycopodium curves are represented by the top three symbols and the glass curves are represented by the bottom three symbols. The copper U curves are everywhere greater than the glass or lycopodium curves; (b) rms streamwise U', wall-normal V', and spanwise W' velocity for copper; (c) rms streamwise, wall-normal, and spanwise velocity for glass; (d) rms streamwise, wall-normal, and spanwise velocity for lycopodium spores. Note for (b) (c) (d), the U' and W' curves are represented by the top three symbols and the V' curves are represented by the bottom three symbols. The U' curves are everywhere greater than the W' curves.

Rouson and Eaton [6] suggest that initial non-uniform loading and an insufficient wind tunnel development length in the Kulick et al. experiments could produce the "check mark" shape in the measurements. The analysis of Graham [1] supports the Rouson and Eaton suggestion that the wind tunnel development length was insufficient. Rouson and Eaton demonstrate that the particle motion of 70 μm copper is collision dominated. Therefore a correct particle-wall interaction model is essential to capturing the subtleties of the near wall measurements. None of the simulation analysis (ODT, DNS, or LES) uses anything but a spectral reflection at the wall, so phenomena which may be important such as inelastic collision, wall roughness,

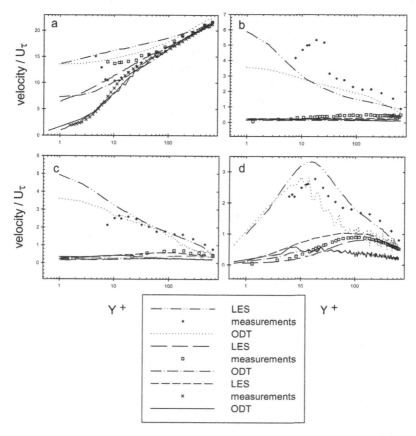

Fig. 2. ODT model wall-normal profiles of wall-normalized velocity for copper ($\tau_p^+ = 2400$), glass ($\tau_p^+ = 350$), and lycopodium spores ($\tau_p^+ = 31$) for $\text{Re}_\tau = 640$, compared to the measurements of Kulick et al. [4] and the LES of Wang and Squires [12]. (a) Mean streamwise velocity U for all three. Note the copper U curves are represented by the top three symbols, the glass are represented by the middle three symbols, and the lycopodium curves are represented by the bottom three symbols; (b) rms streamwise U', wall-normal V', and spanwise W' velocity for copper; (c) rms streamwise, wall-normal, and spanwise velocity for glass; (d) rms streamwise, wall-normal, and spanwise velocity for lycopodium spores. Note for (b) (c) (d), the U' curves are represented by the top three symbols, the V' curves are represented by the middle three symbols, and the W' curves are represented by the bottom three symbols. There are no W' measurements.

particle spin, and Magnus effect, are not represented. Kulick et al. describe the development section of the wind tunnel as made out of particle board, where the walls of the test section was made out of acrylic. The change in material brings up obvious issues on the particle-wall collision dynamics and if it were true that the materials had different collision dynamics the development time for the particles between when the wall material was changed to acrylic was obviously too small to smooth out any transients caused by the material change.

The results for the velocity statistics of copper, glass, and lycopodium compared to the measurements and LES are shown in Figure 2. All velocities are scaled by the friction velocity U_τ. The ODT is about as good a match to the measurements as the LES is, though neither captures the overall shape of the measurement curves of either copper or glass. Figure 2a shows U^+ for all three particles. The rms statistics for copper, glass, and lycopodium are shown in Figures 2b, 2c, and 2d, respectively.

4.3 Conclusions, future work

For the particle simulations at $Re_\tau = 180$ and 640 the ODT simulations are reasonable representations of steady-state statistics for U, U', V', and W' when compared to how the LES compares to the respective DNS and measurements, with the greatest deviation in U' occurring for lycopodium spores at $Re_\tau = 180$, and for copper in the near wall region at $Re_\tau = 640$. This 1D model is less expensive than LES yet compares well with the more complex formulations and can solve higher Re problems.

This two-phase flow model for ODT fails to meet the tracer particle limit. This failure could be a contributing factor to the $Re_\tau = 180$ lycopodium curves not being as good a match to the DNS. These particles have the smallest Stokes number ($\tau_p^+ = 10$) and would be most susceptible to the aforementioned problem. An alternate two-phase flow model for ODT which correctly captures the tracer particle limit has been formulated [8].

Schmidt et al. [10] implement ODT as a sub-grid model for LES. The addition of the ODT particle model would prove fruitful as this would readily give particle sub-grid fluctuations without the need of an additional sub-grid model.

To eliminate the ambiguities associated with the comparison to Kulick et al. [4], Kerstein and Krueger (in progress) are working on a droplet collision representation to compare to Reade and Collins [5].

Acknowledgements

The authors would like to thank Damian Rouson for providing data files from his work and Vebjorn Nilson for providing a FORTRAN code for ODT proper which served as the starting point for the numerical portion of this work.

This work was partially supported by the U.S. Department of Energy, Office of Basic Energy Sciences, Division of Chemical Sciences, Geosciences, and Biosciences. Sandia National Laboratories is a multi-program laboratory operated by Sandia Corporation, a Lockheed Martin Company, for the United States Department of Energy under contract DE-AC04-94-AL85000.

References

1. Graham, D.I., 2004, Development of particle dispersion characteristics from arbitrary initial conditions in isotropic turbulence, *J. Fluid Mech.*, **501**, 149–168.
2. Kerstein, A.R., 1999, One-dimensional turbulence: Model formulation and application to homogeneous turbulence, shear flows, and buoyant stratified flows, *J. Fluid Mech.*, **392**, 277–334.
3. Kerstein, A.R., Ashurst, W.T., Nilsen, V. and Wunsch, S.E., 2001, One-dimensional turbulence: Vector formulation and application to free shear flows, *J. Fluid Mech.*, **447**, 85–109.
4. Kulick, J.D., Fessler, J.R. and Eaton, J.K., 1993, On the interactions between particles and turbulence in a fully-developed channel flow in air, Mech Engng Dept Report MD-66, Stanford University, Stanford, CA.
5. Reade, W.C. and Collins, L.R., 2000, Effect of preferential concentration on turbulent collision rates, *Phys. Fluids*, **12**, 2530–2540.
6. Rouson, D.W. and Eaton, J.K., 2001, On the preferential concentration of solid particles in turbulent channel flow, *J. Fluid Mech.*, **428**, 149–169.
7. Rowe, P.N., 1961, The drag coefficient of a sphere, *Trans. Inst. Chem. Eng.*, **39**, 175–181.
8. Schmidt, J.R., 2004, Stochastic models for the prediction of individual particle trajectories in one dimensional turbulence flows, Ph.D. Thesis, The University of Arizona.
9. Schmidt, R.C., Kerstein, A.R., Wunsch, S. and Nilsen, V., 2003, Near-wall LES closure based on one-dimensional turbulence modeling, *J. Comp. Phys.*, **186**, 317–355.
10. Schmidt, R.C., McDermott, R. and Kerstein, A.R., 2005, ODTLES: A model for 3D turbulent flow based on one-dimensional turbulence modeling concepts, Sandia National Laboratories Report SAND2005-0206.
11. Stock, D.E., 1996, Particle dispersion in flowing gases – 1994 Freeman Scholar Lecture, *Trans. ASME*, **118**, 4–17.
12. Wang, Q. and Squires, K.D., 1996, Large eddy simulation of particle-laden turbulent channel flow, *Int. J. Multiphase Flow*, **22**, 667–683.

Author Index

Subject Index

Mechanics

FLUID MECHANICS AND ITS APPLICATIONS
Series Editor: R. Moreau

Aims and Scope of the Series

The purpose of this series is to focus on subjects in which fluid mechanics plays a fundamental role. As well as the more traditional applications of aeronautics, hydraulics, heat and mass transfer etc., books will be published dealing with topics which are currently in a state of rapid development, such as turbulence, suspensions and multiphase fluids, super and hypersonic flows and numerical modelling techniques. It is a widely held view that it is the interdisciplinary subjects that will receive intense scientific attention, bringing them to the forefront of technological advancement. Fluids have the ability to transport matter and its properties as well as transmit force, therefore fluid mechanics is a subject that is particularly open to cross fertilisation with other sciences and disciplines of engineering. The subject of fluid mechanics will be highly relevant in domains such as chemical, metallurgical, biological and ecological engineering. This series is particularly open to such new multidisciplinary domains.

Mechanics

FLUID MECHANICS AND ITS APPLICATIONS
Series Editor: R. Moreau

Mechanics

FLUID MECHANICS AND ITS APPLICATIONS
Series Editor: R. Moreau

Mechanics

FLUID MECHANICS AND ITS APPLICATIONS
Series Editor: R. Moreau